Developing Charging Infrastructure and Technologies for Electric Vehicles

Mohammad Saad Alam
Aligarh Muslim University, Aligarh, India

Reji Kumar Pillai
India Smart Grid Forum, India

N. Murugesan
India Smart Grid Forum, India

IGI Global
PUBLISHER of TIMELY KNOWLEDGE

A volume in the Advances in Mechatronics and Mechanical Engineering (AMME) Book Series

Published in the United States of America by
IGI Global
Engineering Science Reference (an imprint of IGI Global)
701 E. Chocolate Avenue
Hershey PA, USA 17033
Tel: 717-533-8845
Fax: 717-533-8661
E-mail: cust@igi-global.com
Web site: http://www.igi-global.com

Library of Congress Cataloging-in-Publication Data

Names: Alam, Mohammad S., 1961- editor. | Pillai, Reji, 1958- editor. | Navaneetha Krishnan, Murugesan, 1959- editor.
Title: Developing charging infrastructure and technologies for electric vehicles / Mohammad Saud Alam, Reji Pillai, and Murugesan Navaneetha Krishnan, editors.
Description: Hershey, PA : Engineering Science Reference, an imprint of IGI Global, [2021] | Includes bibliographical references and index. | Summary: "This book reviews and communicates the state of the art in Electric Vehicle charging technologies, their applications, economic, environmental and social impact and integration with the renewable energy". Provided by publisher.
Identifiers: LCCN 2020040297 (print) | LCCN 2020040298 (ebook) | ISBN 9781799868583 (hardcover) | ISBN 9781799868590 (paperback) | ISBN 9781799868606 (ebook)
Subjects: LCSH: Battery charging stations (Electric vehicles) | Electric vehicles--Power supply. | Battery chargers.
Classification: LCC TL220.5 .D48 2021 (print) | LCC TL220.5 (ebook) | DDC 629.28/6--dc23
LC record available at https://lccn.loc.gov/2020040297
LC ebook record available at https://lccn.loc.gov/2020040298

This book is published in the IGI Global book series Advances in Mechatronics and Mechanical Engineering (AMME) (ISSN: 2328-8205; eISSN: 2328-823X)

British Cataloguing in Publication Data
A Cataloguing in Publication record for this book is available from the British Library.

All work contributed to this book is new, previously-unpublished material. The views expressed in this book are those of the authors, but not necessarily of the publisher.

For electronic access to this publication, please contact: eresources@igi-global.com.

Advances in Mechatronics and Mechanical Engineering (AMME) Book Series

J. Paulo Davim
University of Aveiro, Portugal

ISSN:2328-8205
EISSN:2328-823X

Mission

With its aid in the creation of smartphones, cars, medical imaging devices, and manufacturing tools, the mechatronics engineering field is in high demand. Mechatronics aims to combine the principles of mechanical, computer, and electrical engineering together to bridge the gap of communication between the different disciplines.

The **Advances in Mechatronics and Mechanical Engineering (AMME) Book Series** provides innovative research and practical developments in the field of mechatronics and mechanical engineering. This series covers a wide variety of application areas in electrical engineering, mechanical engineering, computer and software engineering; essential for academics, practitioners, researchers, and industry leaders.

Coverage

- Mechanisms and machines
- Micro and nanomechanics
- Vibration and acoustics
- Tribology and surface engineering
- Manufacturing Methodologies
- Computer-Based Manufacturing
- Bioengineering Materials
- Design and Manufacture
- Computational Mechanics
- Sustainable and green manufacturing

IGI Global is currently accepting manuscripts for publication within this series. To submit a proposal for a volume in this series, please contact our Acquisition Editors at Acquisitions@igi-global.com or visit: http://www.igi-global.com/publish/.

The Advances in Mechatronics and Mechanical Engineering (AMME) Book Series (ISSN 2328-8205) is published by IGI Global, 701 E. Chocolate Avenue, Hershey, PA 17033-1240, USA, www.igi-global.com. This series is composed of titles available for purchase individually; each title is edited to be contextually exclusive from any other title within the series. For pricing and ordering information please visit http://www.igi-global.com/book-series/advances-mechatronics-mechanical-engineering/73808. Postmaster: Send all address changes to above address.

Titles in this Series

For a list of additional titles in this series, please visit: www.igi-global.com/book-series

Electric Vehicles and the Future of Energy Efficient Transportation
Umashankar Subramaniam (Prince Sultan University, Saudi Arabia) Sheldon S. Williamson (Ontario Tech University, Canada) Mohan Krishna S. (Alliance University, India) and Febin Daya J. L. (Vellore Institute of Technology, India)
Engineering Science Reference • copyright 2021 • 293pp • H/C (ISBN: 9781799876267) • US $215.00 (our price)

Modeling and Optimization of Solar Thermal Systems Emerging Research and Opportunities
Jagadish (National Institute of Technology, Raipur, India) and Agnimitra Biswas (National Institute of Technology, Silchar, India)
Engineering Science Reference • copyright 2021 • 215pp • H/C (ISBN: 9781799835233) • US $185.00 (our price)

Recent Trends on Electromagnetic Environmental Effects for Aeronautics and Space Applications
Christos D. Nikolopoulos (Department of Electronic Engineering, School of Engineering, Hellenic Mediterranean University, Greece)
Engineering Science Reference • copyright 2021 • 285pp • H/C (ISBN: 9781799848790) • US $265.00 (our price)

Modern Machining Processes and Techniques for Aerospace Materials
Anand Pandey (Manipal University Jaipur, India)
Engineering Science Reference • copyright 2021 • 300pp • H/C (ISBN: 9781799845102) • US $215.00 (our price)

Evaluation and Applications for Ultrasonic Methods of Non-Destructive Testing
Ashish Khaira (Samrat Ashok Technological Institute, India)
Engineering Science Reference • copyright 2021 • 215pp • H/C (ISBN: 9781799845645) • US $195.00 (our price)

Handbook of Research on Advancements in Manufacturing, Materials, and Mechanical Engineering
Leonid Burstein (Independent Researcher, Israel)
Engineering Science Reference • copyright 2021 • 462pp • H/C (ISBN: 9781799849391) • US $295.00 (our price)

Genetic Algorithms and Remote Sensing Technology for Tracking Flight Debris
Maged Marghany (Syiah Kuala University, Indonesia)
Engineering Science Reference • copyright 2020 • 334pp • H/C (ISBN: 9781799819202) • US $195.00 (our price)

Humanizing Cities Through Car-Free City Development and Transformation
Rahma M. Doheim (Assiut University, Egypt & University of Business and Technology, Saudi Arabia) Alshimaa Aboelmakarem Farag (Zagazig University, Egypt) and Ehab Kamel (University of Central Lancashire, UK)
Engineering Science Reference • copyright 2020 • 305pp • H/C (ISBN: 9781799835073) • US $215.00 (our price)

IGI Global
PUBLISHER of TIMELY KNOWLEDGE

701 East Chocolate Avenue, Hershey, PA 17033, USA
Tel: 717-533-8845 x100 • Fax: 717-533-8661
E-Mail: cust@igi-global.com • www.igi-global.com

Dedicated to all the human beings who died or suffered due to transportation pollution.

Table of Contents

Foreword xv

Preface xvii

Chapter 1
Smart Grid-Integrated Electric Vehicle Charging Infrastructure: Future Vision 1
Isa S. Qamber, University of Bahrain, Bahrain
Mohamed Y. Alhamad, GCC Interconnection Authority, Saudi Arabia

Chapter 2
Proposed Power Systems Planning in Indian Scenario for Integrating EV Charging Infrastructure.... 25
Sanchari Deb, School of Engineering, University of Warwick, Coventry, UK
Sulabh Sachan, Department of Electrical Engineering, Indian Institute of Technology, Kanpur, India

Chapter 3
A Comprehensive Review of Blockchain Technology Implementation in the EV Charging Infrastructure 38
Toni Zhimomi, Aligarh Muslim University, India
Mohammad Saad Alam, Aligarh Muslim University, India
Hafiz Malik, University of Michigan, Dearborn, USA

Chapter 4
Integrated Battery Chargers: Challenges and Opportunities in Design and Infrastructure 68
Syed Qaseem Ali, Quanta Technology, Canada
Geza Joos, McGill University, Canada
Chu Sun, Beihang University, Canada

Chapter 5
Impact of Charging on Battery Life and Battery Degradation in Electric Vehicles 96
Mamdouh Ahmed Ezzeldin, Nile University, Egypt
Ahmed Mohsen Ali, Nile University, Egypt
Jomana Ashraf Mahmoud, Nile University, Egypt
Sohaila Ahmed Rabie, Nile University, Egypt
Hossam Hassan Ammar, Nile University, Egypt

Chapter 6
Impact of Electric Vehicle Charging Infrastructure on the Distribution Network Power Quality: Challenges and Opportunities 114
Gagandeep Sharma, Ontario Tech University, Oshawa, Canada
Vijay K. Sood, Ontario Tech University, Oshawa, Canada

Chapter 7
Optimal Charging Management of Microgrid-Integrated Electric Vehicles 133
Giulio Ferro, University of Genova, Italy
Riccardo Minciardi, University of Genova, Italy
Luca Parodi, University of Genova, Italy
Michela Robba, University of Genova, Italy

Chapter 8
Environmental Impact of EV Batteries and Their Recycling 156
Zeeshan Ahmad Arfeen, The Islamia University, Bahawalpur, Pakistan
Rabia Hassan, Institute of Business Management, Karachi, Pakistan
Mehreen Kausar Azam, Institute of Business Management, Karachi, Pakistan
Mohammad Pauzi Abdullah, Universiti Technologi Malaysia, Malaysia

Chapter 9
Multi-Level Inverters Interfacing Electric Vehicle Charging Stations With Microgrid for Vehicle-to-Grid (V2G) Applications 178
Mohd Rizwan Khalid, Aligarh Muslim University, India
Adil Sarwar, Aligarh Muslim University, India
Ibrahim Alsaidan, College of Engineering, Qassim University, Buraydah, Saudi Arabia

Chapter 10
Battery Swapping Station: A Potential Solution to Address the Limitations of EV Charging Infrastructure 195
Furkan Ahmad, Center for Automotive Research and Tribology (CART), Indian Institute of Technology, Delhi, India
Essam A. Al-Ammar, King Saud University, Saudi Arabia
Ibrahim Alsaidan, College of Engineering, Qassim University, Saudi Arabia

Chapter 11
Drivers of EV Charging Infrastructure Entrepreneurship in India 208
Mohd Yasir Arafat, Aligarh Muslim University, India
Imran Saleem, Aligarh Muslim University, India
Thoudam Prabha Devi, University of Buraimi, Oman

Chapter 12
Cryptocurrency-Based Payment Framework for EV Charging 220
Wasif Karim, Aligarh Muslim University, India
Mohammad Saad Alam, Aligarh Muslim University, India
Yasser Rafat, CARET, Aligarh Muslim University, India

Chapter 13
Recent Advancements in PEM Fuel Cell Technologies for Electrified Transportation...................... 233
Mohammad Arqam, Griffith University, Australia
Sameer Usmani, Australian National University, Australia
Sadegh Aberoumand, Griffith University, Australia
Vinay Kumar, Griffith University, Australia

Chapter 14
V2G Pilot Projects: Review and Lessons Learnt 252
Sanchari Deb, School of Engineering, University of Warwick, Coventry, UK
Essam A. Al Ammar,, King Saud University, Saudi Arabia
Hasan AlRajhi, Umm AlQura University, Saudi Arabia
Ibrahin Alsaidan, Qassim University, Saudi Arabia
Samir Shariff, Taibah University, Saudi Arabia

Chapter 15
Implementation and Demonstration of Electric Vehicle-to-Home (V2H) Application: A Case Study 268
Mahdi Shafaati Shemami, CARET, Aligarh Muslim University, India
Marzieh Sefid, Aligarh Muslim University, India

Compilation of References 294

About the Contributors 334

Index 342

Detailed Table of Contents

Foreword xv

Preface xvii

Chapter 1
Smart Grid-Integrated Electric Vehicle Charging Infrastructure: Future Vision 1
Isa S. Qamber, University of Bahrain, Bahrain
Mohamed Y. Alhamad, GCC Interconnection Authority, Saudi Arabia

The movements for any type of electric vehicle (EV) can be powered by wheels or driven by rotary motors. EVs derive their power from various sources, including fossil fuels. In the long term, reducing the cost of electrically powered vehicles (EDV) is seen as an essential ingredient to increase consumer acceptance. In addition, it aims to reduce the weight and volume of EDV. Moreover, the focus is on improving the performance, efficiency, and reliability of the EDV. The development of innovative modules is important when the acceleration of production and marketing needs to be improved. Consumers are looking for the production and transmission of electrical energy. This contributes to a greener environment. One of the most important parts of an EV is its battery. A proposed model presented in this chapter considers several parameters: solar radiation (PV panels), EV backup battery, and main charger. The model allows energy storage to be developed efficiently.

Chapter 2
Proposed Power Systems Planning in Indian Scenario for Integrating EV Charging Infrastructure.... 25
Sanchari Deb, School of Engineering, University of Warwick, Coventry, UK
Sulabh Sachan, Department of Electrical Engineering, Indian Institute of Technology, Kanpur, India

The growing concern about fossil energy exhaustion, air pollution, and ecological deprivation has made electric vehicles (EVs) a practical option in contrast to combustion engine-driven vehicles. In any case, driving extent uneasiness is one of the innate inadequacies related with EVs. Massive integration of EV charging load into the power system may be a threat to the distribution network. Spontaneous situation of charging stations in the distribution system and uncoordinated charging will augment the load demand thereby resulting in voltage instability, deterioration of reliability indices, harmonic distortions, and escalated power losses. This chapter will concentrate on breaking down the effect of EV chargers on the working parameters, for example, voltage dependability, unwavering quality, and force misfortune. The examination will be completed on standard test systems. The discoveries of the proposed part will evaluate the effect of EV charging load on the working parameters of the distribution system and help in proposing a framework for charging station planning.

Chapter 3
A Comprehensive Review of Blockchain Technology Implementation in the EV Charging Infrastructure 38
Toni Zhimomi, Aligarh Muslim University, India
Mohammad Saad Alam, Aligarh Muslim University, India
Hafiz Malik, University of Michigan, Dearborn, USA

Charging infrastructure is a key factor in successful electric vehicle adoption. Charging stations are still a fragmented market in terms of ownership, lack of standards, and charging protocols. The increasing decentralised grid has made energy and communication flow bi-directional. Challenges arise in maintaining the increasing decentralised structure, security, and privacy of the network. Blockchain facilitates the interconnectedness of such a distributed and decentralised network. Blockchain's versatility lies in its transparent and immutable decentralized architecture that enables direct transactions between users without the need of a middleman. It provides powerful safeguards against cyberattacks with its advanced cryptography enabling privacy-preserving authentication. This chapter presents a comprehensive review on the application of blockchain technology in EV charging infrastructure such as facilitating the peer-to-peer energy exchange, increased security and privacy, immutable transactions, and mitigating trust issues among the participants in the charging infrastructure.

Chapter 4
Integrated Battery Chargers: Challenges and Opportunities in Design and Infrastructure 68
Syed Qaseem Ali, Quanta Technology, Canada
Geza Joos, McGill University, Canada
Chu Sun, Beihang University, Canada

Integrated battery chargers (IBCs) have been proposed as a low-weight, low-volume, and high-power solution to conventional conductive chargers. However, the design of such chargers is complicated, requiring special components or control techniques to solve inherent issues (such as galvanic isolation, torque generation, and system reconfiguration) associated with their design. Solutions vary based on charging power, drive topology, and motor technology. This chapter introduces designs for IBCs, including solutions as proposed in the literature. It also presents challenges in their industrial adoption. Finally, the chapter presents opportunities for fleet charging applications using IBCs.

Chapter 5
Impact of Charging on Battery Life and Battery Degradation in Electric Vehicles 96
Mamdouh Ahmed Ezzeldin, Nile University, Egypt
Ahmed Mohsen Ali, Nile University, Egypt
Jomana Ashraf Mahmoud, Nile University, Egypt
Sohaila Ahmed Rabie, Nile University, Egypt
Hossam Hassan Ammar, Nile University, Egypt

Electrical vehicles are the future of the world; hence, there is a necessity to pave the way for the upcoming technology and to ensure its contribution to the society fairly. Nevertheless, if the EVs completely replaced the fuel-based cars, more EV charging stations would be needed which might develop overconsumption of the main grid power causing remarkable instability. Consequently, the micro grids become the solution to this problem, in which they are defined as relatively small networks of energy sources and loads at the distribution level that aim to provide electricity to remote locations where the charging stations are located. In this chapter, the EV is considered as a load to the micro grid indirectly through the EV charging stations. Thus, micro grid loads will be retrieved from experimental data of an actual prototype electric vehicle to reflect on the battery degradation in a micro-grid connected system.

Chapter 6
Impact of Electric Vehicle Charging Infrastructure on the Distribution Network Power Quality: Challenges and Opportunities 114
Gagandeep Sharma, Ontario Tech University, Oshawa, Canada
Vijay K. Sood, Ontario Tech University, Oshawa, Canada

This chapter discusses the available charging systems for electric vehicles (EV) which include battery electric vehicles (BEV) and plugged hybrid electric vehicles (PHEV). These architectures are categorized as common DC bus charging (CDCB) station and common AC bus charging (CACB) station. CACB charging stations are generally used as slow chargers or semi-fast chargers (on-board chargers). CDCB charging stations are used as fast chargers (off-board chargers). These chargers are vital to popularize the electric vehicles (EVs) as a green alternative to the internal combustion engine (ICE) vehicles. Further, this chapter covers the power quality problems related to the grid-connected fast charging stations (FCS), AC-DC converter, control strategies for converters, proposed system of architectures, methodology, system results with comparisons, and finally, a conclusion.

Chapter 7
Optimal Charging Management of Microgrid-Integrated Electric Vehicles 133
Giulio Ferro, University of Genova, Italy
Riccardo Minciardi, University of Genova, Italy
Luca Parodi, University of Genova, Italy
Michela Robba, University of Genova, Italy

The relevance of electric vehicles (EVs) is increasing along with the relative issues. The definition of smart policies for scheduling the EVs charging process represents one of the most important problems. A discrete-event approach is proposed for the optimal scheduling of EVs in microgrids. This choice is due to the necessity of limiting the number of the decision variables, which rapidly grows when a small-time discretization step is chosen. The considered optimization problem regards the charging of a series of vehicles in a microgrid characterized by renewable energy source, a storage element, the connection to the main grid, and a charging station. The objective function to be minimized results from the weighted sum of the cost for purchasing energy from the external grid, the weighted tardiness of the services provided, and a cost related to the occupancy of the socket. The approach is tested on a real case study.

Chapter 8
Environmental Impact of EV Batteries and Their Recycling 156
Zeeshan Ahmad Arfeen, The Islamia University, Bahawalpur, Pakistan
Rabia Hassan, Institute of Business Management, Karachi, Pakistan
Mehreen Kausar Azam, Institute of Business Management, Karachi, Pakistan
Mohammad Pauzi Abdullah, Universiti Technologi Malaysia, Malaysia

Electrifying transportation is one of the biggest keys to solving the looming climate crisis. The demand for electric vehicles (EV) is booming in the last five years and will increase in the coming years. In this modern age, where EV is the finest means of transportation due to null exhaust gases, there is a dire need to think about ways of recycling and reusing those batteries associated with EVs. In this context, it is estimated that post-vehicle battery packs application will be crossed from 1.4 million to 6.8 million by the year 2035. Numerous researches have been done on the re-purposing and safe disposal of EV batteries. However, presently, Lithium-ion batteries (LiBs) are the optimal choice for electric transportation due to greater energy density, compact size, and extended life cycles. Nonetheless, the trade-off between re-purposing and disposal of LiBs is substantial for the protection of the environment and human health. Regrettably, Lithium-ion battery recycling percentage is only 3% currently whereas its revival is negligible.

Chapter 9
Multi-Level Inverters Interfacing Electric Vehicle Charging Stations With Microgrid for Vehicle-to-Grid (V2G) Applications 178
Mohd Rizwan Khalid, Aligarh Muslim University, India
Adil Sarwar, Aligarh Muslim University, India
Ibrahim Alsaidan, College of Engineering, Qassim University, Buraydah, Saudi Arabia

Multi-level inverters (MLI) are power electronic converters that convert DC power to AC power with high power-quality of output voltage waveforms. These MLI are the main streamline converters for integrating dc power of EV with microgrids. Thus, the recent interest of researchers is to investigate the MLI with a lower number of active and passive switch counts which could integrate the DC power of EV to the microgrid with the boosting ability. This chapter discusses various topologies of MLI for the integration of the DC power of EV to the grid for vehicle-to-grid (V2G) applications. MLI converts DC power to AC power with high quality of output voltage waveform. Thus, the recent interest of researchers is to investigate the MLI with a lower number of active and passive switch counts which could integrate the DC power of EV to the microgrid. Also, MLI must be capable of boosting the voltage level to meet the grid requirements. The aim of this chapter is to discuss the various topologies of MLI for the integration of DC power of EV to the grid for vehicle-to-grid (V2G) applications.

Chapter 10
Battery Swapping Station: A Potential Solution to Address the Limitations of EV Charging Infrastructure 195
Furkan Ahmad, Center for Automotive Research and Tribology (CART), Indian Institute of Technology, Delhi, India
Essam A. Al-Ammar, King Saud University, Saudi Arabia
Ibrahim Alsaidan, College of Engineering, Qassim University, Saudi Arabia

State-of-the-art research to solve the grid congestion due to EVs is focused on smart charging and using (centralized, de-centralized, vehicle-to-grid) stationery energy storage as a buffer between times of peak and off-peak demand. On the other hand, the charging of EVs introduces new challenges and opportunities. This can prove to be beneficial for the EV aggregator as well as to consumers, regarding the economy. Also, EV as distributed storage makes the grid more steady, secure, and resilient by regulating frequency and the spinning reserve as backup power. However, the charging time and range anxiety lead to peak challenges for the use of EVs. In this chapter battery swapping station (BSS) as solution to the EV charging station is discussed.

Chapter 11
Drivers of EV Charging Infrastructure Entrepreneurship in India 208
Mohd Yasir Arafat, Aligarh Muslim University, India
Imran Saleem, Aligarh Muslim University, India
Thoudam Prabha Devi, University of Buraimi, Oman

The existing research advocating entrepreneurship as an important way to increase the uptake of electric vehicles (EVs) in developing countries and EV charging business is also playing a crucial role in increasing the adoption of EVs. EV charging is important for EV adoption, and entrepreneurship is also important for EV adoption; therefore, it is important that we must understand what mobilizes or prevents EV charging entrepreneurship. This chapter aimed at explaining drivers of EV charging entrepreneurship. A survey of 121 potential entrepreneurs shows that personal attitude, self-confidence, and opportunity perceptions shape the decision to engage in EV charging entrepreneurship. Policy measures to boost EV charging entrepreneurship have been suggested.

Chapter 12
Cryptocurrency-Based Payment Framework for EV Charging 220
Wasif Karim, Aligarh Muslim University, India
Mohammad Saad Alam, Aligarh Muslim University, India
Yasser Rafat, CARET, Aligarh Muslim University, India

The world is transforming from paper or plastic currency towards digital currency. Everyone is trying to benefit from the decentralized system, the anonymity, and lower transaction cost offered by cryptocurrencies. One of the hurdles in the ascension of electric vehicle is its charging and the transaction for charging. In this chapter, a cryptocurrency-based payment framework is formulated for electric vehicle charging to eliminate the third-party involvement as in the plastic currency system. In this payment framework, smart contracts are deployed which would ensure the conditions to be satisfied, like the amount sent by the consumer. Smart contract includes the transfer function that would transfer the amount of ether and set charging time according to the charging rate, amount of ether sent, and power delivered. Overview of the platform prototype of payment is presented to demonstrate the implementation of the framework in the real world. At last, the authors look at the security issues, challenges ahead, and also suggested some of the improvements that could be further done.

Chapter 13
Recent Advancements in PEM Fuel Cell Technologies for Electrified Transportation...................... 233
Mohammad Arqam, Griffith University, Australia
Sameer Usmani, Australian National University, Australia
Sadegh Aberoumand, Griffith University, Australia
Vinay Kumar, Griffith University, Australia

The interest in developing clean and environmentally-friendly energy devices to be used on vehicles is intensifying because of emissions from conventional internal combustion engines considered as one of the significant contributors to the rapidly changing climate. Fuel cell energy devices, especially the proton exchange membrane (PEM) type, are the solid contender to replace the conventional vehicle propulsion technology in the transport sector. The PEM fuel cell technology needs a lot of efforts to overcome some existing problems such as durability, hydrogen storage, and cost for its successful worldwide commercialisation. This chapter deals with the durability, cost, and performance challenges related to the utilization of PEM fuel cell technology in electrified transportation. Recent advancements concerning the current challenges have been discussed. Moreover, issues of hydrogen storage and infrastructure are outlined.

Chapter 14
V2G Pilot Projects: Review and Lessons Learnt 252
Sanchari Deb, School of Engineering, University of Warwick, Coventry, UK
Essam A. Al Ammar,, King Saud University, Saudi Arabia
Hasan AlRajhi, Umm AlQura University, Saudi Arabia
Ibrahin Alsaidan, Qassim University, Saudi Arabia
Samir Shariff, Taibah University, Saudi Arabia

Electric vehicles (EVs) are not only a viable energy efficient mode of transport, but they have considerable capacity of providing flexible and quick responding storage alternative based on vehicle-to-grid (V2G) scheme. V2G technology facilitates bidirectional flow of energy to and from the vehicle by a power converter. However, there is skepticism regarding the economic profitability of the V2G scheme. Despite the aforementioned challenges, the V2G technology is explored in matured markets. A number of V2G pilot projects across the world have investigated different aspects of V2G integration such as technological readiness, economic feasibility, social benefits, and challenges of V2G. This work aims to review the existing pilot projects on V2G functionality.

Chapter 15
Implementation and Demonstration of Electric Vehicle-to-Home (V2H) Application: A Case Study 268
Mahdi Shafaati Shemami, CARET, Aligarh Muslim University, India
Marzieh Sefid, Aligarh Muslim University, India

This chapter emphasizes the utilization of the plug-in hybrid electric vehicle (PHEV) as a backup power source for residential loads in under-developing and developing countries. It works as a source of energy in residential micro-grid based on the condition of vehicle battery without harming its function as an EV (electric vehicle). The suggested V2H system uses solar PV power to charge vehicle battery; therefore, the entire system works as a residential nano-grid system. The EV is considered as a load of home when its batteries are charged by solar PV or grid. However, the main emphasis is given to use solar PV power to reduce charging from the grid. The key objectives of this work are to minimize the energy cost of a household by reducing the dependency of residential loads on the power grid to enhance the reliability of power supply to residential loads during load shedding and blackouts and to maximize the utilization of power produced by solar PV array mounted on the rooftop.

Compilation of References 294

About the Contributors 334

Index 342

Foreword

It is an honor for me to add my comments to a book which addresses the need of the hour- *Developing Charging Infrastructure and Technologies for Electric Vehicles*, edited by Professor Mohammad Saad Alam, Sri Reji Kumar Pillai, and Sri N. Murugesan.

Climate Change is a pressing issue in today's time. As the world continues to grapple with the aftermaths caused by novel coronavirus disease (COVID-19), we are now being forced to think and act urgently on the next big issue affecting globally- the climate crisis. Nations across the world have to come forward with policies to combat this threat and to accelerate their actions towards the targets set under Paris agreement and United Nations Framework Convention on Climate Change (UNFCCC). There is a need for collective action from the governments and people around the globe on issues threatening the planet as a whole.

The main driver of climate change is the greenhouse gases as emitted by burning of fossil fuels. Transportation followed by production of electricity from conventional means, are the largest contributors of these greenhouse gases. In order to reduce these emissions and their consequential impact on climate change, there has been a huge push worldwide towards transformative clean energy. As we see today, the entire energy sector paradigm is changing in the ways we produce and consume energy. Technology is playing a vital role in bringing this transformation and is also acting as an enabler for shifting control in the hands of consumers. With the advent of solutions like solar roof top, we are now seeing consumers becoming prosumers.

As an effort towards mitigating climate change, India is seen as a leader in terms of its ambitious targets of installing 450 MW of renewable energy by 2030. Further, the government aims to achieve 30% new vehicle sales to be electric by 2030 to decarbonize the transport sector. To achieve these targets, a conducive eco-system needs to be built up. For the transition to electric mobility, one of the essential pre-requisites is a robust network of electric vehicle (EV) charging infrastructure and a reliable grid infrastructure which is what the prime focus of this book.

The book brings out the various aspects of charging infrastructure and discusses state of the art electric vehicle charging technologies, their impact on distribution system, economics and environmental analysis. It also discusses about the impact on battery life of an electric vehicle when charged using state of the art chargers. Furthermore, the book address the challenges of integrating electric vehicle chargers and their impact on the distribution system as well as the impact of charging from a poor power quality distribution system on the battery life of electric vehicle. In addition, the concept of integrating renewable energy sources to reduce dependency on the distribution system and battery swapping stations as a potential solution to the limited electric vehicle charging infrastructure have also been presented.

I believe the contents of this book will be interesting to the readers and could potentially alter the paradigm for developing new charging infrastructure and technologies for electric vehicles. It will also serve as a key guide to new researchers and graduate level students, especially for India and developing countries of the Middle East and South East Asia, to identify various bottlenecks in the state of the art charging infrastructure and finding solutions to them for the success of electric mobility. I would like to congratulate Professor Mohammad Saad Alam, Sri Reji Kumar Pillai, and Sri N. Murugesan for their expertise and keen interest in this emerging area of electrified transportation and compliment them for their efforts on introducing the state of art topics to all of us.

Praveer Sinha
TATA Power, India

Preface

The increase in air pollution and vehicular emissions has led to the development of the renewable energy-based generation and electrification of transportation. Further, the electrification shift faces an enormous challenge due to limited driving range, long charging time, and high initial cost of deployment. Firstly, there has been a discussion on renewable energy such as how wind power and solar power can be generated by wind turbines and photovoltaic, respectively, while these are intermittent in nature. The combination of these renewable energy resources with the available power generation system will make electric vehicle (EV) charging sustainable and viable after the payback period. Recently, there has also been a significant discussion focused on various EV charging types and the level of power for charging to minimize the charging time. By focusing on both sustainable and renewable energy, as well as charging infrastructures and technologies, the future for EV can be explored.

Chapter 1 presents the planning of power system to absorb the electric vehicle load effectively. Chapter 2 covers the concept of integrating electric vehicle load in the power system along with various renewable energy resources in the smart grid. Chapter 3 presents challenges in maintaining the increasing decentralized structure, security and privacy of the network. It shows that Blockchain facilitates the interconnectedness of such a distributed and decentralized network. Chapter 4 introduces the concept of integrated battery charger which is a low-weight, low-volume, and high-power solution to conventional conductive chargers.

Chapter 5 presents the impact of charging on battery life and battery degradation in electric vehicles, while Chapter 6 focuses on the impact of electric vehicle charging on the power quality of distribution network. Chapter 7 is based on optimal charging management of microgrid integrated electric vehicles. Chapter 8 discusses the impact of electric vehicle batteries on the environment. Different multi-level inverters for interfacing electric vehicle charging stations with the microgrid are discussed in Chapter 9. A new concept of a Battery Swapping Station to elevate the limitations of EV Charging Infrastructure is presented in Chapter 10. Chapters 11 and 12 present a new idea for researchers based on the drivers of electric vehicle charging infrastructure entrepreneurship, and cryptocurrency-based payment framework for EV Charging, respectively. Recent advancements in PEM fuel cell technologies for electrified transportation are discussed in Chapter 13. Chapter 14 presents the review on V2G Pilot Projects. A case study based on the implementation and demonstration of electric vehicle to home is shown in Chapter 15.

It is expected that this book will serve as a reference for a larger audience such as power system architects, practitioners, developers, researchers, and graduate-level students, especially in developing countries of the Middle East and Southeast Asia.

The authors wish to express their sincere thanks to the contributors who brought their insights to enhance the content of this book. The authors would also like to thank their families who have given tremendous support. Finally, the authors are extremely grateful to IGI Global Publishers and the editorial staff for the opportunity to publish this book and help in all possible manners.

Mohammad Saad Alam
Reji Kumar Pillai
N. Murugesan

Chapter 1
Smart Grid-Integrated Electric Vehicle Charging Infrastructure:
Future Vision

Isa S. Qamber
University of Bahrain, Bahrain

Mohamed Y. Alhamad
GCC Interconnection Authority, Saudi Arabia

ABSTRACT

The movements for any type of electric vehicle (EV) can be powered by wheels or driven by rotary motors. EVs derive their power from various sources, including fossil fuels. In the long term, reducing the cost of electrically powered vehicles (EDV) is seen as an essential ingredient to increase consumer acceptance. In addition, it aims to reduce the weight and volume of EDV. Moreover, the focus is on improving the performance, efficiency, and reliability of the EDV. The development of innovative modules is important when the acceleration of production and marketing needs to be improved. Consumers are looking for the production and transmission of electrical energy. This contributes to a greener environment. One of the most important parts of an EV is its battery. A proposed model presented in this chapter considers several parameters: solar radiation (PV panels), EV backup battery, and main charger. The model allows energy storage to be developed efficiently.

1 INTRODUCTION

Many hydroelectric vehicles are designed to absorb the energy generated by the wheel sources, in order to minimize power losses. In addition, the charging and distribution equipment of the energy storage unit (ESU) differs from conventional hybrid equipment. At the same time, costs and distribution are not limited to the interaction between the energy storage unit. In some cases of an electric vehicle, the power of the vehicle may be lower than that of the internal combustion engine (ICE). In this case, the excess power generated is stored in the energy storage unit (ESU). Electric vehicles can significantly reduce

DOI: 10.4018/978-1-7998-6858-3.ch001

urban pollution. The objective of reducing pollution is zero emissions. Greenhouse gas savings depend on how electricity is produced. Using the American energy mix with an electric car would reduce CO_2 emissions by 30%. In addition, there are other countries where emissions are expected to drop to 40% in the UK, with China at 19% and Germany at only 1%.

Referring to the load curve, the power consumed varies from one point to another depending on customer demand. All the stored energy can be used to replace the current electricity shortage. At the same time, remember that all stored energy can be used efficiently to improve the quality, efficiency, reliability, profitability, and flexibility of power tool systems. To achieve this, it means someone whose goal is to meet the requirements of superpower and meet basic load generation.

A new EV battery can charge an electric vehicle in 10 minutes, discovered by researchers at Penn State University in the United States. Researchers expect to be able to reach distances up to 322 km (200 miles). The EV battery is a lithium-ion battery. In addition, Formula 1 (F1) will strengthen its plan to become carbon neutral by 2030. F1 aims to achieve carbon neutrality by 2030, which will help modernize the world on one billion internal combustion vehicles. One of the ways F1 can continue to evolve towards carbon neutral fuels such as methane is colorless gas and carbon dioxide.

2 LITERATURE REVIEW

The integration of electric vehicles into the electricity network is explained and discussed in this chapter based on a number of sections and subsections. The different number of electric vehicles is also compared. The smart grid appears in this chapter where it is deemed necessary. The smart grid helps, focuses and strives to trust collectively to provide fuel for electric vehicles. The various objective methods and functions can be described and used to normalize EV performance in smart grid techniques discussed and described in this chapter in the context of a future smart grid environment. Finally, electric vehicles are expected to cause a revolution in the automotive industry due to the current and future benefits of urban pollution.

Tang et al. (Tang et al 2019), in their study, they focus on intelligent energy activity. Tang et al. focusing on the economy of electric vehicle batteries in the photovoltaic (PV) diffusion system. Four cases emerged in their study. These cases are without battery storage and without PV coproduction, with battery storage and without PV coproduction, with battery storage and PV production shared under the same user base, and finally battery storage and PV generation distributed to different users. With reference to a criterion, the first location is selected. They conclude that the third and fourth cases can reduce electricity costs. In addition, the fourth position boosts the rate of self-consumption of PV and improves the operational capacity of batteries in retired electric vehicles. This means that the second life of the batteries is improved.

Vehicle to the Grid is a new technology that Shariff et al (Shariff et al 2019) introduced in their new study. The electric vehicle can be used both as a load and energy storage system. This method is used to support the grid. This study includes a modern and complete overview of the vehicle system of the network, in which it describes the method of power flow selected under the vehicle to the network. In addition, the study describes the main barriers to choosing a commercial electric car. Also, the most recent comparison of network transport, domestic vehicle (domestic transport) and transport (vehicle to vehicle) is provided.

Ahmad et al. Emphasize how extreme air pollution in their study (Ahmad et al 2019) is the result of abstract growth in cities, industry, and an increasing number of vehicles on the road, creating alarming conditions for urban communities. In their research, there is an immediate need to implement the intelligent fuel-based production and transport system. The best framework for an energy management system for a power station of electric vehicles connected to a community micro-network is under study. The study was carried out based on the change of equipment from one trading market to another. The purpose of the energy management system is to reduce the energy costs of recharging the vehicle. In addition, another objective is to meet demand and increase the costs of selling additional energy from the photovoltaic solar system. In addition, it should be noted that the energy management system regulates the energy emitted by electric vehicles. The study was applied to a community of 100 homes connected to the Indian community, if each home drives an electric vehicle. In addition, the study identifies an economic reason for creating a broad and hybrid perspective of electricity consumption on the electricity market.

El-Taweel et al. (El-Taweel et al 2019) in their study, they emphasize the diversity of electronic transport systems which are presented as a promising solution. The study could help reduce environmental pollution. Using gas stations of the right size is an important step in setting the way for exciting transport. The models presented are easy to develop at the end of a globally accessible site where they do not require an optimization solution. In a situation similar to market or transportation prices, there are different zoom settings in different places. El-Taweel et al. In their study, they present analytical methods to assess their performance on the size of electric and hydrogen service stations. The scales of the various components are configured according to the percentage of operation of the system. In addition, the estimation error of the model is analyzed. The proposed model was then compared to artificial intelligence models for validation.

Xiang et al. (Xiang et al 2018) in their study, they highlight electric vehicles where electric vehicles have evolved rapidly in recent years due to the low-carbon industry and smart grid initiatives. The effect of large-scale integration of electric vehicles on the reliability and security of the network is becoming more visible. One of the objectives is to solve the problems and challenges of advanced charging of electric vehicles. Vehicles that charge load for cost modeling are taken into account to improve the required accuracy. During the study, the impact characteristics of recharging electric vehicles were studied.

Kurdve et al (Kurdve et al 2019) their research focuses on the electric vehicle market, which is expected to grow in the coming years. The electric car imposes new requirements at the end of life. In addition, recycling systems are requested. It should be noted that the ecological footprint of vehicles is involved in the extraction and production of raw materials. Use a template to show the design and development of the system. The main contribution of their research is to identify the factors of the design system and to deduce the associated factors. Kurdve et al (Kurdve et al 2019) achieved this objective in their research using an adaptive analysis composed of several factors (political, economic, social and technological factors). At the same time, these factors are distinguished. Their study allows previous discoveries on the functioning of the vehicle's battery systems.

In their research, Al-Shariff et al (Al-Shariff et al 2019) highlight smart cities in need of an intellectual transport system (ITS). In smart cities, transport must be hassle-free, environmentally friendly, and connect and share vehicles for better public transport services. Through better transportation, ITS ensures better traffic, energy, and fleet management. Electric vehicles with the necessary information need smart city coverage. These help to communicate relevant transport information to other relevant recipients. The authors of their research framework rely on communications between the electric car and Internet of Things network applications. Public information for travelers is shared via a mobile app. One of the objectives of the study is to optimize and manage traffic and public transport. Recent developments are under consideration regarding the design of innovation and the use of centers as a smart grid for cost levels at all levels, second and third costs.

Mohamed et al (Mohamed et al 2019) reviewed their research on wireless power transfer. This technology offers new technology to charge electric vehicles without physical warning. This system should solve part of the problem of electric vehicles and buses. Many researchers have considered two categories, these two groups being non-stop charging systems and dynamic wireless charging. A recent solution is the wireless charging method. The problem is therefore related to the state of the vehicle. These standards are examined in two cases, for example if it moves and if it is in execution mode. The article (Mohamed et al 2019) explains and discusses the mathematical model of the wireless dynamic charging system in terms of several components related to the location of the vehicle. The authors developed the mathematical model and identified the correct conditions. This compensation method is deemed to be subject to vehicle traffic.

In their research on electric vehicles, Un-Noor et al (Un-Noor et al 2017) have become more and more widespread in the transport sector in recent years. Means of transport will probably soon replace combustion engines. Electric vehicles can have a major impact. This has consequences for the environment, the electrical system, and other related sectors. Electric vehicles can make an important contribution to the smart network. At the same time, electric vehicles can significantly reduce greenhouse gas emissions. It should be noted that electric vehicles face major obstacles. Un-Noor et al (Un-Noor et al 2017) aimed to examine all the useful data available on the configuration of electric vehicles, energy sources, motors, charging and optimization methods, trends and possible directions for future development. The trends and planned approaches are future developments, with the results of the study summarizing the full text and providing a clear picture of this kind.

In their research, Khan et al (Khan et al 2017) highlight electric vehicles that are popular in different countries of the world. Electric vehicles are more energy efficient. In addition, it is more environmentally friendly than motor vehicles for internal combustion. The installation of electric vehicles on the existing electricity network, powered by a fossil fuel production system, has implications for the circulation system. Solar energy has great potential for generating electricity. The energy produced comes from a PV panel where electric vehicles are charged via PV panels. Photovoltaic panels are an excellent solution as a sustainable step towards the environment. The study includes a comprehensive analysis of electric vehicle charging and handling systems around the world.

The study carried out by (Ashique et al 2017) assesses the important characteristics of a rapidly integrated DC charger for photovoltaic networks. During the review, particular attention was paid to the elements of the cost system. These components are architecture, operating modes and control. The examination focuses on the interaction of different components, namely the photovoltaic energy source, the electrical network, the energy storage unit and power electronics for chargers. From the point of view of power electronics, the study assesses the gentle conversion of non-external DC transformers to DC, where they can be used as electric chargers in the future. In addition, the cost effect of a PV network on the network and the circulation system and their processing efforts were briefly discussed. During the study, the Vehicle to Gris concept was explained. It has proven to be useful as a one-stop document for engineers, scientists and others. The study is well suited to the rapid cost of direct current for electric vehicles with a renewable energy source.

The costs and distribution of electric vehicles are becoming increasingly important at home. Therefore (Shemami et al. 2019), in their study, they aim for good load management and use solar energy. Based on the fuzzy logic controller designed for the home delivery system, the authors (Shemami et al. 2019) proposed an optimized approach of fuzzy rules and the member function of the delivery system with the fuzzy system. The authors in their study designed the vehicle for the home system. In addition, the study was conducted based on the results of the experimental configuration. The functionality of product members and server entry increased as the optimization system was tested and performance improved.

The Ferdowsi study (Ferdowsi 2007) focuses on plug-in hybrid vehicles that improve the efficiency, fuel consumption and reliability of power and transportation systems. Energy storage systems improve the efficiency and reliability of energy production, distribution and recycling. This helps balance the load.

Roche et al. (Roche et al. 2015) are studies on the design of electrically mobile power transmission systems, which leads to a wide variety of configurations. The development of an electric vehicle is envisaged from the start. In addition, a strong relationship between the size of the battery and the weight, range and performance of the vehicle. The result is a series of interconnected dependencies where they can lead to multiple design loops to simultaneously achieve goals and rules. Integrating key relationships into vehicle goals, markets and regulations will facilitate vehicle development. The vehicle is a cost-effective vehicle assembly that can achieve the objectives. It provides a function and an analysis of sensitivity to the price of the vehicle weight and the size of the battery. The precision model engine avoids vehicles when vehicle functions are loaded at an early stage. According to the study, the unit was assessed by formulating one of its travel solutions. Finally, there are two similar control strategies, leading to a total of four simulations. Simulation results compared to a preconceived tool solution that assesses loyalty assessments and trends in finished products. The results depend on the architecture, components and management strategy.

In their study, Morais et al. (Morais et al. 2014) have evaluated smart grids that intensively deploy useless energy sources. This idea will play an important role in future adjustments to the electrical system. The nature of renewable energy sources brings new challenges. The repeated use of electric vehicles, especially the main vehicles, is becoming increasingly important. The study focuses on the effects of electric vehicles on system operating costs and the distribution network efficiency curve. A cost-effective management method increases costs and distribution to electric vehicles. The main objective of the study is to reduce the operating costs of the system and to reduce the difference between the lowest and the highest. As a case study, the circular network of 32 buses is used as a case study to consider different scenarios for electric vehicles.

3 POWER GENERATION AND TRADING

Increase the security and reliability of changes in the electricity supply necessary for electricity exchanges. These changes are limited and are provided to improve capacity capabilities (Al-Hamad 2018). According to international practice, many obstacles prevent markets from functioning. Renewable Energy Resources (RER) compete for production. Challenges for the categorization of legislative organization, market model, market design, regulation of the energy market and economic and financial events are classified. To better understanding the obstacles to the functioning of the energy market, it is necessary to assess the organizational level of national energy markets, market equipment and understand the history of the energy trade. These will assist and understand potential barriers to the commercial energy market, if any. Qamber highlights on the reserve markets, where it was defined as the electricity reserves playing a key role in stabilizing the power system (Qamber, 2020). As a definition it means that the market frequency remains within a stable band.

Increased cooperation in the energy trade between countries can bring several benefits for the energy trade (Al-Hamad 2018). Some of them will bring economic benefits, reduce the cost of supplying interconnected countries, allow more efficient and effective use of infrastructure for additional production capacity, the impact of carbon emissions from more energy renewable and other factors available (Al-Hamad 2018) are reduced.

Charging is cost effective. It uses a standard 110V unit, so EV drivers can use the charging cost provided by the most EVs almost anywhere. This cost will take the longest time and will be primarily used as an additional cost, emergency, or security solution. The electrical load varies slightly but is generally between 12 amps and 16 amps continuous power. At these output levels, a Level 1 charger is estimated to provide between 3.5 and 6.5 miles of charging range per hour. These loading distances can be satisfactory for drivers who do not drive more than 30 to 40 miles per day and who can use the charger at night. There are different levels (Reed 2011) of voltage and electricity (Center for Sustainable Energy 2020):

Level 1: This is a 120-volt charging standard (AC, this is available in some countries), and the power source is at home, through which the electric car is charged.

Level 2: Loading units installed in consumer garages, compared to conventional equipment, they are very slow. They can charge the car's electric battery in 4 to 10 hours. Second level chargers are often placed in shopping malls / workplaces. This helps drivers to charge the car during work or shopping.

Surprisingly, the British government has set a future target to ban the sale of petrol and diesel cars by 2040 (EDF 2020). This is in line with the speech of the British mayor of London since road transport in London accounts for about half of the capital's air pollution.

The introduction of electric vehicles on the general market can change the structure of the market and the electric vehicle charging group adds new customers to the electricity market. Electric vehicles can be used as "virtual power plants" to participate in the active energy market, an emergency services market (Xiang et al. 2018). Renewable energy includes many forms, such as wind, sun, water, biomass, and a variety of geothermal energy sources. Solar energy production is one of the most dynamic today and is the most promising renewable energy source in the world. Today, electrical energy is becoming one of the most important things in our daily life and it is in increasing demand. Therefore, the energy generated by solar energy is the conversion of solar radiation into electricity by solar photovoltaic energy. This transformation takes place in the solar cell with photovoltaic energy. Producing solar energy is of course the most efficient way to generate electricity, since fewer steps are required to generate electricity compared to other production methods.

4 SMART GRIDS AND CHARGING STATIONS

Deza et al in their study (Deza et al 2020) concentrate on the diversion of vehicles to one-way shared networks for electric cars is an operational burden. It has been proven that the number of stations and vehicles is increasing and the best place to enjoy it. Therefore, it is important to focus on finding the best locations for the cost stations for one-way electric car sharing in order to correct network imbalances. An electric car charging station (Deza et al 2020) or (electronic charging station) is considered an electric charging station and is a type of electric car power equipment that provides electric power for rechargeable cars or wireless.

Napoli et al in their study (Napoli et al 2019) highlight on the electric vehicle market should be analyzed first, and the input variables related to the vehicle system should be properly filtered. For this purpose, registration data and technical specifications provided by manufacturers have been used to meet the requirements which satisfy the goal. The road system, infrastructure and driver behavior are considered and studied. To charge at work or at home, some electric vehicles have built-in transformers that can be plugged into a wall outlet or a high-capacity device (Napoli et al 2019). Other types of electric vehicles require a station switch with an electric switch, a monitor, or stations with safety devices. In some street countries, there are facilities provided by electricity companies or in shopping malls, restaurants and car parks managed only by private companies.

Mehrjerdi and Hemmati (Mehrjerdi Hemmati 2019) highlight on an electric vehicle charging station is connected to the distribution network and has a battery energy storage system. The cost levels are very well implemented and planned to buy a suitable EV. The power and the capacity of the battery energy storage system are considered for the model that are looking for and targeted. The capacity of the charging station (i.e. the number of parking spaces) and the network upgrade speed are important to be considered for modeling. For electric cars, there are four basic categories. These four types (Mehrjerdi and Hemmati 2019) are:

1. Residential charging stations: The owner of the electric car connects his/her car to sockets when he/she returns home. Some portable charging devices can also be mounted on the wall as charging stations.
2. The public charging terminal, which is available during the parking, is often a private or commercial project for a fee or free of charge. This terminal can sometimes be offered in partnership with the owners of the public park. In summary, the electric charge can be slow or fast and often motivate electric car owners to rebuild their cars and take advantage of nearby facilities.
3. Fast charging stations, often more than 4,000 watts, can provide more than 60 miles (97 km) of range in a time estimated to be 10-30 minutes as a charging period.
4. In some charging stations, the battery exchanger can be done in less than 15 minutes. One of the specific objectives is to make an electric car emit up to zero, where it can run for approximately 200 miles (about 320 km).

Gradual electrification is widely seen as an operational strategy to reduce dependence on oil and Carbon dioxide emissions from road transport (Napoli et al 2020). For these purposes, electric vehicles (EVs) have become increasingly important. The objective of Napoli et al in their study (Napoli et al 2020) is to estimate the number of cost items and to determine the correct distribution on the road network by analyzing the supply and taking into account the psychological demand and the behavior of the driver. The battery capacity (Dixon and Bell 2020) and charging speed developed according to the different charging methods mentioned earlier (Napoli et al 2020). Different approaches and solutions from different manufacturers have adopted a standard approach, and in 2015, it is strongly recognized that consolidation is necessary.

The electric vehicle (EV) market is constantly evolving rapidly, both in terms of battery capacity and charging power brought to market by vehicle manufacturers and the increasingly widespread penetration of charging infrastructure (Dixon and Bell 2020). The charge level being responsible for managing the electric voltage in the battery car (Dixon and Bell 2020), where the battery charging time depends on the battery capacity and the charging capacity, which simply means that the time required depends on the cost level of the battery used. The charging period (Electrical Technology 2020) can be calculated by the following equation:

$$Charging\ Time(h) = \frac{Battery\ Capacity(kWh)}{Charging\ Power(kW)} \tag{1.1}$$

For normal charging (up to 7.4 kW), the car manufacturer has designed a battery charger (Dixon and Bell 2020) in the car, which is used by through a charging cable. The charging cable is connected to the electrical grid to supply 230-volt AC (Electrical Technology 2020). Most battery manufacturers recommend around 25% of battery capacity charger size (Electrical Technology 2020). Larger chargers can be used to reduce charging time but might reduce battery life. Small chargers are good for long term liquidity (Electrical Technology 2020). For fast charging (22 kW, up to 43 kW and more), manufacturers have chosen two solutions:

1. Using the built-in charger from the moment of manufacture, the built-in car is designed to charge 3 to 43 kW in 230 V Single-Phase or 400 V Three-Phase.
2. Use an external charger, which converts AC to a continuous current, and charges the car at 50 kilowatts (e.g. Nissan Leaf) or more than 120-135 kW (Tesla Model S).

Charging schedules are derived from the travel diaries for different battery capacities, charging powers and charge access level to examine the likely effect of changing the charged behavior and the resulting power demand. However, to make sure that this process takes place in complete safety, the charging system must perform many safety functions for a car Praise charging (Electrical Technology 2020).

In the battery exchange station (known as swap station), the car battery or accumulator can be fully charged (Dixon and Bell 2020). This procedure eliminates the waiting time for replacing a car battery.

The benefits of battery swap for the electric car (Dixon and Bell 2020):

1. Battery change is quickly changed in less than five minutes
2. There are no limits to the electric car battery, as battery swap is available
3. Driver does not have to get out of the car while swap the battery
4. Battery swap can support electric car at a lower price than gasoline equivalent cars

Mehrjerdi and Hemmati (Mehrjerdi Hemmati 2019) highlight on an electric vehicle charging station is connected to the distribution network and has a battery energy storage system. The cost levels are very well implemented and planned to buy a suitable EV. The power and the capacity of the battery energy storage system are considered for the model that are looking for and targeted. The International Electrotechnical Commission (IEC) is a worldwide standard organization comprising all national electrotechnical committees (IEC national committees). The aim of the IEC (IEC 62196-1:2014 2014) is to encourage international cooperation on all questions of standardization in the electrical and electronic fields. In addition, the IEC publishes international standards, technical specifications, technical reports, publicly available specifications, and guidance. Sub-committee of the IEC developed International

Standard IEC 62196-1 belong to the plugs, sockets, and twins for industrial applications, and for electric vehicles. The International Electrical Technical Committee, defines media charging (IEC 62196), and divides the types of charging into:

Type 1: Slow charging from a normal power socket (single or three-phase).
Type 2: Slow charging from a normal connection, but with some special protection arrangements for electric cars.
Type 3: Slow or fast charging using a special multi-pin socket with control and protection functions (e.g., SAE J1772 and IEC 62196).
Type 4: Fast charging using some special charger technology.

Hanauer in the study carried out (Hanauer 2018) provides an analysis of the current IEC 62752. Establishing protective measures against electric shock is one of the key tasks in the development of safe electrical equipment. The electric shock is also very important for electric vehicles as they are mobile and in a public place most of the time. Even when the batteries are recharged, the mobility device is unlikely to be accessible to children or other players. IEC 62752 is the standard for products that connect the electric vehicle to a standard household outlet. This connection requires charging electricity on board. The complexity of the worldwide standardization of eMobility results are developed of the products continues, where the IEC 62752 describes several specific requirements and several specific tests to help design engineers to design a safe, reliable and stable product. There are three cases of connections (Hanauer 2018):

1. The first case is a charger connected to the power supply (the power supply cable is usually attached to the charger).
2. The second case is a car charger on board with a power supply cable that can be separated from both the width and the car.
3. The third case is a dedicated charging station with a fixed current supply for the vehicle, and the power supply cable may be permanently attached to the charging station.

To charge electric vehicles, there are four types of plugs used (IEC 62196-1:2014 2014):

1. The first type is a single-phase vehicle pairing.
2. The second type is a single and three-phase compound pairing.
3. The third type is a single and three-phase car interface equipped with safety shutters
4. The fourth type is a quick-delivery interface for special systems

The capacity of the charging station (i.e. the number of parking spaces) and the network upgrade speed are important to be considered for modeling. Charging stations (Napoli et al 2020) are usually connected to the electrical grid (Mehrjerdi and Hemmati 2019), which often means that electricity originates from fossil fuel power plants or solar power plants and is also suitable for electric vehicles. These stations usually use the marketing method of solar systems with the electric car charging station.

Electric vehicles use lithium-ion-eye batteries, like those used in mobile phones and laptops, on a much larger scale (Reed 2011). The choice of electric car at a high value of kilowatt-hours is classified as buying a car with a larger gasoline tank (Reed 2011). This means that the car will be able to travel more miles (kilometers) before the need for filling the tank (full), or electrical charging. But be aware that because of the way private vehicles work, you won't really have access to full battery capacity, so that the car's circuit system prevents the battery from becoming 100 percent charged, i.e. it cannot be fully charged.

As with conventional powered cars, electric vehicles are tested and selected based on operating range and power consumption under controlled conditions in the laboratory (Reed 2011). A start testing of the electric car will be carried-out with a fully charged battery.

Electric cars (Reed 2011) are usually charged at home, by traditional 120-volt electric circuits (known as the first level for charging, in some countries) or 240 volts (level 2 for charging). It is recommended to install 240 volts in the private garage (Reed 2011). This type of charging procedure is reducing the charging period.

There is a faster alternative to charging, which is called the third level of charging (Reed 2011). Although, the charging period is limited and is considered a general charging station. This charge is called fast charging (DC current), where the electric car battery can be brought up to 80 percent of its capacity in a period up to 30 minutes, depending on the vehicle's type and specifications.

The combined vehicle known as hybrid vehicles which is driven with a traditional fuel engine which produce during driving some emission of carbon dioxide (EDF 2020). Therefore, it is important to think that the hybrid vehicle use the solar for production of electricity. This production is helping to charge the battery of the plug-in hybrid, where the fuel will be reduced and carbon dioxide will be reduced, too. In this case the environment will be saved from pollution.

The charge and discharge cycle of the EV batteries occurred when driving (EDF 2020). This process repetition over time affects the charge amount of the battery that can hold. In addition, this decreases the range and time needed between each journey to charge. The warranty of the battery is five to eight years issued by the most manufacturers. Also, it is predicated that the future EV battery becomes between 10 – 20 years before they need to be replaced.

One type of the rechargeable batteries is known as a Lithium-ion (Li-ion) battery. This type of battery has ions, but not Lithium metals. Even though, this type of battery used for the EV (EDF 2020). In addition, this type of battery when compared by typical lead-acid or nickel-cadmium rechargeable batteries it has higher energy density. At the same time, the manufacturers can save space by reducing the overall size of the battery pack. It should be noted that Lithium-ion batteries are also safer than many alternatives batteries. The manufacturers equip EV with charging safeguards to protect the batteries during repeated rapid charging in a short period of time.

The electric vehicles are charged overnight at home or during the day whilst the working time. This means when the car is not in use (EDF 2020). The EV convenience of charging at home. Another factor is affecting. This factor is the temperature, that may impact the number of times you need to charge your EV. Lithium-ion batteries perform better in warm weather, so you might notice a slight drop in the range your EV can travel in the colder weather.

The time to charge the EV battery depends on (EDF 2020) as shown in Figure 1.

1. Car's battery size
2. Number miles (km) you do between charges
3. Charging behavior, i.e. topping up often vs charging from low to full less often
4. Power rating of the charger has different types of chargers and their kWh ratings

Figure 1. Dependency of EV Battery Charging Time

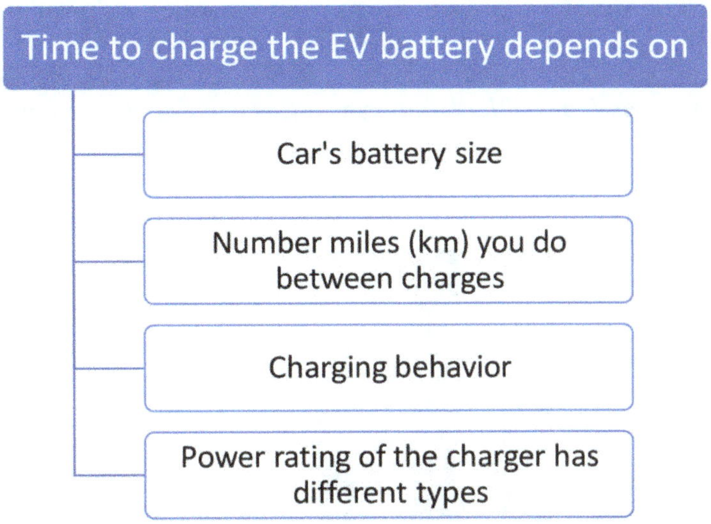

5 ELECTRIC VEHICLES (EV) MEANING AND TYPES

The electric vehicle's battery capacity is measured in kilowatt-hours, the same unit used in household electric meter records to determine the monthly electricity bill (Reed 2011). In the electric car system, the kilowatt-hour used for batteries as equivalent system to the gallons of gasoline tank used in regular cars. But the full battery can't be fully equal to a full fuel tank. In order to maintain the battery efficiency and battery life, the "Charging Mode" management system never allows the battery to become 100% fully charged or 100% empty. The most appropriate measurement may be as much as a usable battery.

The most important factor to consider when shopping for an electric car is the estimated range of the battery on full charge (Reed 2011). Fixing a sticker on the side window is helping to obtain the car specifications. This system of posting sticker could be applied in all countries, because of the benefit to the customer. There are different types of electric cars (Reed 2011). A car with an electric battery, PHEV and HEV. There are three main types of electric vehicles (vehicles), classified to the extent that electricity is used as a source of energy. BEVs, or battery electric vehicles, PHEVs of plug-in hybrid electric vehicles, or hybrid electric vehicles.

The actual range of the electric car can vary according to several factors (Reed 2011). The heavier the vehicle, i.e. the higher the load, the more energy required to reach the desired location. The battery will also be drained at a faster rate when operating with full load of passengers and cargo. The heavier the vehicle, i.e. the higher the load, the more energy required to reach the desired location. Maintaining a certain and specific speed will also be less consumption.

The electric vehicle's charging position can continue for longer period in cold or extremely hot weather (Reed 2011). Gasoline engines generate large amounts of heat that can be harvested to warm up inside the car, and the electric vehicle's climate control system relies solely on battery power. In use, the average electric vehicle is reduced by 41 percent. The same study (Reed 2011) determined that when external temperatures reach 95 degrees Fahrenheit (35 degrees Celsius) and air conditioning is used, the electric vehicle's range will be reduced by 17 percent.

Figure 2 illustrates some of the full and hybrid electric vehicles type versus their ranges in kilometers. The full and hybrid electric vehicles type versus their ranges in miles is shown in Figure 3.

Figure 2. Ranges in km versus Types of Full and Hybrid Electric Cars (Based on JOJU Solar 2018)

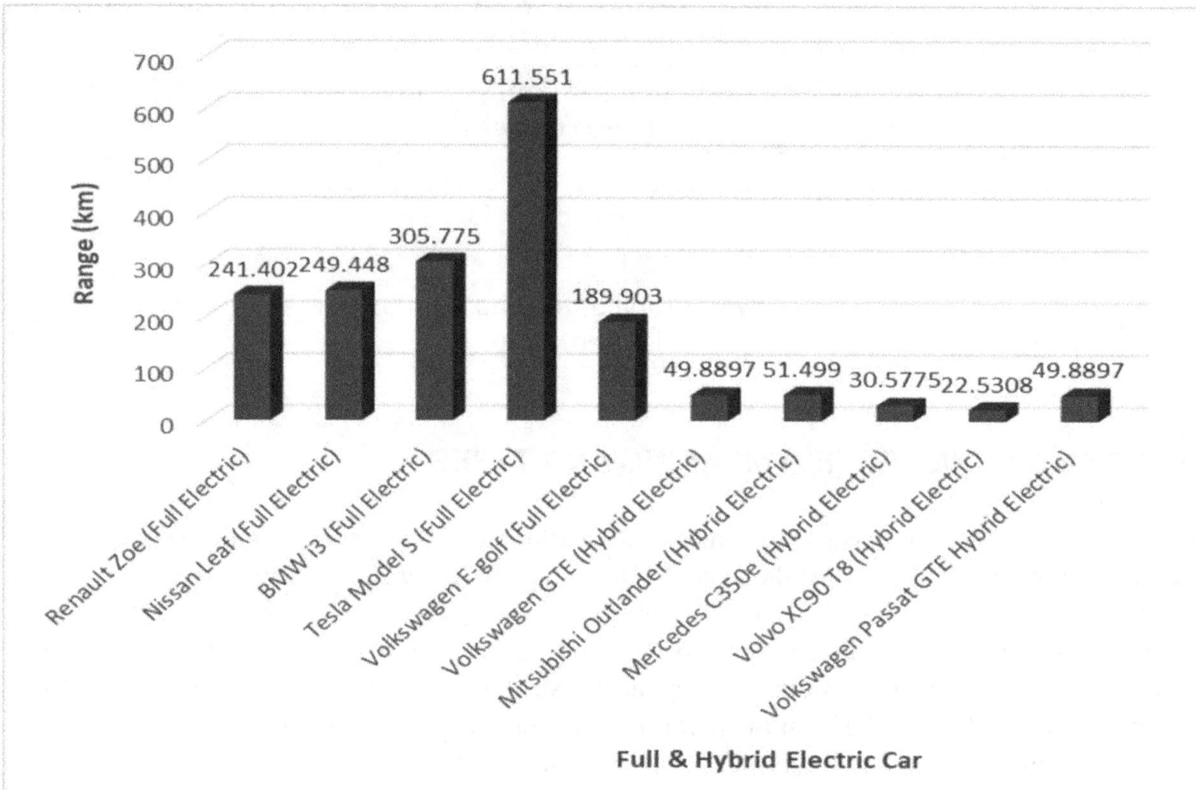

The EV depends, as well, on number of sub-components such:

1. The vehicle ranges
2. The vehicle prices
3. The vehicle weights
4. The vehicle motor torque

The Electric Drive Vehicle is a type of partial electric power or entire propulsion electric power. Furthermore, there are number of EV. These are:

1. EV known as Electric Drive Vehicle (EDV) which includes battery electric vehicle
2. Hybrid Electric Vehicle (HEV)
3. Fuel Cell Electric Vehicle (FCEV)
4. Plug-in Hybrid Electric Vehicle (PHEV)
5. Photo Voltaic Electric Vehicle (PVEV)

Figure 3. Ranges in miles versus Types of Full and Hybrid Electric Cars (Based on JOJU Solar 2018)

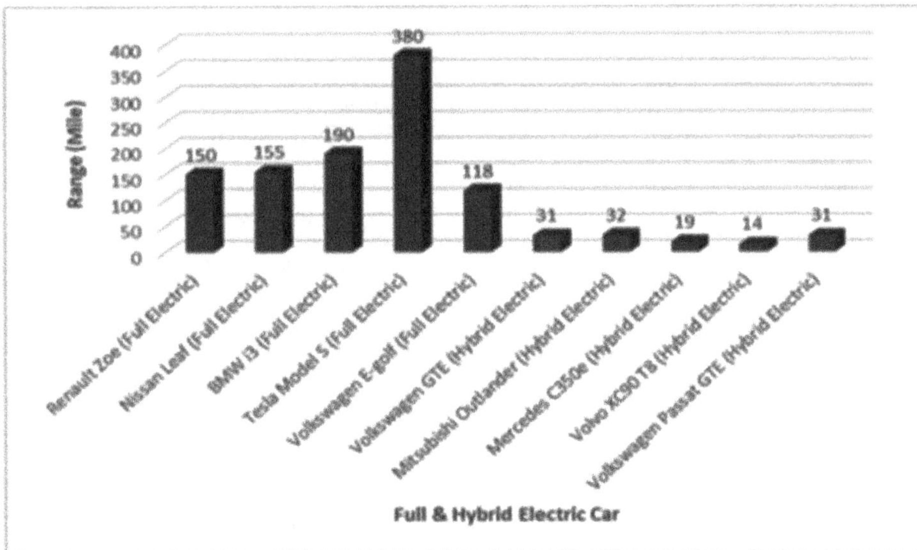

6 COMPARISON BETWEEN THE TYPES OF ELECTRIC VEHICLES

The Tesla (JOJU Solar 2018), the first company to offer high-production electric vehicles and a longer extension, was initially launched model S, with battery capacity of 40 kilowatt hours, 60 kWh and 85 kWh, where it depends on the model The latter is a total of about 480 km (300 miles). As for hybrid vehicles between electric and gasoline, they have as much as 3 to 5 kilowatt hours, as far as electricity is concerned, they can travel between 20 and 40 kilometers, but the gasoline engine guarantees a full range of conventional cars. For example, the Nissan Leaf can be approached, as the first-generation usable battery (JOJU Solar 2018), such as the Nissan Leaf, is about 20 kilowatt hours, giving it an estimated distance of about100 miles (160 km).

The cost of charging an electric car will depend on several factors (JOJU Solar 2018):

1. Battery capacity
2. Range for car
3. The cost of the electricity you buy.

Both Tables 1a and 2 illustrate the types of full and hybrid electric vehicles versus the battery capacity. In addition, these Tables transferred to two figures, respectively. Figure 4 represents Table 1, where Figure 5 represents Table 2.

Table 3 shows a comparison between different electric car's models. Some of these cars are 2020 model and others 2021 model (Electric Vehicle Database 2020).

Table 1. Full Electric Car and their Battery Capacity (Source: JOJU Solar 2018)

Full Electric Car	Battery Capacity (kWh)
Renault Zoe	22
Nissan Leaf	30
BMW i3	33
Tesla Model S	100
Volkswagen E-golf	24.2

Table 2. Hybrid Electric Car and their Battery Capacity (Source: JOJU Solar 2018)

Hybrid Electric Car	Battery Capacity (kWh)
Volkswagen GTE	8
Mitsubishi Outlander	12
Mercedes C350e	6.2
Volvo XC90 T8	9.2
Volkswagen Passat GTE	9

Figure 4. Full Electric Car and their Battery Capacity (Based on JOJU Solar 2018)

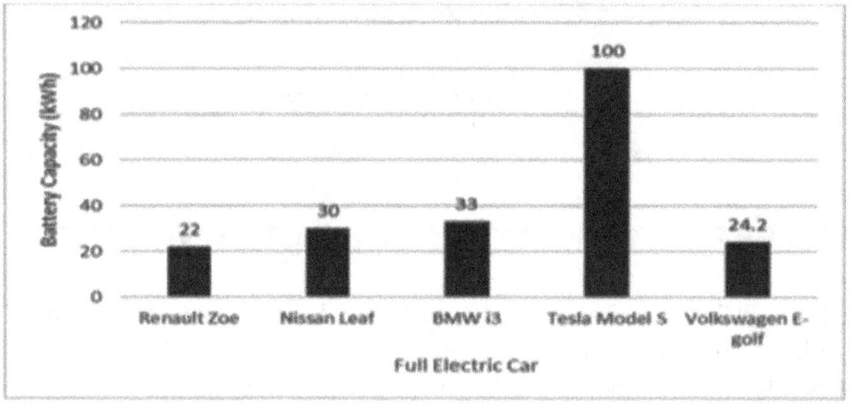

Figure 5. Hybrid Electric Car and their Battery Capacity (Based on JOJU Solar 2018)

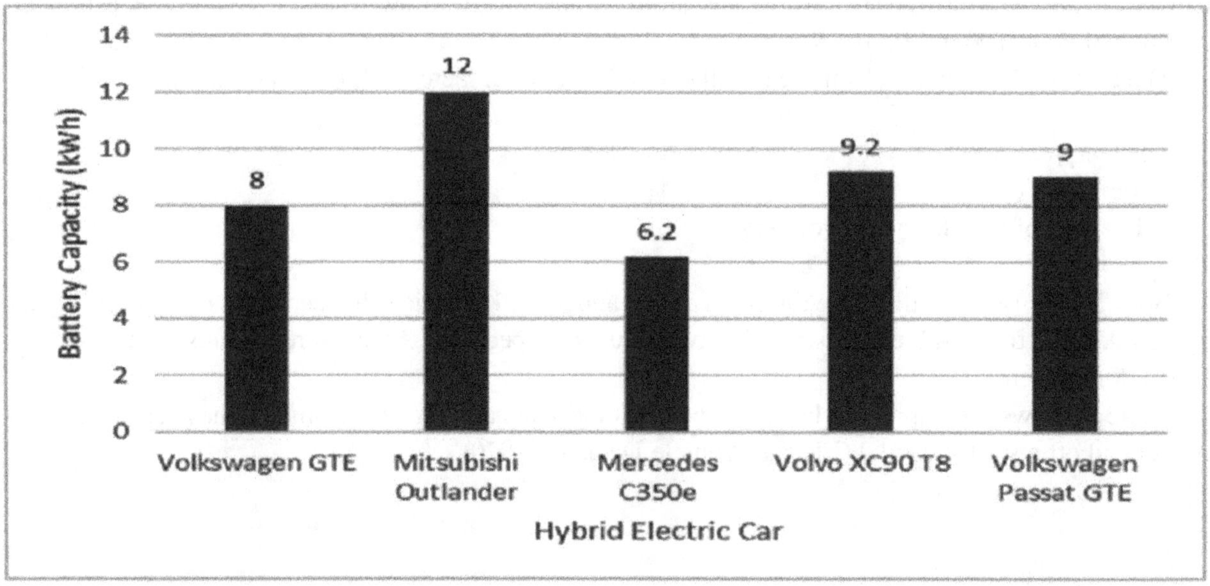

Table 3. Full Electric Car and their Battery Capacity (Source: Electric Vehicle Database 2020)

Car Type/Model	Raise Speed in 0 - 100	Top Speed	Range distance	Efficiency	Fastcharge
Tesla Model 3 Long Range Dual MotorTesla Model 3 Long Range Dual Motor Battery Electric Vehicle - 75 kWh	4.6 sec	233 km/h	450 km	161 Wh/km	940 km/h
Volkswagen ID.3 Pure (later this year) Battery Electric Vehicle - 48 kWh	10.0 sec	160 km/h	270 km	167 Wh/km	250 km/h
BMW iX3 (later this year) Battery Electric Vehicle - 80 kWh	6.8 sec	180 km/h	360 km	206 Wh/km	560 km/h
Polestar 2 Battery Electric Vehicle - 78 kWh	4.7 sec	210 km/h	400 km	181 Wh/km	620 km/h
Audi Q4 e-tron (concept) (from 2021) Battery Electric Vehicle - 82 kWh	6.3 sec	180 km/h	400 km	193 Wh/km	540 km/h
Honda e Battery Electric Vehicle - 35.5 kWh	9.5 sec	145 km/h	170 km	168 Wh/km	190 km/h
Peugeot e-208 Battery Electric Vehicle - 50 kWh	8.1 sec	150 km/h	275 km	164 Wh/km	420 km/h
Volkswagen e-Golf Battery Electric Vehicle - 35.8 kWh	9.6 sec	150 km/h	190 km	168 Wh/km	220 km/h
Tesla Model 3 Standard Range Plus Battery Electric Vehicle - 50 kWh	5.6 sec	225 km/h	310 km	153 Wh/km	650 km/h
Mercedes EQC 400 4MATIC Battery Electric Vehicle - 85 kWh	5.1 sec	180 km/h	370 km	216 Wh/km	440 km/h
Hyundai Kona Electric 64 kWh Battery Electric Vehicle - 67.5 kWh	7.9 sec	167 km/h	400 km	160 Wh/km	380 km/h
Nissan Leaf Battery Electric Vehicle - 40 kWh	7.9 sec	144 km/h	220 km	164 Wh/km	230 km/h
Porsche Taycan Turbo S Battery Electric Vehicle - 93.4 kWh	2.8 sec	260 km/h	375 km	223 Wh/km	780 km/h
BMW i4 (concept) (from 2021) Battery Electric Vehicle - 80 kWh	4.0 sec	200 km/h	450 km	178 Wh/km	650 km/h

7 FUTURE VISION OF ELECTRIC VEHICLES (EV) AND PROPOSED MODEL

The fuzzy logic system might be applied to the present case of EV under normal driving circumstances. There are some rules (Al-Hamad and Qamber 2019) need to go through before going through several parameters that need to be taken under consideration. The rule-based neural network applies a multi-layer network structure. The transfer functions with links represent a fuzzy logic system. Minimizing the neuro-fuzzy weight and updating vector are needed to be investigated as:

$$w^x = \frac{w_r}{\sum_{i=0}^{m} w_i} \tag{1.2}$$

finding a step size and weight's direction are driving a cost function through its minimum. The symbol w^x is known as firing strengths of the fuzzy rules. Furthermore, the cost function JN (.) can be approximated using the Taylor's expansion by the quadratic function:

$$J_N(w+\Delta w) = J_N(w) + \Delta w \frac{dJ_N(w)}{dw} + \frac{1}{2}\Delta w^T \frac{d^2 J_N(w)}{dw^2}\Delta w \tag{1.3}$$

where Δw is the weight vector update. Obtaining the differentiation of equation (1.3) with respect to Δw, where the result set to zero. By setting equation (1.3) to zero will minimize the equation to:

$$\frac{dJ_N(w)}{dw} = \frac{d^2 J_N(w)}{dw^2}\Delta w \tag{1.4}$$

$$g = -H\Delta w \tag{1.5}$$

where g and H represent the gradient and Hessian of $J_N(w_k)$, respectively.

Reducing conjugate gradient algorithm is helping to find the required solution of the equation. Therefore, the weight vector lies in a subset of the region. Therefore, the obtained algorithm is summarized as follows:

1. $k = 0.$
2. Set initial weight vector, w_k.
3. Determine gradient $g(w_k)$ and Hessian $H(w_k)$.
4. If $\|g(w_k)\| <\in$ then terminate.
5. Solve $H(w_k)\ \Delta w_k + g(w_k)$ using a conjugate gradient, ensuring $\| w_{k+1}\| \leq D_k$.
6. Starting with $\lambda_k = 1$, find λ_k where $w_{k+1} = w_k + \lambda_k \Delta w_k$.
7. Adjust the subset of the trust region radius (D_{k+1}) with the following heuristic:

$$\boldsymbol{D}_{k+1} = \begin{cases} 2\boldsymbol{D}_k & \boldsymbol{if}\ \lambda_k \geq 1 \\ \dfrac{\boldsymbol{D}_k}{3} & \boldsymbol{if}\ \lambda_k < 1 \end{cases} \tag{1.6}$$

8. Go to step (iv).

Implementing the previous algorithm, where the gradient and Hessian of the cost function $J_N(w_k)$ are needed. The weight vector w to be considered, where both first and second derivatives of the cost function $J_N(w_k)$ can be obtained. These derivatives can be formed as follows:

$$\frac{\partial J_N(w)}{\partial w_p} = \frac{1}{N}\sum_{k=1}^{N} 2\frac{\partial \hat{y}(w(k),w)}{\partial w_p}\left[y(k) - \hat{y}(w(k),w)\right] \tag{1.7}$$

$$\frac{\partial^2 J_N(w)}{\partial w_p \partial w_q} = \frac{1}{N}\sum_{k=1}^{N}\left[2\frac{\partial^2 \hat{y}(y(k),w)}{\partial w_p \partial w_q} y(k) - \hat{y}(w(k),w) + 2\frac{\partial \hat{y}(w(k),w)}{\partial w_p}\frac{\partial \hat{y}(w(k),w)}{\partial w_q}\right] \quad (1.8)$$

The model-gradient output with respect to the weights represented by the following:

$$\frac{\partial \hat{y}(x,w)}{\partial w_p} = \begin{cases} \mu^u_{A^i} \\ \left[\prod_{u=1,u\neq k}^{U} \sum_{i=1}^{P_u} \mu^u_{A^i} w^u_i \right] \mu^k_{A^j} \end{cases} \quad (1.9)$$

w_p refers to the ith weight of the uth tensor model.
w_q refers to the jth weight of the kth sub-model.

$$\frac{\partial^2 \hat{y}(x,w)}{\partial w_p \partial w_q} = \begin{cases} \left[\prod_{u=1,u\neq m\neq k}^{U} \sum_{i=1}^{P_u} \mu^u_{A^i} w^u_i \right] \mu^k_{A^j} \mu^m_{A^l} \\ 0 \end{cases} \quad (1.10)$$

w_p and w_q refer to the jth weight of the kth sub-model and l^{th} weight of the mth sub-model.

where the dependency of $\mu^u_{A^i}(\mathbf{x}_i)$ on the input vector $\boldsymbol{x}_i$ has been dropped to reduce the notation.

The parameters need to be considered are the sun irradiance, EV battery and battery size from the energy point of view. These parameters might be recovered by the solar PV (ON-OFF charging procedure), EV backup battery (ON-OFF battery) and grid availability charger. These will help the EV charging procedure, where if satisfied and solved the problems facing the EV. These points are shown and represented by the proposed model illustrated in Figure 1.6 under three layers. In case of emergency, the ON-OFF notation helps recover the previous layer to be replaced by the second layer.

As future vision, the presented model (Figure 6) is constructed. At the same time, in future the researchers might consider the charging case as follows:

1. Low charge
2. Medium charge
3. High charge

These three cases are taken into account when using the EV battery. In this case, the fuzzy criteria following the proposed rule in this section is applied to calculate the charging behavior of the EV battery. This phenomenon might be known as smart EV batter charged/replaced by one of the boxes in the second layer. The same phenomena might be applied to the battery size represented from the energy point of view. The solar PV is charged through the sun irradiance.

Figure 6. Proposed Fuzzy Logic Controller Evolution Model

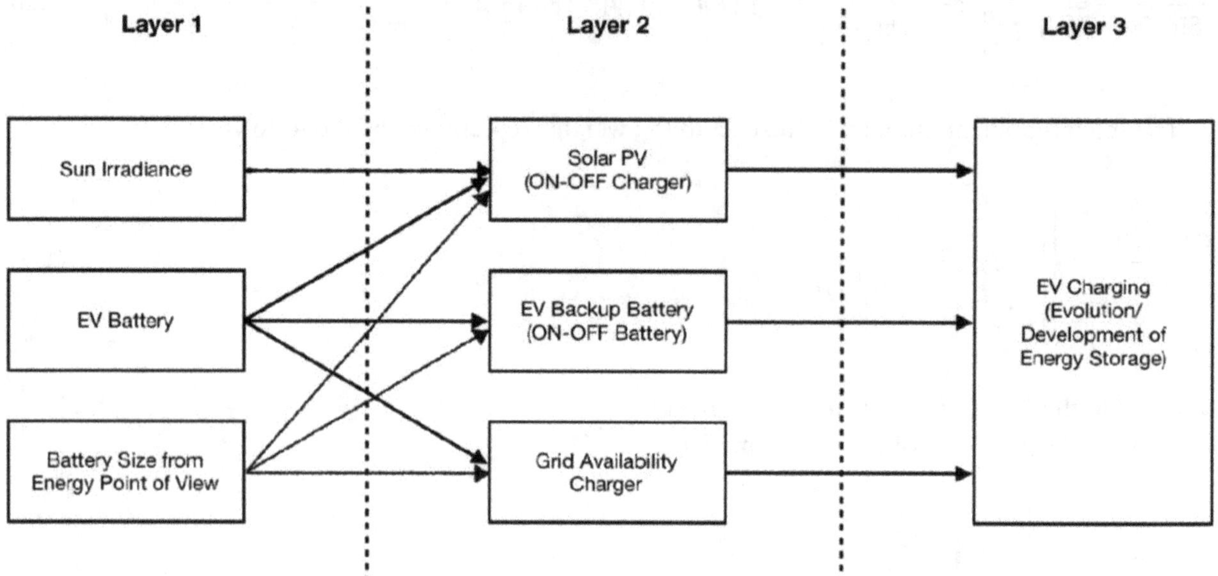

The fuzzy sets are initialized according to the known data of fuzzy variables in the first layer, where the weights (w) representing the truth-values of the rules in layer (2). The firing strengths of the fuzzy rules are normalized according to equation (1.2) and represented in layers (1) and (2). The quadratic function represented by equation (1.3) and derived from Newton Algorithm using Taylor's expansion to find the Neuro-fuzzy weight update vector. The result of differentiating equation (1.3) is set to zero to minimize equation (1.4). Both equations (1.9) and (1.10) represent the model's output with respect to the weights and represented in layer (3). With reference to the solar charge controlling, the Fuzzy Inference System (FIS) is applied. Three inputs are shown in layer (1) which are the sun irradiance, EV battery, and battery size from the energy point of view. The output of the model (Layer 3) is the EV charging which is the evolution per development of energy storage. Layer 2 is the obtained results module specifications required which is called defuzzification. This process of converting an output fuzzy set for a solution variable into a single value known as output, where this output is the EV charging module.

8 CASE STUDY

The ability of a vehicle to slip and flow through the air without creating much resistance is crucial when it comes to fuel economy (Douglas, J., 2019). The solar battery boasts and is impressive with approximately 34% efficiency rating, compared to an average 15% to 20% efficiency rating of most of its solar panels, where efficiency is the amount of sunlight that falls on the solar panel, through which it can be converted into electricity. Solar energy provides a great opportunity to harness the power of the sun to recharge the batteries. Therefore, combining both electricity and solar (as electric-solar vehicle) helping much the electric-solar vehicle in reducing the waiting time for battery-recharging. In addition, the lifetime of the battery is taken in consideration.

Therefore, the proposed Fuzzy Logic Controller Evolution model shown in Figure 6 represented by the first layer, where (Sun Irradiance block) is representing the amount of sunlight that falls on the solar panel. The Solar Panel block represented by the (Solar PV, the second layer), where the amount of sunlight (from the first layer) will pass to. This will result to charge the battery (Third layer block).

In case, it is started from the second block (EV Battery, First Layer) will be charged through (Solar PV, second layer) or replaced by (EV Backup Battery, Second Layer). Then, it will lead to the (Third layer block). Otherwise, it might be charged through the (Grid Availability, Second Layer) which leads to the third layer.

The next case, the (Battery Size, first layer) considered. The battery whatever its size considered, it will be charged through (Solar PV, second layer) which leads to the third layer, or replaced by (EV Backup Battery, Second Layer) and leads to the third layer. Otherwise, it might be charged through the (Grid Availability, Second Layer). Then it will lead to the (Third layer block).

9 GREEN ENVIRONMENT AND SMART GRID

Although a car engine can be a source of electric driving, the battery is considered to be the virtual source and soul of the meaning of an electric car (Reed 2011). If you are considering purchasing a new or refurbished electric car, consider the battery specifications and be aware of the other factors that ultimately affect performance. For a fixed charge system, this charge requires a power cable with the addition of two additional connectors at the bottom and the car outlet, as well as connectors to adjust the high voltage current of the fixed current with the charging stations to the car battery (Reed 2011). The electric vehicle is reducing CO_2 emissions and reducing pollution (EDF 2020). This is one of the benefits of EV as it is better for a clean environment. In addition, the EV of the production process contributes to creating a better environment. Electric vehicles can help reduce noise where it reaches the cities. This makes the EV much quieter than normal vehicles. Several number of electric vehicles types, as an example, were proposed in both sections 1.5 and 1.6.

In addition, the pure EV has no exhaust pipes. This means that it will significantly contribute to the reduction of air pollution (EDF 2020). As a result, carbon dioxide emissions disappear and are not emitted. This helps in creating a clean environment, clean street and allows cyclists and pedestrians to breathe fresh air.

Driving with an electric vehicle reduces CO_2 emissions by around 17% to 30% less than driving with petrol or diesel (EDF 2020). This has been confirmed by studies from the European Energy Agency.

At night, the sunlight reduced. This allows the battery to be recharged at home. At the same time, renewable energies can bring benefits, and energy is stored all day (EDF 2020). This approach can be useful when it saves additional time on utility bills and reduces the amount of energy required in the network.

For intelligent charging and discharge, the network manager manages the time and power of electric vehicles in the charge and discharge of batteries. There were costs and limits on the amount of electricity (Morais et al. 2014). Recording costs and discharge, reducing labor costs, reducing greenhouse gas emissions, or reducing costs can serve several purposes. The electric vehicle user determines the energy required to store the battery at specific times. This intelligent control is also integrated into the smart grid to help control the cost of charging an electric vehicle in the system.

The "smartness" of the power grid can be measured by browsing smart meters because smart meters provide information about the state of the local power grid and how EV chargers enter. An example of power generation to supply the EV is the PV panels, where the solar and electric vehicles depend on solar array that convert the sunlight into electricity. In addition, it is clear that the global market for electric vehicles (EVs) is growing. Therefore, the utilities and other power stations generators need to prepare at this stage for the increased loads. The reason behind this increment is the growing of Electric Vehicles.

The basis of how electricity is supplied to the grid, it always takes the shortest route to the point where it is needed (EV Charging Technology 2020). A mains charger charges the device and draws electricity from the car battery and simply pass it back to the mains, where it follows its path to the nearest point where it is needed.

In 2013, the Finnish energy sources discussed future transport (EV Charging Technology 2020). The future of electric transport is clear. Together, the 18 energy utilizations have created a company to build a national charging network called Verta where it was born. Before launching the services in Finland, the first international customers found, and the solutions are designed to support EV charging vehicles worldwide. Virta allows the customers to take charge of electric vehicles using renewable energies as well.

Currently, Nissan is outperforming other automakers by marketing models of V2G-compatible cars (EV Charging Technology 2020). Any Nissan Leafs on the market can be discharged with net vehicle stations. The ability to support V2G for vehicles is hopefully join the vehicle-to-grid, and later many more manufacturers will soon join the Auto Network Compatibility Club. For example, Mitsubishi has also announced plans to market the V2G with the Outlander PHEV. In the United States, the EV charging company Smart electric vehicle charging experiences are powered by Virta.

The abbreviation EVSE stands for Electric Vehicle Supply Equipment (Nikki Gordon-Bloomfield 2010). This is the name of the company for the charging station of electric cars. The terms "EVSE", "EV Charging Station" and "Station Charging" mean the same thing and can be replaced.

There was a time when an electric car could be connected to almost all the electricity discovered by the owner (Nikki Gordon-Bloomfield 2010). With a certain ingenuity and sophisticated design, an electric car can be connected almost anywhere.

Several numbers of stories heard from members of the National Electric Drag Racing Association (NEDRA) which connected to facility's 240V, 50A business centers for a quick charging, and Toyota RAV4 EV owners bought special chargers to access cars (Nikki Gordon-Bloomfield 2010). The ability of Toyota RAV4 EV owners have the chance to plug in J1772 Charging stations and even owners of Tesla companies that carry almost all viable electrical converters so that they can charge from 110 V, 15 A to 240 V, 70 A.

In 2017, Dubai police added an electric car - the BMW i3 - fully electric to their fleet (Leyal Khalife 2020). The fact is, Tesla opened Dubai's first trade fair in the Middle East and has more than 200 charging stations across the United Arab Emirates. If Dubai can guarantee that 75% of its energy will come from clean sources by 2050, it would have the smallest carbon footprint in the world.

10 CONCLUSION

It can be concluded that the EV helps for a better environment. In addition, it is recommended to increase the production of the EV in future and using the PV to produce the electricity to charge the EV. The EV helps improve the air quality in cities, towns, and villages. This means it helps significantly in improving the air quality in the world. One EV on the road might save an average of 1.5 million grams of carbon dioxide in one-year time. The EV driving creates a more peaceful environment for the world.

The process of accelerating the deployment of renewable energy and the electric vehicle will fuel economic growth. In addition, the renewable energy will create new employment opportunities, enhance human welfare, and contribute to a safe future climate.

The proposed model is investigated where it can be applied with the considered parameters. The solar energy helps significantly in the model, where it saved time in charging the battery of the EV. For this reason, the charger has ON-OFF switch. This type of switch helps the continuation of driving the EV without stopping for charging the EV battery. At the same time, the EV backup battery (as ON-OFF battery switch) will reduce the waiting time of charge by the time of replacement which is reduced to approximately five minutes.

Using the PV to recharge the EV battery (ON-OFF charge) is considered as self-consumption battery. It will also improve the economic achievement of the retired EV battery sufficiently, where it helps to upgrade the second life use of the EV battery and protect the environment.

The EV large scale to power system becomes a challenge as presented to deal with EV charging. It will be considered and used in an optimal way as a flexible resource. As a future development of EV, the EV battery becomes as a required charging key requirement. Furthermore, the EV in the coming future expected to lead and to be a challenge in manufacturing the vehicles.

ACKNOWLEDGMENT

The art works and the modeling of the data on the initial manuscript satisfied by Ms. Aysha Isa Qamber. The authors would like to thank her for the great help during the preparation of this chapter.

REFERENCES

Ahmad, F., Alam, M., Shariff, S., & Krishnamurthy, M. (2019). *A Cost-efficient Approach to EV Charging Station Integrated Community Microgrid. IEEE Transactions on Transportation Electrification*. doi:10.1109/TTE.2019.2893766

Al-Hamad & Qamber. (2019). GCC electrical long-term peak load forecasting modeling using ANFIS and MLR methods. *International Arab Journal of Basic and Applied Sciences, 26*, 269–282. doi:10.1080/25765299.2019.1565464

Al-Hamad, M., & Qamber, I. (2018). The GCC Countries Power Trading Modeling and Challenges. In *Saudi Arabia Smart Grid Conference on Smart Grids and Green Energy (SASG 2018)*, Jeddah, Saudi Arabia.

Al-Shariff, S., Alam, M., Ahmad, Z., & Ahmad, F. (2019). Smart Transportation System: Mobility solution for Smart Cities. *Second Smart Cities Symposium (SCS 2019, University of Bahrain).*

Ashique, R., Salam, Z., Bin Abdul Aziz, M., & Bhatti, A. R. (2017). Integrated photovoltaic-grid dc fast charging system for electric vehicle: A review of the architecture and control. *Renewable & Sustainable Energy Reviews, 69*, 1243–1257. doi:10.1016/j.rser.2016.11.245

Center for Sustainable Energy. (2020). *Electric Vehicle Charging 101(Learn more about different charging options for electric vehicles (EVs)).* https://calevip.org/electric-vehicle-charging-101#:~:text=One%20distinction%20between%20these%20three,between%20200%20and%20600%20volts

Charging Technology, E. V. (2020). *Vehicle-to-Grid: Everything you need to know, V2G technology.* Liikennevirta Oy (Ltd.). https://www.virta.global/vehicle-to-grid-v2g

Deza, A., Huang, K., & Mete, M. R. (2020). Charging station optimization for balanced electric car sharing. *Discrete Applied Mathematics.* Advance online publication. doi:10.1016/j.dam.2020.01.042

Dixon & Bell. (2020). Electric vehicles: Battery capacity, charger power, access to charging and the impacts on distribution networks. *eTransportation, 4*, 1-18. doi:10.1016/j.etran.2020.100059

Douglas, J. (2019). *Tesla's Musk says solar panels on cars make little sense, but that's not stopping Toyota, Hyundai.* CNBC. https://www.cnbc.com/2019/09/28/teslas-musk-rejects-solar-on-cars-its-not-stopping-toyota-hyundai.html

EDF. (2020). *Electric cars and the environment.* Électricité de France S.A. https://www.edfenergy.com/electric-cars/environment

El-Taweel, N., Khani, H., & Farag, H. (2019). *Analytical Size Estimation Methodologies for Electrified Transportation Fueling Infrastructures Using Public–Domain Market Data.* IEEE Transactions on Transportation Electrification. doi:10.1109/TTE.2019.2927802

Electric Vehicle Database. (2020). *2020 EV Database - v4.2.* https://ev-database.org/#sort:path~type~order=.rank~number~desc|range-slider-range:prev~next=0~1200|range-slider-acceleration:prev~next=2~23|range-slider-topspeed:prev~next=110~450|range-slider-battery:prev~next=10~200|range-slider-eff:prev~next=100~300|range-slider-fastcharge:prev~next=0~1500|paging:currentPage=0|paging:number=9

Electrical Technology. (2020). *How to Calculate the Battery Charging Time & Battery Charging Current – Example.* https://www.electricaltechnology.org/2013/03/easy-charging-time-formula-for.html

Ferdowsi, M. (2007). Plug-in Hybrid Vehicles - A Vision for the Future. *Proceedings of the IEEE Vehicle Power and Propulsion Conference*, 457-462. 10.1109/VPPC.2007.4544169

Hanauer, D. (2018). Technical Note: Mode 2 Charging—Testing and Certification for International Market Access. *World Electric Vehicle Journal*, *9*(2), 26. doi:10.3390/wevj9020026

IEC 62196-1:2014, Plugs, socket-outlets, vehicle connectors and vehicle inlets - Conductive charging of electric vehicles - Part 1: General requirements. Retrieve Web Page: https://webstore.iec.ch/preview/info_iec62196-1%7Bed3.0%7Db.pdf

Khalife. (2020). *Brighter Future: UAE is investing in electric cars, green oasis & solar parks.* Retrieve Web Page: https://stepfeed.com/brighter-future-uae-is-investing-in-electric-cars-green-oasis-and-solar-parks-6928

Khan, S., Ahmad, A., & Ahmad, F. (2017). *A Comprehensive Review on Solar Powered Electric Vehicle Charging System.* Smart Science. doi:10.1080/23080477.2017.1419054

Kurdve, M., Zackrisson, M., Johansson, M., Ebin, B., & Harlin, U. (2019). Considerations when Modelling EV Battery Circularity Systems. *Batteries*, *5*(2), 1–20. doi:10.3390/batteries5020040

Mehrjerdi, H., & Hemmati, R. (2019). Electric vehicle charging station with multilevel charging infrastructure and hybrid solar-battery-diesel generation incorporating comfort of drivers. *Journal of Energy Storage*, *26*, 1–9. doi:10.1016/j.est.2019.100924

Mohamed, N., Aymen, F., Ben Hamed, M., & Lassaad, S. (2019). *Analysis of battery-EV state of charge for a dynamic wireless charging system*. Energy Storage. doi:10.1002/est2.117

Morais, H., Sousa, T., Vale, Z., & Faria, P. (2014). Evaluation of the electric vehicle impact in the power demand curve in a smart grid environment. *Energy Conversion and Management*, *82*, 268–282. doi:10.1016/j.enconman.2014.03.032

Napoli, G., Polimeni, A., Micari, S., Andaloro, L., & Antonucci, V. (2020). Optimal allocation of electric vehicle charging stations in a highway network: Part 1. Methodology and test application. *Journal of Energy Storage*, *27*, 1–9. doi:10.1016/j.est.2019.101102

Napoli, G., Polimeni, A., Micari, S., Dispenza, G., & Antonucci, V. (2019). Optimal allocation of electric vehicle charging stations in a highway network: Part 2. The Italian case study. *Journal of Energy Storage*, *26*, 1–23. doi:10.1016/j.est.2019.101015

Nikki Gordon-Bloomfield. (2010). *What Is EVSE And Why Does Your Electric Car Charger Need It?* https://www.greencarreports.com/news/1050948_what-is-evse-and-why-does-your-electric-car-charger-need-it

Qamber, I. (2020). Power System Control and Reliability: Electric Power Design and Enhancement. AAP & CRC, Taylor & Francis.

Reed, P. (2011). *Electric-Vehicle Battery Basics*. https://www.myev.com/research/ev-101/electric-vehicle-battery-basics

Roche & Sabrià. (2015). An Accessible Pre-design Calculation Tool to Support the Definition of EV Components. *World Electric Vehicle Journal, 7*, 1-13.

Shariff, S., Iqbal, D., Alam, M., & Ahmad, F. (2019). A State-of-the-Art Review of Electric Vehicle to Grid (V2G) technology. *IOP Conf. Series: Materials Science and Engineering, 561*. 10.1088/1757-899X/561/1/012103

Shemami, M., Alam, M., Asghar, M., & Shariff, S. (2019). Adaptive Neuro-Fuzzy Inference System (ANFIS) for Optimization of Solar Based Electric Vehicle-to-Home (V2H) Fuzzy Inference System (FIS) Controller. *IEEE Conference Paper*. 10.1109/ITEC.2019.8790579

Solar, J. O. J. U. (2018). *EV Facts and Figures, Batteries and Smart Grids*. https://www.jojusolar.co.uk/ev-facts-and-figures/

Tang, Y., Zhang, Q., Li, H., Li, Y., & Liu, B. (2019). Economic Analysis on Repurposed EV batteries in a Distributed PV System under Sharing Business Models. *Energy Procedia*, *158*, 4304–4310. doi:10.1016/j.egypro.2019.01.793

Un-Noor, F., Padmanaban, S., & Mihet-Popa, L. (2017). *A Comprehensive Study of Key Electric Vehicle (EV) Components, Technologies, Challenges, Impacts, and Future Direction of Development*. doi:10.20944/preprints201705.0090.v1

Xiang, Y., Hu, S., Liu, Y., Zhang, X., & Liu, J. (2018). Electric vehicles in smart grid: a survey on charging load modelling. IET Smart Grid. doi:10.1049/iet-stg.2018.0053

Chapter 2
Proposed Power Systems Planning in Indian Scenario for Integrating EV Charging Infrastructure

Sanchari Deb
School of Engineering, University of Warwick, Coventry, UK

Sulabh Sachan
https://orcid.org/0000-0003-0309-5001
Department of Electrical Engineering, Indian Institute of Technology, Kanpur, India

ABSTRACT

The growing concern about fossil energy exhaustion, air pollution, and ecological deprivation has made electric vehicles (EVs) a practical option in contrast to combustion engine-driven vehicles. In any case, driving extent uneasiness is one of the innate inadequacies related with EVs. Massive integration of EV charging load into the power system may be a threat to the distribution network. Spontaneous situation of charging stations in the distribution system and uncoordinated charging will augment the load demand thereby resulting in voltage instability, deterioration of reliability indices, harmonic distortions, and escalated power losses. This chapter will concentrate on breaking down the effect of EV chargers on the working parameters, for example, voltage dependability, unwavering quality, and force misfortune. The examination will be completed on standard test systems. The discoveries of the proposed part will evaluate the effect of EV charging load on the working parameters of the distribution system and help in proposing a framework for charging station planning.

DOI: 10.4018/978-1-7998-6858-3.ch002

1. INTRODUCTION

Electric Vehicles are a sustainable and environmentally friendly substitute to grab the problems of air pollution as well as energy crisis. There is need of charging the EVs in the charging stations in the wake of voyaging a specific separation and hence, EV adoption calls for development of charging stations. Massive integration of EV charger load to the power distribution system may be a threat to the smooth operation of the network (Leou et al., 2013)-(Zhang et al., 2019). The most visible and pronounced effect of EV charger integration is increase in evening peak load as most of the EV drivers charge their EVs after returning home in the evening. Research works concluded that location of charging spot at feeble points of the network will increase the power losses by 85% (Deb et al., 2018a). Ola Mobility concluded that Indian power grid is not equipped to support EVs and overhaul of the power system is required for adoption of EVs (QZ, n.d.). In this context, this work attempts to study the effect of EV charger integration on various operating constraints of power system such as peak power, voltage profile, reliability, power losses in a systematic approach supported by case studies. Further, some recommendations for charger placement are also presented in this work.

Apart from that, Section 2 depicts the evaluation of prevailing research works on effect of EV charger load on power grid. Section 3 presents an overview of distribution network operating parameters. Section 4 presents some case studies on effect of charger placement in distribution grid. Section 5 presents numerical analysis thereby reporting the effect of charger integration on 33 bus distribution grid. Section 6 present recent trends of charging infrastructure planning in India. Conclusion work is depicted in section 7. An outline of the chapter is presented in Figure 1.

Figure 1. Outline of the chapter

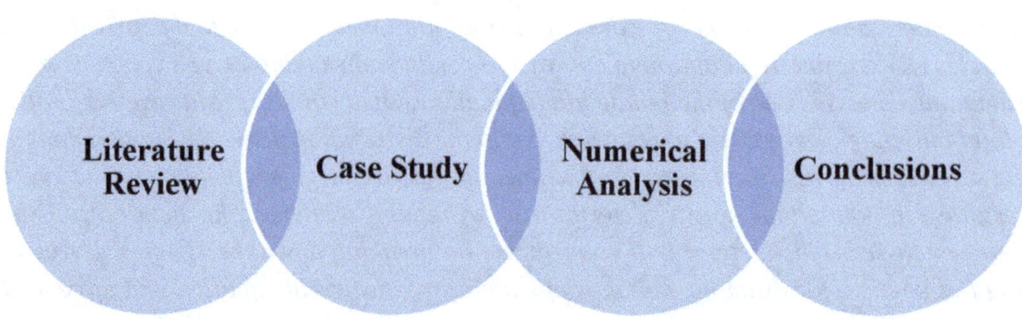

2. LITERATURE REVIEW

In recent centuries, a lot of research works have analysed the effect of EV charger integration on grid. In (Sachan & Kishor, 2016), a study has been conducted regarding effect of EV charger on distribution grid of California considering real time charging profiles of 64 EVs at changed perception rate of EVs. The research work concluded that most of the EV charging occurred between 3 PM-5 PM and suggested that the charging request during daytime can be fed utilizing solar power. In (Sachan & Kishor, 2018), researchers have studied effect of EV charger on practical grid of Ostrava considering charging profiles of different EV models. The research work showed that power loading and harmonics increased after introduction of EV charger load. In (Gómez et al., 2017), the authors have studied the effect of EV load on grid of Cuenca. The analysis showed that in the worst situation, the voltage harmonic distortion changed by 1.36%. Thus, re-dimensioning of the distribution network is necessary because the capacity of the system could not withstand these new burdens. In (Gunes et al., 2018), (Sachan & Adnan, 2018) the effect of EV charger load on reliability indices of power grid is evaluated. In (Deb et al., 2018a), (Deb et al., 2019) the researchers studied the effect of EV charger on 33 bus and 69 bus grid. In (Karmaker et al., 2019), the researchers studied the effect of EV charger load on a practical power grid of Bangladesh. Simulation results presented in the aforesaid work clearly showed that power quality degraded after introduction of charger load. In (Manríquez et al., 2020), the authors have considered the effect of EV load while planning the expansion of power grid. In (Pinto et al., 2018), the authors have analyzed the impact of EV charger on power quality of a low voltage urban grid considering Brazilian standard. In (Susuki et al., 2017), the researchers have studied the effect of EV charger considering spatial temporal charging pattern. In (Pothinun & Premrudeepreechacharn, 2018), the researchers have studied the effect of EV charger on THD of the Thailand grid. The effect of EV charger in loading of transformer is reported in (Satarworn & Hoonchareon, 2017)-(Dong et al., 2018).

In comparison to the ref (Sachan & Kishor, 2016)-(Dong et al., 2018) the contributions of this chapter are listed below:

- This chapter reports global case studies related to impact of EV charger on power grid
- This chapter quantifies the impact of EV charger load on voltage profile, reliability and power loss of standard test network
- Further, some recommendations for charger placement are provided

3. OVERVIEW OF DISTRIBUTION NETWORK OPERATING PARAMETERS

The working parameters of distribution grid must be within breaking point for smooth activity of the framework. An overview of the functioning constraints of the grid is presented in Table 1.

Table 1. Functioning Indices of Distribution network

Parameter	
Voltage Steadiness	Power losses
SAIFI	Harmonics
SAIDI	AENS
CAIDI	

4. CASE STUDIES

Globally, a number of case studies have quantified the effect of EV charger burden positioned at power distribution grid. Selected case studies are reported in this section. The system methodology adopted in the case studies is as shown in Fig. 2. The term target parameter in the step 4 of the methodology refers to the parameter which is under investigation. The target parameter can be load, reliability indices, voltage profile. *etc*.

Figure 2. System Methodology adopted in the case studies

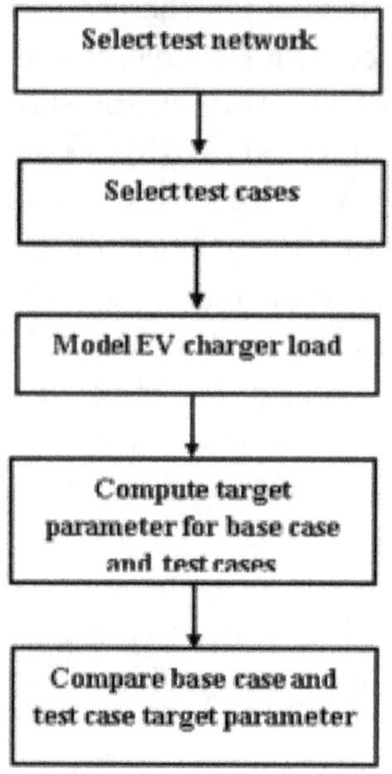

4.1. Case Study on Effect of EV Charger on Sub-Regions of South East Reliability Council (SERC) by Advanced Meteorological Systems

A gathering beneath the direction of Baron Advanced Climatic Systems analyzed the impact of EV charger load on different sub regions of SERC in the year 2006 (Hadley, 2006). The group quantified the effect of Vehicle charger on generation level reliability indices for different scenarios. Power system engineers' practices an index termed as Loss of Load Probability (LOLP), which characterizes the probability that aggregated load won't be adequate to fulfill need (Hadley, 2006), (Ghosh et al., 2017). The effect of EV charger load on LOLP for various situations is accounted for in Table 2. Table 2 depicts that the LOLP increments with the presentation of EV charger load. Additionally, the LOLP for late evening charging is more than night charging.

4.2. Impact of EV charger on Residential Feeder by Forum of Regulators (FOR)

Forum of Regulators (FOR) quantified the impact of fast EV charger on residential feeder for different EV charging scenario (Forum of Regulators, n.d.). The details of the test system on which analysis was performed can be found in (Forum of Regulators, n.d.). The different test cases considered in the study are listed in Table 3. It is realized that the test cases considered different values of EV load with different battery capacity. Table 4 reports the voltage drops and increase in maximum demand across scenarios compared to the baseline scenario for Residential Feeder. It is observed that case 3 is most critical for smooth operation of power system.

Table 2. Effect of EV charger load on LOLP of changed sub regions of SERC (Hadley, 2006)

Scenario	LOLP
Base case	0.166
6-8 PM (1.4 kW)	0.189
8-12 PM (1.4 kW)	0.171
6-8 PM(2 kW)	0.194
8-12 PM (2 kW)	0.178
6-8 PM (6 kW)	0.217
8-12 PM (6 kW)	0.185

Table 3. Test cases considered by FOR (Forum of Regulators, n.d.)

Case	Description
1	Base case with baseline load (10 MW)
2	Baseline Load + 4 MW EV load (85 kWh battery capacity)
3	Baseline Load + 8 MW EV load (85 kWh battery capacity)
4	Baseline Load + 8 MW EV load (24 kWh battery capacity)
5	Baseline Load + 4 MW EV load (24 kWh battery capacity)

Table 4. Voltage drops and increase in Maximum Demand across scenarios compared to the baseline scenario for Residential Feeder (Forum of Regulators, n.d.)

Case	Voltage drop (%)	Increase in maximum demand (%)
2	1.6	13
3	2	36
4	1.8	35
5	1.5	11

4.3. Effect of EV charger on Peak load of five Indian cities

A research group of IIT Kanpur studied the effect of EV load on peak load of five large metropolises of India such as Kolkata, Chennai, Delhi, Mumbai, and Bengaluru (Sharma et al., 2019). They forecasted the number of different EV segments for the year 2025 and 2030 for the five cities as shown in Table 5. Further, the effect of EV charger on peak demand of five cities for changed circumstances described in Table 6. The peak load of the five cities for the year 2025 and 2030 for the test circumstances described in Table 6 is presented in Table 7.

5. NUMERICAL ANALYSIS

The effect of EV charger load on 33 bus distribution grid is tested for different circumstances reported in Table 8. There is necessity of thoroughly studying the test network before planning the placement of charging stations. The 33 bus distribution network is depicted in Fig 3. There are logical reasons for selection of circumstances cited in Table 8. It is agreed that Distribution Company would not approve location of charging stations at feeble node of the distribution grid. However, such cases are considered in order to quantify the severity of citing fast charging stations at the feeble node. From the perspective of power system engineers it may seem that some of the cases are impractical. However, system planners need to have a complete 360 degree overview of the network and hence, all the cases considered are necessary.

Table 5. Forecasted number of different EV segments for the year 2025 and 2030 (Sharma et al., 2019)

City	Year	Type of EV segment					
		2W	E rickshaw	3 W	Commercial 4W	Heavy Commercial	Non Commercial 4W
Delhi	2025	2621226	122965	34485	25264	9797	661408
	2030	7015582	128765	88026	64488	25008	1688286
Mumbai	2025	1441024	No data	79005	23044	12497	434155
	2030	3678306		183177	58823	27295	1217851
Kolkata	2025	639110	No data	6015	4094	51012	67313
	2030	1579019		13458	15172	96333	166310
Chennai	2025	1363641	No data	30648	14286	12903	156589
	2030	3480781		78232	34763	28493	345775
Bengaluru	2025	2292800	No data	30648	19986	14269	279012
	2030	7055515		78232	51015	36422	712196

Table 6. Different test cases (Sharma et al., 2019)

Case	Description for charging
1	90% of total EVs @ home and rest 10% @ charging station (Sharma et al., 2019)
2	70% of total EVs @ home and rest 30% @ charging station (Sharma et al., 2019)
3	50% of total EVs @ home and rest 50% @ charging station (Sharma et al., 2019)
4	All EVs @ charging station (Sharma et al., 2019)

It is considered that the fast charging stations consume 50 kW power (Deb et al., 2018a). The impact of EV charger load on various working parameters of distribution system, for example, voltage steadiness, power quality, losses, and VRP index is comprehensively analysed. Fig. 4 represents the voltage profile of each buses of 33 bus network for the test cases reported in Table 8 It is observed that the voltage profile degraded after introduction of EV charger load. For case 5, where charger is placed at bus 14, the voltage fall below the critical value off acceptable voltage. Fig 5-8 shows the impact of EV charger load on SAIFI, SAIDI, CAIDI, and AENS respectively. The degradation of reliability indices is obvious from the aforementioned figures. Further, it is perceived that the degradation of reliability indices is more for case 3 as compared to case 4 confirming that distributing the charging stations between two buses is advantageous. Fig 9 quantifies the effect of EV charger load on loss of the network.

Table 7. Impact of EV charger load on peak load of five Indian cities (Sharma et al., 2019)

Case	Peak without EV load (MW)		Peak with EV load (MW)	
	Year			
	2025	2030	2025	2030
Delhi				
1	8196	9734	11092	17865
2			10828	19553
3			10917	19481
4			12153	23992
Mumbai				
1	4514	5248	6333	10537
2			6159	10291
3			6309	11356
4			7584	16075
Kolkata				
1	2629	3126	4321	6896
2			4314	6829
3			4287	6763
4			4220	6596
Chennai				
1	4188	4869	5877	8962
2			5815	8824
3			5752	8686
4			5596	8511
Bengaluru				
1	5009	5824	7034	12204
2			6922	11919
3			7020	11745
4			7834	13208

Figure 3. 3 bus Distribution Network

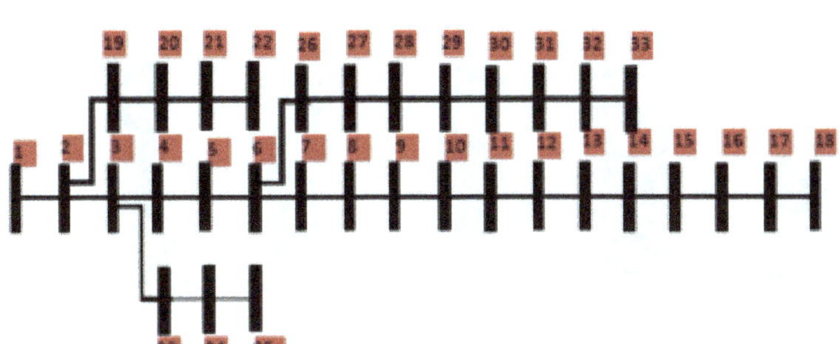

Table 8. Test cases deliberated for exploration

Case	Explanation
1	Base case with no EV load
2	1 FCS is cited at bus 2
3	2 FCS is cited at bus 2
4	1 FCS is cited at bus 2 and bus 19
5	1 FCS is cited at bus 14

Figure 4. Impact of EV charger load on voltage profile of 33 bus distribution network

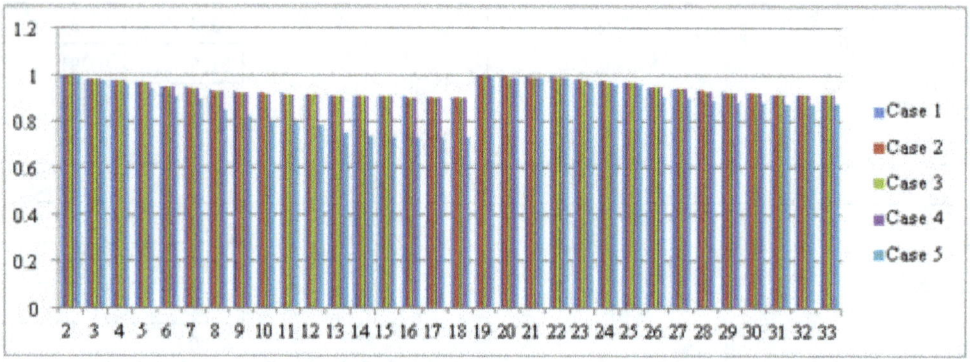

Figure 5. Effect of EV charger load on SAIFI of 33 bus distribution network

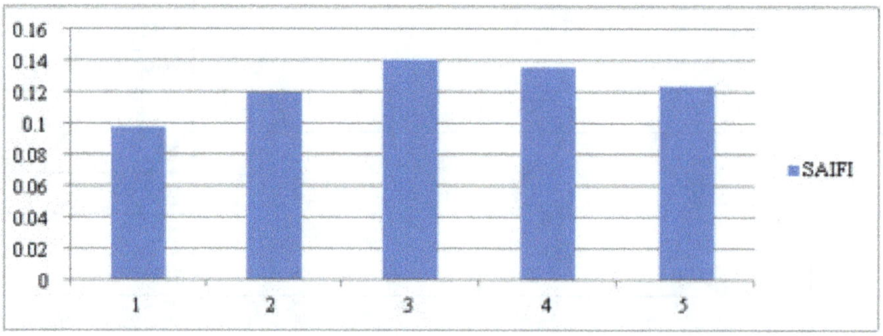

Figure 6. Impact of EV charger load on SAIDI of 33 bus distribution network

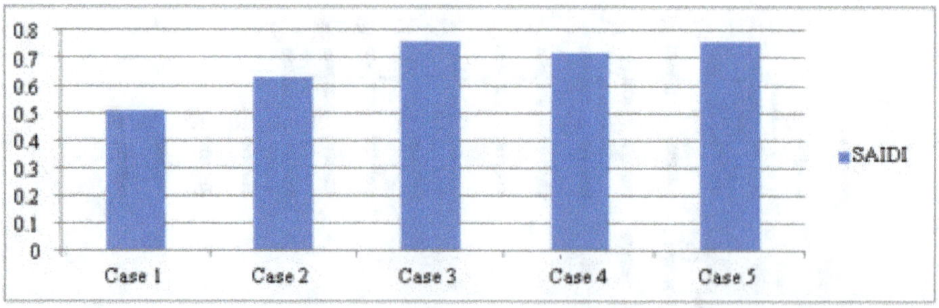

Figure 7. Impact of EV charger load on CAIDI of 33 bus distribution network

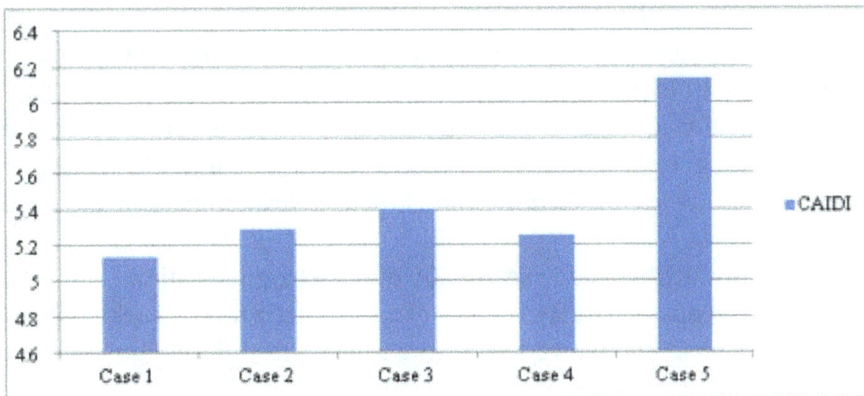

Figure 8. Impact of EV charger load on AENS of 33 bus distribution network

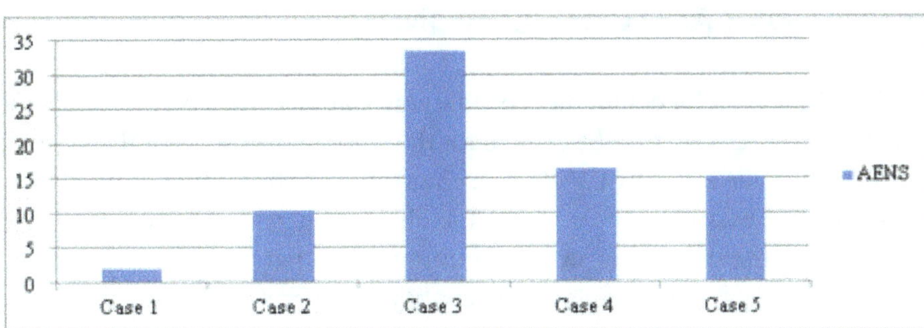

Figure 9. Impact of EV charger load on power loss of 33 bus distribution network

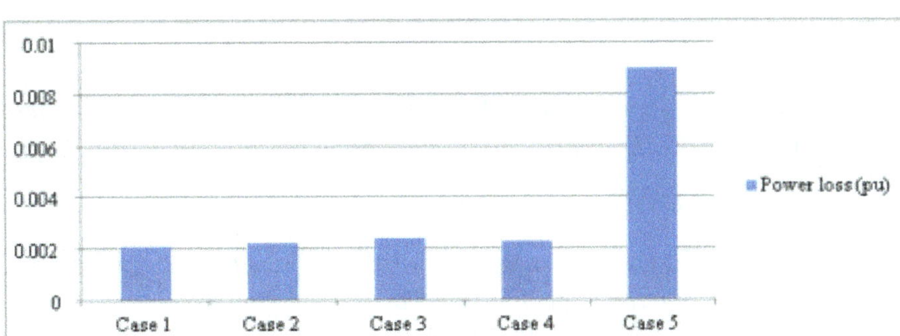

6. RECENT TRENDS AND DEVELOPMENTS IN INDIA

The shift from ICE vehicles to EVs is still under process in India. Despite having a low EV penetration (0.1% of total cars sold) many companies have plans to spend in charging station development because of the following - (Deb et al., 2018b)

1. India released the 'National Electric Mobility Mission Plan (NEMMP) 2020' in 2013 to address national energy security, vehicular pollution, and growth of domestic manufacturing capabilities. Further, as a consequence of its commitment to the Paris Agreement, the Government of India has plans to make a 30% shift to EVs by 2030.
2. Absence of charging infrastructure is a barrier abstaining Indians from purchasing EVs. Thus, development of charging infrastructure is needed to boost the Indian EV market.
3. The milestones in the history of charging infrastructure development in India are- (Deb et al., 2018b)
4. Central government has deliberated to set up 206 public charging stations.
5. NTPC is planning to set up of charging stations at a subsidized rate of 1 lakh rupees (1300 Euros).
6. TATA Power Delhi is planning to invest 100 crore rupees (13 million Euros) for establishment of 1000 charging stations across Delhi.
7. Ola has invested 50 crore rupees (6.7 million Euros) for EV and charging infrastructure in Nagpur.

7. RECOMMENDATIONS AND CONCLUSIONS

Wide spread adoption of EVs call for deployment of charging stations. Massive and unplanned integration of EV charger load to the power grid may be critical for the smooth operation of the system. This chapter made an attempt to quantify the effect of EV load on changed operating constraints of the grid. Selected global case studies that have examined the effect of EV charger on grid are also reported in this chapter. Simulation results indicate the severity of improper citing of CS particularly at the feeble buses of the grid. It is expected that findings of the chapter will quantify the effect of EV load on working constraints of grid and help in grid planning in existence of CS load. Our future work will deal with some other aspects associated with EV charging such as V2G scheme, smart charging, and EV as component of VPP.

REFERENCES

Anastasiadis, A. G., Kondylis, G. P., Polyzakis, A., & Vokas, G. (2019). Effects of Increased Electric Vehicles into a Distribution Network. *Energy Procedia, 157*, 586–593. doi:10.1016/j.egypro.2018.11.223

Arrillaga, J., & Watson, N. R. (2004). *Power system harmonics*. John Wiley & Sons.

Bhadra, J., & Chattopadhyay, T. K. (2015). Analysis of distribution network by reliability indices. *Proceedings of the International Conference on Energy, Power, and Environment: Towards Sustainable Growth (ICEPE)*, 1–5. 10.1109/EPETSG.2015.7510094

Billinton, R., & Alen, R. (1994). *Reliability Evaluation of Power Systems* (2nd ed.). Plenum.

Chowdhury, A., & Koval, D. (2015). *Power Distribution System Reliability: Practical Methods and Applications*. John Wiley & Sons.

de MattosAffonso, CKezunovic, M. (2018). Technical and economic impact of PV-BESS charging station on transformer life: A case study. *IEEE Transactions on Smart Grid, 10*(4), 4683–4692.

Deb, S., Kalita, K., & Mahanta, P. (2017, December). Review of impact of electric vehicle charging station on the power grid. In *2017 International Conference on Technological Advancements in Power and Energy (TAP Energy)* (pp. 1-6). IEEE. 10.1109/TAPENERGY.2017.8397215

Deb, S., Kalita, K., & Mahanta, P. (2019). Distribution Network planning considering the impact of Electric Vehicle charging station load. In *Smart Power Distribution Systems* (pp. 529–553). Academic Press. doi:10.1016/B978-0-12-812154-2.00022-5

Deb, S., Tammi, K., Kalita, K., & Mahanta, P. (2018a). Impact of electric vehicle charging station load on distribution network. *Energies*, *11*(1), 178. doi:10.3390/en11010178

Deb, S., Tammi, K., Kalita, K., & Mahanta, P. (2018b). Review of recent trends in charging infrastructure planning for electric vehicles. *Wiley Interdisciplinary Reviews. Energy and Environment*, *7*(6), e306. doi:10.1002/wene.306

Dong, M., Nassif, A. B., & Li, B. (2018). A data-driven residential transformer overloading risk assessment method. *IEEE Transactions on Power Delivery*, *34*(1), 387–396. doi:10.1109/TPWRD.2018.2882215

Dubey, A., & Santoso, S. (2015). Electric vehicle charging on residential distribution systems: Impacts and mitigations. *IEEE Access: Practical Innovations, Open Solutions*, *3*, 1871–1893. doi:10.1109/ACCESS.2015.2476996

Forum of Regulators. (n.d.). http://www.forumofregulators.gov.in/Data/study/EV.pdf

Ghosh, D., Deb, S., & Mohanta, D. K. (2017). Reliability evaluation and enhancement of microgrid incorporating the effect of distributed generation. In *Handbook of Distributed Generation* (pp. 685–730). Springer. doi:10.1007/978-3-319-51343-0_20

Gómez, J. R., González, L. G., Vide, P. C., Gameiro, N. S., & Guerron, G. (2017, September). Impact of electric vehicle charging on the distribution system in cuenca, ecuador. In 2017 IEEE PES Innovative Smart Grid Technologies Conference-Latin America (ISGT Latin America) (pp. 1-6). IEEE. doi:10.1109/ISGT-LA.2017.8126756

González, L. G., Siavichay, E., & Espinoza, J. L. (2019). Impact of EV fast charging stations on the power distribution network of a Latin American intermediate city. *Renewable & Sustainable Energy Reviews*, *107*, 309–318. doi:10.1016/j.rser.2019.03.017

Gunes, D., Tekdemir, I., Karaarslan, M., & Alboyaci, B. (2018). *Assessment of the impact of electric vehicle charge station loads on reliability indices*. Academic Press.

Hadley, S. W. (2006). *Impact of plug-in hybrid vehicles on the electric grid.* ORNL Report.

Karmaker, A. K., Roy, S., & Ahmed, M. R. (2019, February). Analysis of the Impact of Electric Vehicle Charging Station on Power Quality Issues. In *2019 International Conference on Electrical, Computer and Communication Engineering (ECCE)* (pp. 1-6). IEEE. 10.1109/ECACE.2019.8679164

Kundur, P., Balu, N. J., & Lauby, M. G. (1994). *Power system stability and control* (Vol. 7). McGraw-Hill.

Leou, R. C., Su, C. L., & Lu, C. N. (2013). Stochastic analyses of electric vehicle charging impacts on distribution network. *IEEE Transactions on Power Systems*, *29*(3), 1055–1063. doi:10.1109/TPWRS.2013.2291556

Ma, G., Jiang, L., Chen, Y., Dai, C., & Ju, R. (2017). Study on the impact of electric vehicle charging load on nodal voltage deviation. *Archives of Electrical Engineering*, *66*(3), 495–505. doi:10.1515/aee-2017-0037

Manríquez, F., Sauma, E., Aguado, J., de la Torre, S., & Contreras, J. (2020). The impact of electric vehicle charging schemes in power system expansion planning. *Applied Energy*, *262*, 114527. doi:10.1016/j.apenergy.2020.114527

Pinto, D. R., Arioli, V. T., Hax, G. R., Taniguchi, F. K., Torquato, R., Teixeira, W. W., & Eduardo, P. L. (2018, May). Field investigation of the power quality impact of electric vehicles in secondary residential systems. In *2018 18th International Conference on Harmonics and Quality of Power (ICHQP)* (pp. 1-6). IEEE. 10.1109/ICHQP.2018.8378904

Pothinun, T., & Premrudeepreechacharn, S. (2018, September). Power quality impact of charging station on mv distribution networks: A case study in pea electrical power system. In *2018 53rd International Universities Power Engineering Conference (UPEC)* (pp. 1-6). IEEE. 10.1109/UPEC.2018.8541921

QZ. (n.d.). https://qz.com/india/1603051/ola-says-electric-vehicles-need-a-whole-new-indian-power-sector/

Rupa, J. A. M., & Ganesh, S. (2014). Power flow analysis for radial distribution system using backward/forward sweep method. *Int. J. Electr. Comput. Electron. Commun. Eng.*, *10*, 1540–1544.

Sachan, S., & Adnan, N. (2018). Stochastic charging of electric vehicles in smart power distribution grids. *Sustainable Cities and Society, 40,* 91-100.

Sachan, S., & Kishor, N. (2016). Optimal location for centralized charging of electric vehicle in distribution network. *2016 18th Mediterranean Electrotechnical Conference (MELECON)*, 1-6. 10.1109/MELCON.2016.7495326

Sachan, S., & Kishor, N. (2018). A Probabilistic Approach for DG with Electric Vehicles. *2018 IEEE International Conference on Power Electronics, Drives and Energy Systems (PEDES)*, 1-4. 10.1109/PEDES.2018.8707451

Satarworn, S., & Hoonchareon, N. (2017, June). Impact of EV home charger on distribution transformer overloading in an urban area. In *2017 14th International Conference on Electrical Engineering/Electronics, Computer, Telecommunications and Information Technology (ECTI-CON)* (pp. 469-472). IEEE. 10.1109/ECTICon.2017.8096276

Sharma, A., Kapoor, M. A., & Chakrabarti, S. (2019). *Impact of Plug-in Electric Vehicles on Power Distribution System of Major Cities of India: A Case Study*. Academic Press.

Susuki, Y., Mizuta, N., Ishigame, A., Ota, Y., Kawashima, A., Inagaki, S., & Suzuki, T. (2017, October). A continuum approach to assessing the impact of spatio-temporal EV charging to distribution grids. In *2017 IEEE 20th International Conference on Intelligent Transportation Systems (ITSC)* (pp. 1-6). IEEE. 10.1109/ITSC.2017.8317625

Utakrue, M., & Hongesombut, K. (2018, June). Impact analysis of electric vehicle quick charging to power transformer life time in distribution system. In 2018 IEEE Transportation Electrification Conference and Expo, Asia-Pacific (ITEC Asia-Pacific) (pp. 1-5). IEEE. doi:10.1109/ITEC-AP.2018.8433281

Wamburu, J., Lee, S., Shenoy, P., & Irwin, D. (2018, June). Analyzing distribution transformers at city scale and the impact of evs and storage. In *Proceedings of the Ninth International Conference on Future Energy Systems* (pp. 157-167). 10.1145/3208903.3208925

Zhang, Y., Dai, X., Han, X., & Jin, X. (2019, June). Analysis of Impact of Electric Vehicle Charging Load on City Grid. *IOP Conference Series. Materials Science and Engineering*, *486*(1), 012139. doi:10.1088/1757-899X/486/1/012139

Chapter 3
A Comprehensive Review of Blockchain Technology Implementation in the EV Charging Infrastructure

Toni Zhimomi
Aligarh Muslim University, India

Mohammad Saad Alam
Aligarh Muslim University, India

Hafiz Malik
University of Michigan, Dearborn, USA

ABSTRACT

Charging infrastructure is a key factor in successful electric vehicle adoption. Charging stations are still a fragmented market in terms of ownership, lack of standards, and charging protocols. The increasing decentralised grid has made energy and communication flow bi-directional. Challenges arise in maintaining the increasing decentralised structure, security, and privacy of the network. Blockchain facilitates the interconnectedness of such a distributed and decentralised network. Blockchain's versatility lies in its transparent and immutable decentralized architecture that enables direct transactions between users without the need of a middleman. It provides powerful safeguards against cyberattacks with its advanced cryptography enabling privacy-preserving authentication. This chapter presents a comprehensive review on the application of blockchain technology in EV charging infrastructure such as facilitating the peer-to-peer energy exchange, increased security and privacy, immutable transactions, and mitigating trust issues among the participants in the charging infrastructure.

DOI: 10.4018/978-1-7998-6858-3.ch003

1. INTRODUCTION

The history of Electric vehicles dates back to the mid-19th century (Guarnieri, 2012). Cheap abundant gasoline along with the continued improvement in the internal combustion engine contributed to the premature decline of electric vehicles. The energy crisis of 1970's renewed interest in sustainable technological development including alternative means of transport (Behrens & Glover, 2012). However, the technology of the time hampered its successful revival. The advancement in the semiconductor and battery technology have revived the electric vehicle industry (Larminie & Lowry, 2012). The transition towards low carbon energy economy entails adoption of renewable energy resources, energy storage systems, Electric Vehicles (EV's), corresponding charging infrastructure, smart buildings, and prosumers to engage actively in the electricity markets (Kobashi, 2020). E-mobility plays a crucial role in reducing the greenhouse gas (GHG) emissions (International Energy Agency (IEA), 2020). Faster adoption of electric vehicles is only possible with an equal emphasis on EV charging infrastructure. While there are many charging technologies available (Ahmad et al., 2017) (Patil et al., 2017), the commercial success will be one that is accessible, easy to use, and relatively inexpensive.

The challenges of EV charging infrastructures are high cost, multiple charging protocols, disagreements over infrastructure ownership, mistrust and lack of coordination in its fragmented market (Lee & Clark, 2018), (Das et al., 2020). The automation of charging infrastructure adds to these challenges with issues concerning privacy and security of data exchanged. Moreover, the information and communication flow in energy network is becoming increasingly distributed and complex, overwhelming the current centralized structure. Blockchain helps facilitate transaction of data and energy through peer-to-peer(P2P) trading, increased security and privacy via cryptography, provides trust and transparency with its distributed and decentralised database and allows anyone to become a part of the network system (Brilliantova & Thurner, 2019). Their applications have moved beyond the financial sector (Underwood, 2016)–(Casino et al., 2019). Blockchain has been explored as part of the energy transition (Ahl et al., 2020), (Creyts & Trbovich, 2018).The Brooklyn microgrid serves as the first successful pilot project for blockchain enabled peer-to-peer trading of energy (Mengelkamp et al., 2018). Blockchain facilitates the interconnectedness of distributed energy systems through IoT devices in the energy internet ecosystem (Sun et al., 2018), (Hosseinian et al., 2020). In the EV charging infrastructure, the peer-to-peer energy trading through blockchain platform enables prosumers and private owners to share their surplus energy generation for EV charging without a third-party intermediary. This enables autonomy to the users in choosing from multiple energy suppliers at a competitive price, secure platform and real-time EV power demand (Kim, 2019), (Zaheer, 2018).

eMotorWerks is among a host of companies, in the U.S and elsewhere, using blockchain to facilitate electric vehicle charging to alleviate charging anxiety (eMotorWerks, 2017). Meanwhile, the power company Innogy implemented blockchain technology to authenticate and manage the billing process in Germany (Shepard, 2017). Blockchain can facilitate the authentication and validation of billing through its smart contract function. Smart contracts are "lines of code written in the blockchain and automatically execute when predetermined terms and conditions are met" (Buterin, 2014), (Watanabe et al., 2016). Its role in the charging infrastructure is rapidly increasing with its implementation in battery management (Florea & Taralunga, 2020), secure communication with its environment (Kobashi, 2020), (Gai et al., 2020), energy trading from distributed and decentralised energy sources (Kim, 2019), (Qian et al., 2020), (Zhou et al., 2019) to initiatives to track the cobalt mined for EV batteries from the Democratic Republic of Congo (Anonymous, 2019).

This chapter presents a comprehensive review of blockchain technology implementation in the electric vehicle charging infrastructure. To understand what aspect of the technology is being used, we have reviewed extensively and categorized Blockchain's distinctive function into six sub-sections namely: Peer-to-Peer (P2P) energy trading, charging coordination, security issues, mitigating trust issues, Billing system and preserving privacy.

2. BACKGROUND: BLOCKCHAIN TECHNOLOGY

Blockchain technology is amongst the most promising technological trends in the field of information technology. Since its first application in the financial sector with Bitcoin in 2008, blockchain technology has been used in a variety of non-financial sectors. The global blockchain has a market value of USD 3.84 billion and is expected to grow to USD 57.64 billion by 2025 (GrandView Research, 2019). The applications of blockchain are wide ranging: Health (Kuo et al., 2017), education (Chen et al., 2018), energy (Andoni et al., 2019), governance (Ølnes et al., 2017), supply chain management (Saberi et al., 2019), IoT integration (Conoscenti et al., 2016), e-mobility (Kim, 2019), (Florea & Taralunga, 2020). Blockchain can be defined as "a distributed ledger technology that holds immutable data in a secure and encrypted way. This peer to peer, decentralized technology has no central authority to govern and provides complete transparency, privacy, security, and ensures that transactions are not corrupted". Blockchain is regarded as the fifth evolution of computing.

The three core components of blockchain are: blocks, chain, and network. Each block contains a digital signature, timestamp and information. The block includes the cryptographic hash of the previous block in the blockchain which links them forming a chain of blocks (Narayanan et al., 2016). A hash is a unique combination of letters and numbers. It is like a digital fingerprint for the data in a block. Every block in the Blockchain has a unique hash generated. Any change of data in a block changes the hash as well. The bitcoin blockchain uses the SHA-256 algorithm which is a unique fixed size 256-bit hash. The network is composed of nodes located all over the world and can be operated by anyone. Each node contains a complete record of all the transactions recorded in that blockchain. Nodes which participate to validate the transaction are called "full nodes". The validation mechanism is called consensus algorithm and depending on the blockchain type, different consensus algorithms exist. Each consensus algorithm has its advantages and disadvantages. Table 1 shows the comparison among some of the consensus algorithms, Proof of Stake (PoS), Proof of Work (PoW), Practical Byzantine Fault Tolerance (PBFT), Delegated Proof of Stake (DPoS), and RAFT (Du et al., 2017), (Zheng et al., 2017).

Blockchain is classified into three categories: public blockchain, private blockchain, and consortium blockchain (Buterin, 2015), (Pilkington, 2020). Anyone can participate as a user or validator on Public blockchain network without any restrictions, for example, bitcoin. Private blockchain is a permissioned network where participation and validations process is allowed only for authorised participants of a single organization example Ripple. Consortium or federated blockchain is similar to a permissioned platform but multiple organizations govern the platform example Quorum and Hyperledger. Table 2 summarises the differences among the different blockchain types.

Figure 1. Blockchain is a digital data structure that stores data in blocks. Each block contains a hash, timestamp and details of the transaction. Blocks are cryptographically linked to previous blocks forming a chain of records that determines the sequencing order of events.

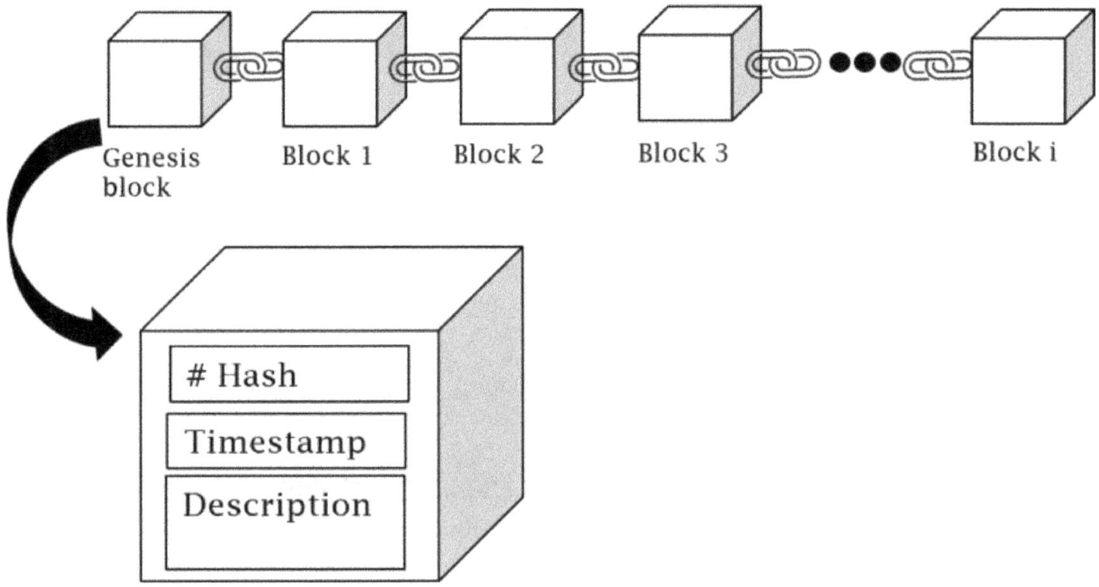

Table 1. Comparison of Consensus Algorithms (Du et al., 2017), (Zheng et al., 2017)

Characteristics	Consensus Algorithms				
	PoW	PoS	DPoS	PBFT	RAFT
Byzantine fault tolerance	50%	50%	50%	33%	N/A
Crash fault tolerance	50%	50%	50%	33%	50%
Verification speed	>100s	<100s	<100s	<10s	<10s
Throughput (TPS)	<100	<1000	<1000	<2000	>10k
Scalability	Strong	Strong	Strong	Weak	Weak
Energy saving	No	Partial	Partial	Yes	
Example	Bitcoin [40]	Peercoin [41]	Bitshares [42]	Hyperledger Fabric [43]	Quorum

Table 2. Difference among the types of blockchain

Features	Public Blockchain	Private Blockchain	Consortium Blockchain
Network type	Decentralized	Partly decentralized	Decentralized
Access	Open to all	Permissioned platform for a single organisation	Permissioned platform for multiple organisation
Transaction speed	Slow	Fast	Fast
Scalability	Good	Good	Good
Efficiency	Low	High	High
Finality	Poor	Good	Good
Transaction cost	High (Fluctuates)	Cheap (Fixed)	Cheap (Fixed)
Consensus mechanism	PoW, PoS	Voting/Multi-party consensus	Voting/Multi-party consensus
Energy consumption	High	Low	Low
Immutability	Difficult to tamper	Partial	Partial
Example	Bitcoin	Ripple	Quorum, Hyperledger, Corda

3. BLOCKCHAIN IMPLEMENTATION IN EV CHARGING INFRASTRUCTURE

Guidehouse Insight report states that the "combined market for blockchain-based EV charging and grid integration is expected to exceed $1 billion by 2029 at a 77.9% compound annual growth rate (CAGR)" (Guidehouse Insight, 2020). The charging infrastructure is an integral aspect in the uptake of e-mobility. While range anxiety issue may be overcome with the advancement in the battery technology (Beresford, 2020), charging anxiety may be the new challenge (Autovista Group, 2020). Simply put, charging anxiety is the fear of insufficient charging stations to charge the vehicle. To address charging anxiety, expansion of charging infrastructure is a must. Several countries have invested millions in setting up charging infrastructure to promote electric vehicle adoption (Olson, 2020)–(Hanley, 2020). Multifaceted and collaborative efforts would be required in promoting the establishment of charging infrastructures. Government utilities, automakers, energy companies, and residential prosumers may collectively work as service providers, creating an ecosystem for charging electric vehicles. However, collaboration among the many service providers may run the risk of distrust among them. Blockchain integrated with the charging infrastructure has significant potential. It enables controlling and sharing of electric charging stations among the EV users (Aljabri et al., 2019), eliminates billing manipulation (Jeong et al., 2018), facilitates EV as a service (EVaaS) for P2P trading to maintain the demand-supply balance (Kobashi et al., 2020), enables automated payment and charging by executing smart contracts with the charging stations (Akin et al., 2019). It also enables a new revenue stream through its peer-to-peer charging applications and grid services (V2X).

Several companies and initiatives have explored the significant benefits of blockchain-enabled solutions in the charging infrastructure. Share&Charge Foundation employs decentralized blockchain system to supports payment and contracts for EV charging. It uses the Open Charge Point Interface protocol (OCPI) to establish connections between the charge point operators and service providers. According to Share&Charge, "the combination of OCPI with blockchain technology can result in secured contracts and connections between parties and improve payment and settlement" (Share & Charge, 2020). Share&Charge partnered with eMotorWerks to establish the first P2P network in the US (eMotorWerks, 2017). Other examples are Alliander, Aerovironment, Chargemap, Easy Park, Charg, Oxygen Initiative, etc.

Figure 2. In the EV charging infrastructure, Blockchain facilitates P2P energy trading, charging coordination, more secure transaction, helps mitigate trust issues among the various entities, provides an immutable billing system and preserves privacy by using pseudonyms.

3.1 Blockchain Enabled Peer-to-Peer Energy Trading for EV Charging

The success of electric vehicle adoption is contingent upon the availability of charging infrastructure that is accessible, easy to use, and cheap. To meet the growing requirement, cost-effective charging infrastructure models are required. Peer-to-peer (P2P) model is an energy trading business model where consumers can trade energy at their agreed price without an intermediary. P2P model came about as a result of increasing adoption of distributed energy resources (DER) and the intention to incentivize its use (IRENA, 2020). P2P trading is a truly decentralised energy market model. Blockchain technology is a key enabler for effective implementation of P2P energy models. The Brooklyn microgrid project implemented by LO3 was the first successful P2P energy trading using blockchain. Many P2P energy trading projects using blockchain have followed since (Zhang et al., 2017).

While the uptake of renewable energy has been significant, the challenge lies in their intermittent nature and storage of energy. Large battery storage facility is currently used to store the energy. The steep fluctuation in the power demand curve between day and night use, the phenomenon widely described as the duck curve, has been observed with higher adoption of renewable energy (Denholm et al., 2009). Electric vehicles can help mitigate the duck curve problem (Pal, 2020). Electric vehicles provide an interesting solution, in that it can work as a movable energy storage unit and trade their energy through vehicle to grid (V2G) or vehicle to vehicle (V2V) mechanism (Mao et al., 2018). Electric vehicles can, therefore, work as a distributed energy resource (DER) and trade their energy to other vehicles or return the stored energy back to the grid to balance the peak demand supply (Alvaro-Hermana et al., 2016).

Table 3. P2P energy trading projects

P2P platform	Country	Year
Brooklyn Microgrid, LO3 energy [16]	US	2016
Centrica Plc [59]	UK	2018
Eemnes energie [60]	Netherlands	2019
Electrify [61]	Singapore	2017
Electron [62]	UK	2015
Grid+ [63]	US	2017
Lichtblick Swarm Energy [64]	Germany	2010
Lumenaza [65]	Germany	2020
PeerEnergyCloud [66]	Germany	2012
Piclo [67]	UK	2014
Power Ledger [68]	Australia	2016
Smart Watts [69]	Germany	2011
Sun Contract [70]	Slovenia	2018
SOlshare [71]	Bangladesh	2013
SonnenCommunity [72]	Germany	2015
TransActive Grid [73]	US	2015
Vandebron [74]	Netherland	2014
Yeloha, Mosaic (Shut down) [75]	US	2015

The blockchain technology provides a secure and trusted platform for P2P energy trading where electric vehicles can participate, both as a consumer and prosumer (as DER, V2X) (Xue et al., 2020), (Sun et al., 2020). The blockchain–based system integrates smart contracts to trade energy for electric vehicle charging have been explored in (Mannaro et al., 2017)–(Umoren et al., 2020). Zhou et al (Zhou et al., 2018) develops a consortium blockchain to securely trade energy among EV's, Local energy aggregator (LEAG), and edge computing service providers (ESP). In (Umoren et al., 2020), an EVaaS (EV as a Service) energy trading prototype is designed which uses a graphical user interface (GUI) application to remotely control energy trading between EV and consumer. The experimental work done in Paisley shows the effectiveness of the energy prototype using blockchain.

The work in (Kang et al., 2017) details the implementation for energy trading among plug-in hybrid electric vehicles (PHEV) in Texas. The results in (Silva et al., 2019) shows that blockchain-enabled P2P trading has been beneficial in generating revenue as well as balancing the load demands for university campus parking lots. In Japan, surplus power generated from solar PV's have been traded on the blockchain platform to charge electric vehicles. This not only charges the vehicle but also maintains the supply and demand balance flattening the peak demand curve (Nagatsuka et al., 2018). To encourage the use of renewable energy, a market mechanism to incentivize the EV users and EV charging providers have also been proposed (Chen et al., 2020). The V2G application of electric vehicles helps maintain the load demand on the grid by injecting or drawing energy during energy shortage or excess (Islam et al., 2020), (Alskaif et al., 2020). Large scale energy trading faces potential security threats like the Sybil attack. In Sybil attack, an adversary creates multiple fake identities to manipulate and control the decision of the entire network (Douceur, 2002). Blockchain can counter the Sybil attack with the proof-of-work (PoW) consensus mechanism (Asfia et al., 2019). The consensus requires each node to solve an expensive cryptopuzzle to participate in the mining process. This makes creation of multiple fake identities difficult and expensive. The P2P model does not discriminate among the participants in the network. Residential and commercial EV owners can profit by sharing their charger. This creates flexibility to the drivers and it allows them to charge their vehicles at different locations.

Figure 3. Increased renewable energy integration to the grid fosters bi-directional flow of energy and communication with the grid. Electric Vehicles can participate as an energy storage system and feed the stored energy back to the grid (V2G)

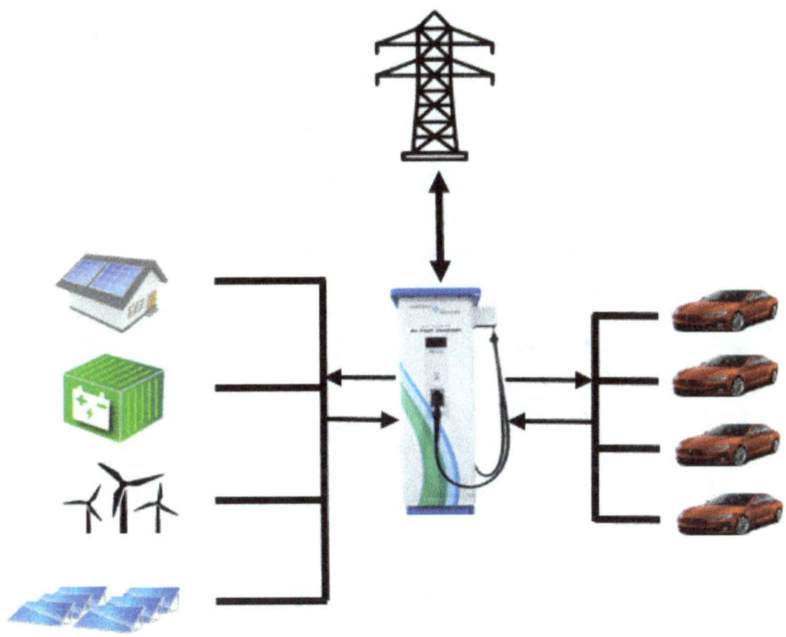

Unlike the legacy wholesale energy markets, where retailers sell the energy to the consumer, a model where EV's (as prosumers) trade energy directly to the consumer is made possible through the blockchain platform (Umoren et al., 2020). Waseda et al (Waseda & Tanaka, 2020) designs an automatic bidding agent for EV charging taking into account the uncertainties such as the demand and supply, and market price. The results show that most surplus energy during the day was absorbed by the EV at low price and deficient energy was sold at higher price at night. In (Xia et al., 2020), a Bayesian game pricing is proposed for electricity trading on the blockchain-enabled Internet of vehicles (BIoV). On the blockchain, the smart contracts act as pricing agents. Transactions among participants on the P2P network are securely recorded on the blocks. The smart contract automatically executes the transaction when the conditions (energy price, power to be traded, or time of charging) are met (Tai et al., 2016). Table 4 below shows the projects using blockchain technology for EV charging application

Table 4. Projects using blockchain technology for EV charging application

Company	Platform	Consensus algorithm	Location
AeroVironment	Ethereum	n/a	US
Alliander (Charge Ledger)	Ethereum	Proof of Work	Netherlands
BlockCharge	Ethereum	n/a	Germany
Car eWallet	Hyperledger Fabric	Practical Byzantine Fault Tolerance	Germany
Charg	Ethereum	n/a	US
Chargemap	Ethereum	n/a	France
eMotorwerks (EnelX)	Ethereum	Proof of Work	US
Energo Labs	Qtum	Proof of Stake (public Qtum), Proof of Time and Raft agreement (consortium Qtum)	China
Everty	Ethereum	Proof of Work	Australia
Innogy Motionwerk (Share&Charge)	Ethereum, Energy Web (Ethereum-based)	Proof of Work, Proof of Authority	Germany
Oxygen initiative	Ethereum	Proof of Work	US
Power Ledger	Ethereum	Proof of Work	Australia
PROSUME	Proprietary	n/a	Switzerland
Slock.it	Ethereum, Energy Web (Ethereum-based)	Proof of Work, Proof of Authority	Germany

3.2 EV Charging Coordination with Blockchain

The increasing adoption of electric vehicles (EV) in distribution systems causes burden on the grid: increased peak power demand, transformer overloading, voltage instability, increased power losses and power quality issues (Deb et al., 2018)–(Rutherford & Yousefzadeh, 2011). Large-scale uncoordinated EV charging may threaten the distribution network's safety operation. Therefore, a charging strategy is pertinent in adoption of electric vehicles. A comparison between coordinated and uncoordinated charging at different EV penetration percentages is drawn in (Zalnidzam et al., 2018). The paper proposes an optimal coordination schedule using particle swarm optimization (PSO) algorithm. The objective function to minimize the active power loss is achieved using the said optimization method for every level of EV penetration. Celli et.al in (Celli et al., 2012) also uses PSO to maximize revenue for the vehicle owners, through its V2G application, and to reduce energy losses in the network. Apart from technical considerations, coordinated charging stations can also offer economic benefits through its V2G services. A multi-objective optimization using whale optimization algorithm is studied in (Adetunji et al., 2020). The multi-objectives are to maximize the grid performance viz minimization of the load variance and power loss, and minimization charging cost. Similarly, various metaheuristic methods have been formulated to provide optimal coordination charging strategy addressing different objective functions and constraints (Das et al., 2020)–(Underwood, 2016). Mukherjee et. al in (Mukherjee & Gupta, 2014) explores the various scheduling algorithms and broadly classifies charging optimization into unidirectional or bidirectional power flow, which is further classified into the type of control (decentralised or centralised) and mobility aspect (static or mobility-aware). Most of the coordination and scheduling of charging stations are centralised where all the charging demands are collected by a central aggregator. The central aggregator then through iterations with the charging station optimizes the charging schedule while considering the system constraints. However, considering the imminent increase in electric vehicle penetration, centralized optimization becomes impractical due to its sheer volume. Optimal charging schedule taking both decentralised and centralised control schemes under consideration is studied in (Xu, Yan, Feng et al, 2019), (Vaya & Andersson, 2012). The study shows that though the objective function is achieved for both the centralised and decentralised control method. The decentralised control scheme appears to be the better control scheme with reduced computational time, better computing efficiency, reduced communication cost, and privacy protection. The decentralised control scheme has the advantage in scalability, an important factor considering the anticipated EV adoption.

The authors in (Pustišek et al., 2018) present a concept using blockchain to select the most convenient charging station depending on different parameters such as lowest price for energy/kWh, user preferences, waiting time etc. Thakur et al. in (Thakur & Breslin, 2018) proposes a blockchain enabled EV queue management procedure. The study shows that the proposed blockchain-maintained solution performs better than random allocation with reduced congestion level, efficient load balance between charging stations, and reduced waiting time for EV charging. In (Ping et al., 2019), the authors design a coordination method taking into account the permissible charging power quota trading platform on blockchain. The trading platform is implemented on Ethereum. The EV coordination is done in a two-step process. The first step involves the allocation of required charging demand to charging stations according to their charging power quota. The second step involves double auction where charging stations trade their charging power quotas. The smart contracts designed in Ethereum platform enables this two-step mechanism and helps make Pareto improvement without the need of a central coordinator. The study conducted in (Liu, Chai, Zhang et al, 2018) proposes a novel adaptive blockchain-based electric vehicle participation (AdBEV) scheme which uses the Iceberg order execution algorithm to obtain an improved schedule to charge and discharge which reduces the overall charging costs and power fluctuation. The iceberg order algorithm is used to mitigate the fluctuations in the power load caused by large volume trading. In (Javed & Javaid, 2019), the authors present Mobile charging vehicle-to-vehicle (M2V) charging strategy. The study shows that the proposed scheduling using mobile charging reduces the charging time as well as charging cost. The authors in (Fu et al., 2020) address the coordination among different energy companies by proposing a novel EV charging system on the consortium blockchain. The paper constructs a mathematical model, Bio-Objective Mixed-Integer Programming model (BOMILP), with the objective (i) to minimize the imbalance between revenue and operation costs and (ii) to minimize the distance between the charging station and the EV user. The smart contract is designed using the Limited Neighbourhood Search with Memory (LNSM) algorithm to solve the BOMILP problem. To evaluate the performance of the proposed charging system, the method is implemented in Beijing, China. The proposed novel LNSM algorithm makes the smart contract run faster and have better performance.

3.3 Blockchain Enabled Security in EV Charging Infrastructure

The EV charging infrastructure has a number of actors participating such as the EV owner, private and public charging station (CS) owners, electricity suppliers and communication companies. Each participating actor shares a large number of critical data between them including its location, EV identity, battery state of charge (SoC), charging information, payment details, and price. These data become vulnerable points for the entire network and are exploited through cyber-attacks. An instance of such a possible security threat was demonstrated on Nissan leaf, where the knowledge of the Vehicle identification number (VIN) enabled the hack into the system and control of the vehicle (Hunt, 2016). The attack could manipulate the climate control, drain the battery life, access the private driving data, and control the car's driving system. The risk of a demand-side cyberattacks using plug-in electric vehicles and EV high-wattage charging station infrastructure is investigated in (Acharya et al., 2020). Dvorkin et al. demonstrated a plausible scenario in which multiple charging EVs can be hacked at the same time disrupting operations of the grid such as stability of grid frequency causing blackout. A detailed study in (Antoun et al., 2020) presents various potential cyber-attacks to the participating entities in the EV charging ecosystem and categorizes the security gaps as: Availability, Confidentiality and privacy, Integrity, and authenticity and nonrepudiation.

- Availability: Cyber-attack such as Denial of service (DOS) and Distributed DOS (DDOS) where the attacker hijacks the charging station and denies service (availability) to other users
- Confidentiality and Privacy: An attacker can access data such as user ID, location, charging preferences, payment methods etc.
- Integrity: Attacks like false data injection, tampering of data where the integrity of the data communicated is compromised and corrupted
- Authenticity and nonrepudiation: Attacks such as impersonation, identity theft threatens the authenticity of the owner

Blockchain has piqued the interest of security experts as it provides transparency and security in tracking the transfer and ownership of data without the use of a trust entity. A smart contract also enables the development of completely automated security systems, such as automated obligation enforcement systems (Lee & Kim, 2017). Blockchain secures the data by maintaining data integrity and authenticity of the owner. It maintains data integrity by making sure that the data recorded on blockchain cannot be altered or removed. The data integrity is ensured with the use of consensus mechanisms. All parties in the network verify the information being added by agreeing to the consensus mechanism. Blockchain uses public key infrastructure (PKI) technology which secures the ownership of the account. This ensures a secure account unless the private key is exposed. The underlying principles of blockchain enables authenticity of the sender and keeps data tamper-proof.

In (Li, Lu, Misic et al, 2018), Blockchain-inspired data coins and energy coins are proposed as the distributed consensus. Depending on their data contribution frequency and energy contribution amounts, the coins are awarded to EV's. The data and energy coins serve as authentication operators to ensure secure interaction for resource allocation. Secure integration of smart communities with EV's and distributed energy resources (DER's) can be facilitated using permissioned blockchain and the delegated Byzantine fault-tolerance (DBFT) consensus algorithm (Wang et al., 2018). The authors in (Su et al., 2019) propose a secure charging scheme for EVs using a delegated Byzantine fault-tolerance (DBFT) consensus algorithm on a permissioned blockchain for achieving secure and optimal consumption. (Chen & Zhang, 2019) proposes a novel secure energy trading for electric vehicles based on China's electricity market rules. To ensure reliability, the digital signature technology employs elliptic curve bilinear pairing and practical Byzantine fault-tolerant (PBFT) algorithm to increase transaction throughput and reduce transmission delay. Another novel decentralized security model called Lightning network and smart contract (LNSC) is presented in (Huang et al., 2018). Lightning Network is a second layer technology that uses micropayments channels that take away transactions from the main blockchain as a way to address the bitcoin scalability issue (Poon & Dryja, 2016). The hash functions are calculated by the LNSC using elliptic curve encryption. The new proposed security model can be easily integrated with existing scheduling mechanisms to improve the security of trading between EVs and charging piles. To securely allocate DER's to EV's, a blockchain based secure incentive scheme (BSIS) through a permissioned network that reaches a consensus based on proof-of-reputation (PoR) is presented in (Wang et al., 2019). The authors in (Choubey et al., 2019) propose a permissioned blockchain to enhance secure energy trading among electric vehicles. They propose an EnergyTradingRank (ETR) algorithm where they incentivize EV's with high ETR score. High ETR scores indicate active users with maximum, truthful and committed participation. This results in a more secure and transparent environment. The blockchain-based proof-of-concept (POC) is implemented using IBM's hyper-ledger fabric and composer to analyse the effectiveness of the new ETR algorithm.

Figure 4. Visual representation of a blockchain transaction. Blockchain uses public key infrastructure (PKI) to secure the transaction. It consists of two different keys, a public key and a private key, that encrypts and decrypts to secure the data.

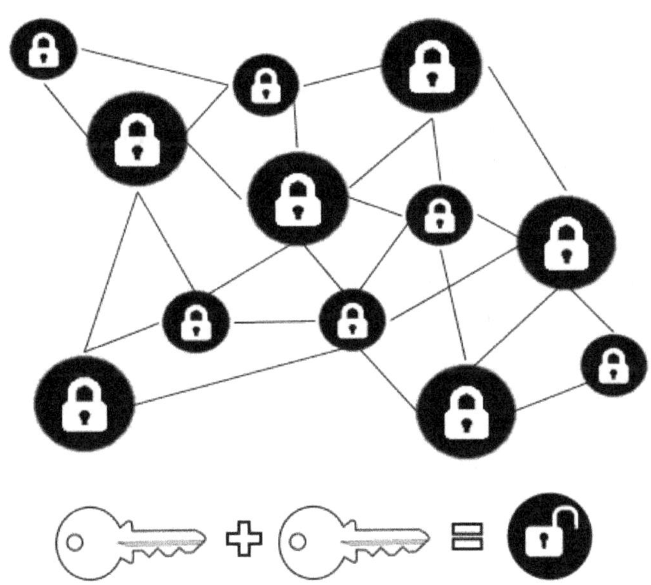

3.4 Mitigating Trust Issues with Blockchain

Establishing trust among multiple entities has always been a challenging task. There is a significant number of parties in providing e-mobility: both from the energy supply side and e-mobility provision. With the imminent electrification of the transportation sector, interaction and transaction among the entities becomes inevitable. The problem arises with a significant amount of parties and data flows. How would the parties share the data? Who would be in charge of the charging management? And Will the other parties trust it? Hence, a need arises for a technological solution to solve the trust issue.

Blockchain is known as 'The Trust Machine' as it can perform reliable transactions in the future without the need for human intervention (The trust machine, 2020). Blockchain increases trust between trading parties while also reducing costs by allowing direct transactions without the need for intermediaries. A lack of trust was observed among the various charging service providers (CSP), property owners, and EV owners (University of Waterloo, 2019). A study in (Gorenflo et al., 2019) identifies trust issues between charging service providers (CSP) and customers, and between property owners and charging service providers (CSP). Customers must not be overcharged by the CSP, while property owners must not be undercharged by the CSP. The authors use blockchain to address these trust issues and outlines a proof-of-concept Hyperledger Fabric implementation.

3.5 Billing System

Electric vehicle users will need to make payments to the charging service providers. A transparent and immutable billing system will be required to facilitate such a transaction. The blockchain enables secure decentralised peer-to-peer transactions, which will form the foundation of the future private transportation system's financial structure. Centralized third parties will not be required to facilitate payments thereby reducing transfer costs by eliminating commissions, and the administrative and bank fees. The United Nations' World Food Program (WFP) is an example which uses Ethereum-based blockchain payments system. A pilot test in Jordan's Azraq refugee camp facilitated cash transfers for over 10,000 refugees. The agency eliminated 98 percent of the bank-related transfer fees by saving $150k per month (Aoun, 2019a). Current energy financial systems are not planned for a bidirectional energy market. The existing method of collecting and processing financial settlement data are susceptible to error and highly ineffective. This results in increased delay time in value settlement and necessitates costly reconciliation processes.

Figure 5. Billing process proposed in (Jeong et al., 2018) shows the steps required in authenticating a transaction between the EV and the Charging station.

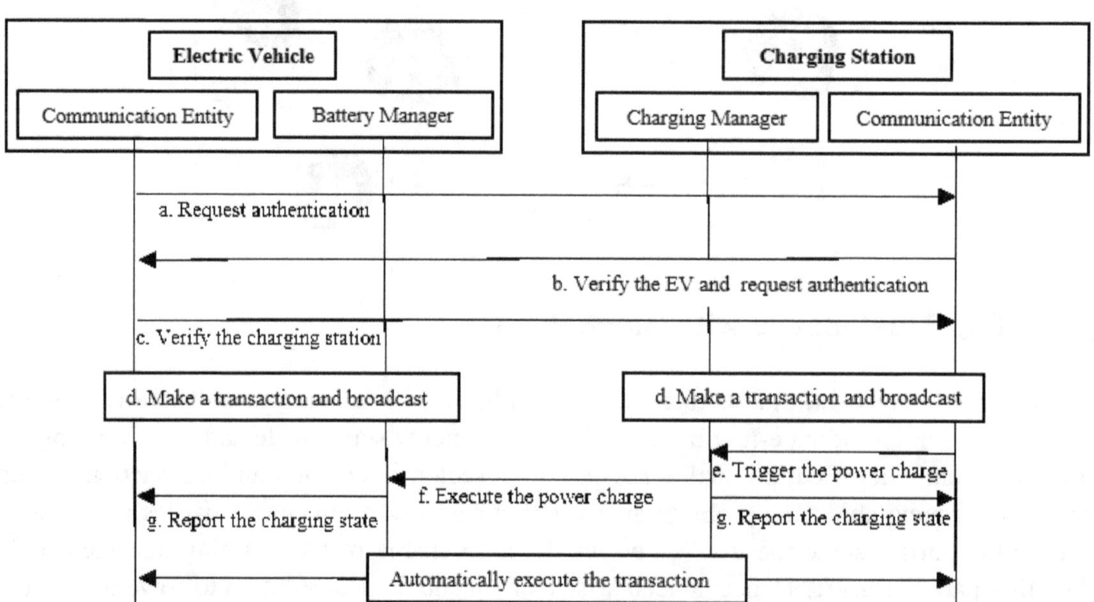

Blockchain can function as an automated decentralised energy trading system that collects energy consumption automatically from thousands of smart meters and ensure a process assured billing system for energy transactions in a bidirectional way. BlockCharge facilitates charging of EV batteries by utilising the Ethereum blockchain. It communicates with the Ethereum blockchain to negotiate the price, record the charging data, and manage the payment process (Aoun, 2019b). Similar billing systems have been proposed using the concept of blockchain in (Jeong et al., 2018), (Zielinska et al., 2019), (Matsuda & Tanaka, 2020). UBS, a Swiss bank, has collaborated with German energy company Innogy SE and automotive technology company ZF to develop blockchain-based eWallets for electric vehicles. If they choose to participate in the sharing economy, the eWallets will be able to make payments for parking, charging, highway tolls, and car-sharing fees (ZF, 2017).

A billing process for mobile charging proposed in (Kim et al., 2017) analyses the challenges of data size in blockchain and proposes a lightweight scheme. For each transaction, existing blocks must include a signature and a secret key. However, the proposed method only includes the idTag, the charging amount, and the hash of the most recent transaction, significantly reducing the data size. Ethereum smart contract has been implemented to ensure secure automated payment for charging (Kohlbrenner et al., 2019), (Chanthong et al., 2020). (Erdin et al., 2018) highlights the issue of Bitcoin-based payments systems transactional fees being comparable to the cost of charging. The study addresses this issue by proposing an overlay payment channel network (PCN) that utilises Bitcoin's off-chain mechanism based on optimization model, multi-commodity flow problem structure. The proposed topology in the study can significantly reduce the transaction fees.

Blockchain offers several advantages in the payment system. It eliminates intermediaries thereby eliminates or reduces the transactional fees allowing for a safe and quick Cross-Border Payments in shorter time. It also offers data (billing) transparency and immutability due to its distributed and cryptographically linked structure. Automation with Smart Contracts also reduces the payment time, helps implement instant payments, and automate payment flows on the Ethereum blockchain.

3.6 Preserving Privacy on the Blockchain

With the rise of the internet, the huge repository of digital footprint left on cyberspace has become a contentious issue of privacy (Shaw, 2009). The massive databases have accrued a wide range of confidential information about organizations and individuals. The search history and multiple transactions made online everyday by unassuming users can and have been used by third party to profile them. The dangers of such social profiling from personally identifiable information (PII) range from gathering information for market analysis for targeted advertising to more malicious activity such as identity theft (Newman & Mcnally, 2005), credit card fraud (Goodin, 2009), (McCurry, 2016), and influencing public opinions creating echo chambers (Unver, 2017). The inconspicuous encroachment over the privacy of individuals and institutions can lead to the very collapse of free society (Shahbaz, 2018), (Westin, 1966). The consequences of privacy breach may seem far-fetched to the unassuming populace. However, with the immense proliferation of the internet-based economy taking over all aspects of transactions and communication, the potential to access and misuse personal information is only magnified and exacerbated by artificial intelligence's unprecedented levels of power and speed (Kerry, 2020).

Electric vehicle users communicate with the charge service providers or smart grid to charge or discharge, in the process leaving behind a trail of personal data. Sensitive material such as identity of the user and vehicle, vehicle preference, the charging and discharging pattern, driving pattern, and geographical location of charging stations, banks, hospitals, schools etc. visited must be protected. Privacy of personal data must be preserved. The risks on privacy increases significantly for EV integration with the smart grid (Jouvray et al., 2013), (Saxena et al., 2017). Various techniques to anonymise data to protect privacy have been proposed. K-anonymity is "a model to protect identity disclosure in which sensitive data of an individual in the dataset is indistinguishable from at least k-1 other individuals in the dataset" (Sweeney, 2002). This model may be described as a 'hiding in the crowd' model. Privacy models to protect attribute disclosure such as l-diversity (Machanavajjhala et al., 2006) and t-closeness (Ninghui et al., 2007) are both *K*-anonymity extensions. Differential privacy is another privacy preserving method where randomised "noise" is added to an aggregate query result to preserve individual input without significantly altering the result (Dwork & Roth, 2013). Homomorphic encryption (Gentry, 2009) is also used to protect data confidentiality.

To allay privacy concerns, (Zyskind et al., 2015) constructs a decentralized personal data management system that protects the data. Personal data can be deduced from financial transactions conducted even on using pseudonyms. To strengthen privacy during financial transactions, a digital currency called Zerocash is constructed using decentralized anonymous payment schemes (DAP schemes) based on zero-knowledge Succinct Non-interactive arguments of Knowledge (zk-SNARKs) (Ben-Sasson, 2014). Zerocash enables direct payment between users without disclosing the amount transferred or the payment's location. The transaction size is below 1kB and take less than 6 ms to verify. The authors in (Aitzhan & Svetinovic, 2018) have similarly implemented PriWatt, a token based trading system, which enables exchange of energy anonymously on the blockchain platform. Privacy preserving protocol using blockchain in EV charging has been proposed in (Knirsch et al., 2018). The protocol enables the EV users to find the lowest priced charging station while protecting the customer's location and energy needs. To build a reliable and effective vehicular announcement network, a novel Blockchain-based incentive announcement network called CreditCoin has been proposed (Li, Liu, Cheng et al, 2018). The incentive mechanism motivates users to share traffic information while protecting the identity of the users. A similar privacy-preserving blockchain incentive mechanism has been proposed in (Praba Devi & Miraclin Joyce Pamila, 2019) to motivate users to report and/or endorse a genuine accident alert message and send the accident location to a nearby emergency service without revealing their identity. Another anonymous payment system has been proposed in (Radi et al., 2019) which utilizes untraceable digital coins and generates random pseudonyms for each transaction on the consortium blockchain. Privacy is preserved as the coins are untraceable and pseudonyms expire after each transaction thereby unlinking the transaction to the identity of the user.

4. EXISTING STANDARDS ON BLOCKCHAIN

Even as a nascent technology, Blockchain has come a long way in terms of the many variations taking form. Each variation has its distinct characteristics wrt scaling solutions, degree of decentralization, consensus mechanisms. This may result in fragmentation of the industry making collaboration between participants difficult which may consequently slow adoption of the technology. Thus, the lack of standardization has been listed as one of the challenges in the implementation of blockchain in section 4. Several renowned standards developing organizations ('SDOs'), such as Institute of Electrical and Electronics Engineers Standards association (IEEE SA), the National Institute of Standards and Technology (NIST), the International Organization for Standardization ('ISO'), Enterprise Ethereum Alliance ('EEA'), German Institute for Standardization (DIN), the World Wide Web Consortium ('W3C'), the Society for Worldwide Interbank Financial Telecommunication ('SWIFT') and the standardization sector of the International Telecommunications Union ('ITU-T') (Anjum et al., 2017), (Aristidou & Marcou, 2019) have recognised the issue of interoperability. A standard is a "published document providing detailed information that saves manufacturers enormous amounts of time and money when developing products" (Institute of Electrical and Electronics Engineers, 2011). Mobility Open Blockchain Initiative (MOBI), a global multi-stakeholder project, released the first standard for blockchain-based EV grid integration (Mobility Open Blockchain Initiative (MOBI), 2020). The standard covers the technical design specification for core use cases: Tokenized Carbon Credits (TCC), Peer to Peer (P2P) applications, and Vehicle to Grid Integration (V2G).

5. CHALLENGES AND FUTURE OUTLOOK

Blockchain has emerged as a promising technology beyond the financial industry by transforming the way industries operate. The electric transportation sector is no exception to its influence. Various companies have implemented blockchain in the EV charging infrastructure such as eMotorWerks and Easy Park. The companies implementing the technology have grown rapidly over the years, steadily increasing its presence in various cities across the globe. However, to realise its full potential, challenges remain to be overcome. These challenges arise due to both the blockchain technology and the electric mobility industry being in its early stages of development. Both being an immature technology, the long-term projection cannot be ascertained due to limitations in the research conducted and published until now. The selected literatures (Casino et al., 2019), (Andoni et al., 2019), (Pilkington, 2020), (Zyskind et al., 2015)–(Radi et al., 2019), (Kshetri, 2017)–(Li et al., 2020) have been analysed to assess the challenges faced in implementing blockchain in the EV charging infrastructure, the following limitations have been ascertained:

Technological Barriers: Blockchain is an immature technology. The system lacks the transactional capacity and the high latency to be adopted on a larger scale. Though advancements have been achieved, with Ripple and Ethereum having a latency of seconds. There is also the "scalability trilemma" termed by Vitalik Buterin which presents the trade-off among the three central components of blockchain – decentralisation, security and scalability. The trade-off is application specific. The Proof-of-Work consensus used in Bitcoin is energy-intensive. The technology can also be inaccessible to the average person due to lack of wallet support and non-user-friendly interfaces. The EV charging infrastructure is also an immature industry consisting of many entities. Though the trust can be established on the blockchain platform, the behavioural change and societal acceptance in using cryptocurrency may take time.

Regulatory Barriers: Regulatory frameworks and energy laws need amendments to support new applications and services in different countries and states. Regulatory uncertainties and related issues such as intellectual property concerns and compliance are critical barriers in its adoption (PwC, 2018). Legacy utilities may feel threatened by its disruptive decentralised nature. However, Blockchain can aid them in creating new business models and underpin economic, business, and social infrastructure.

Table 5. Existing Standards on Blockchain and its use in mobility services (Aristidou & Marcou, 2019), (König et al., 2020)

Organization	Publication	Details	Year
NIST	NISTIR 8202	Blockchain Technology Overview	2018
ANSI	ASC X9	Distributed Ledger and Blockchain Technology Study Group report	2018
ISO	ISO/TR 23455:2019	Blockchain and distributed ledger technologies—Overview of and interactions between smart contracts in blockchain and distributed ledger technology systems	2019
DIN	DIN SPEC 16597	Terminology for Blockchains	2018
	DIN SPEC 3103	Blockchain and Distributed Ledger Technologies in Application Scenarios for Industry 4.0	2019
	DIN SPEC 3104	Blockchain-based validation of data	2018
	DIN SPEC 4996	Blockchain-based approach to the transfer of software licenses	2020
	DIN SPEC 4997	Privacy by Blockchain Design: A standardized model for processing personal data using blockchain	2020
ENISA	ENISA	Distributed Ledger Technology & Cybersecurity - Improving information security in the financial sector	2018
BSI	BSI	Towards Secure Blockchains	2019
BSI/ANSSI	BSI/ANSSI	Franco-German IT-Security Situation Overview, dedicated to cryptocurrency crime	2019
ITU	FG DLT D1.1	Technical specification about terminology and definitions for distributed ledger technology	2019
	FG DLT D1.2	Distributed ledger overview, concepts, ecosystem	2019
	FG DLT D2.1	Technical report about distributed ledger technology use cases	2019
	FG DLT D3.1	Reference architecture for distributed ledger systems	2019
	FG DLT D3.3	Assessment criteria for distributed ledger platforms	2019
	FG DLT D4.1	Distributed ledger technology regulatory framework	2019
	FG DLT D5.1	An overview of future predictions for DLT on the subjects governance and legal regulation, computation networks, identity and privacy, security and resilience, and risk and audit	2019
European Committee for Electrotechnical Standardization (CENELEC)	CENELEC	Recommendations for Successful Adoption in Europe for Emerging Technical Standards on Distributed Ledger/Blockchain Technologies	2018
Mobility Open Blockchain Initiative (MOBI)	MOBI	Technical specification on Blockchain-based EV grid integration for three core use cases: V2G, TCC and P2P	2020
IEEE	P2418.5	Standard for Blockchain in Energy sector	2020
	2142.1-2021	IEEE Recommended Practice for E-Invoice Business Using Blockchain Technology	2021
	2143.1-2020	Standard for General Process of Cryptocurrency Payment	2020
	2144.1-2020	Standard for Framework of Blockchain-based Internet of Things (IoT) Data Management	2020
	2418.2-2020	Standard for Data Format for Blockchain Systems. This standard addresses data structures, data types, and data elements.	2020

Lack of Standardization and Frameworks: Any new technology would benefit by establishing standards and frameworks. Standards can play a vital role in fostering its adoption and ensuring interoperability. Establishing standards to address the issues concerning security, privacy, and governance of data could help build trust in the technology. Using standards could help reduce the risk of a fragmented ecosystem by establishing a stronger consensus on fundamental understanding of the technology as well the technical considerations to be made. There exists a lot of scope in investigating viable and effective standards and frameworks for Blockchain adoption.

6. CONCLUSION

To conclude, blockchain technology can clearly benefit the EV charging infrastructure. Blockchain technology can facilitate the increasing distributed and decentralised EV charging infrastructure. It offers transparency and trust among the various entities in the EV charging infrastructure, offers immutable and secure billing transactions, and enables prosumers to be a part of the energy market and thereby generate revenue even for micro payments. They enable a sharing economy. As detailed in the sections, we have categorized Blockchain's distinctive function into six sub-sections namely: Peer-to-Peer (P2P) energy trading, charging coordination, security issues, mitigating trust issues, Billing system and preserving privacy. The paper contributes in helping understand how blockchain can be implemented in facilitating EV charging infrastructure. Several challenging barriers remain, such as developing resiliency against cyber-attacks, energy-efficient mining, immature technology, and regulatory barriers. Research is ongoing on these key improvement areas and additional initiatives, projects, and collaboration will be critical in reaching the technology's full potential.

REFERENCES

Acharya, S., Dvorkin, Y., & Karri, R. (2020). Public Plug-in Electric Vehicles + Grid Data: Is a New Cyberattack Vector Viable? *IEEE Transactions on Smart Grid, 11*(6), 5099–5113. doi:10.1109/TSG.2020.2994177

Adetunji, K., Hofsajer, I., & Cheng, L. (2020). A coordinated charging model for electric vehicles in a smart grid using whale optimization algorithm. *Proc. 2020 23rd Int. Conf. Inf. Fusion, FUSION 2020.* 10.23919/FUSION45008.2020.9190284

Ahl, A., Yarime, M., Goto, M., Chopra, S. S., Kumar, N. M., Tanaka, K., & Sagawa, D. (2020, January). Exploring blockchain for the energy transition: Opportunities and challenges based on a case study in Japan. *Renewable & Sustainable Energy Reviews, 117*, 109488. Advance online publication. doi:10.1016/j.rser.2019.109488

Ahmad, A., Alam, M. S., & Chabaan, R. (2017). A Comprehensive Review of Wireless Charging Technologies for Electric Vehicles. *IEEE Trans. Transp. Electrif., 4*(1), 38–63. doi:10.1109/TTE.2017.2771619

Aitzhan, N. Z., & Svetinovic, D. (2018). Security and Privacy in Decentralized Energy Trading Through Multi-Signatures, Blockchain and Anonymous Messaging Streams. *IEEE Transactions on Dependable and Secure Computing, 15*(5), 840–852. doi:10.1109/TDSC.2016.2616861

Akin, Y., Dikkollu, C., Kaplan, B. B., Yayan, U., & Yolacan, E. N. (2019). Ethereum Blockchain Network-based Electrical Vehicle Charging Platform with Multi-Criteria. *Decision Support Systems*, 1–5. Advance online publication. doi:10.1109/UBMYK48245.2019.8965557

Aljabri, Aldhaheri, Mohammed, Alkaabi, Abdella, & Shuaib. (2019). *Facilitating Electric Vehicle Charging Across the UAE Using Blockchain*. . doi:10.1109/ICECTA48151.2019.8959705

Alonso, M., Amaris, H., Germain, J. G., & Galan, J. M. (2014). Optimal charging scheduling of electric vehicles in smart grids by heuristic algorithms. *Energies*, *7*(4), 2449–2475. doi:10.3390/en7042449

Alskaif, T., Holthuizen, B., Schram, W., Lampropoulos, I., & Van Sark, W. (2020, December). A blockchain-based configuration for balancing the electricity grid with distributed assets. *World Electr. Veh. J.*, *11*(4), 1–17. doi:10.3390/wevj11040062

Alvarez, J. G., González, M. Á., Vela, C. R., & Varela, R. (2018). Electric Vehicle Charging Scheduling by an Enhanced Artificial Bee Colony Algorithm. *Energies*, *11*(10), 1–19. doi:10.3390/en11102752

Alvaro-Hermana, R., Fraile-Ardanuy, J., Zufiria, P. J., Knapen, L., & Janssens, D. (2016). Peer to Peer Energy Trading with Electric Vehicles. *IEEE Intell. Transp. Syst. Mag.*, *8*(3), 33–44. doi:10.1109/MITS.2016.2573178

Andoni, M., Robu, V., Flynn, D., Abram, S., Geach, D., Jenkins, D., McCallum, P., & Peacock, A. (2019). Blockchain technology in the energy sector: A systematic review of challenges and opportunities. *Renewable & Sustainable Energy Reviews*, *100*, 143–174. Advance online publication. doi:10.1016/j.rser.2018.10.014

Anjum, A., Sporny, M., & Sill, A. (2017). Blockchain Standards for Compliance and Trust. *IEEE Cloud Comput.*, *4*(4), 84–90. doi:10.1109/MCC.2017.3791019

Anonymous. (2019). Partnership for Traceable Cobalt for Batteries: Engineering, Geology, Mineralogy, Metallurgy, Chemistry, etc. *Engineering and Mining Journal*.

Antoun, J., Kabir, M. E., Moussa, B., Atallah, R., & Assi, C. (2020). A Detailed Security Assessment of the EV Charging Ecosystem. *IEEE Network*, *34*(3), 200–207. doi:10.1109/MNET.001.1900348

Aoun, A. (2019a). Using Blockchain for Smart Metering and Billing in Developing Countries. *The Blockchain Journal*. https://blockchainjournal.blog/2019/12/13/using-blockchain-for-smart-metering-and-billing-in-developing-countries/

Aoun, A. (2019b). Blockchain for Electric Vehicle Charging. *The Blockchain Journal*. https://blockchainjournal.blog/2019/12/05/blockchain-for-electric-vehicle-charging/

Aristidou, C., & Marcou, E. (2019). Blockchain Standards and Government Applications. *Blockchain Standards and Government Applications*, *7*(3), 287–312. doi:10.13052/jicts2245-800X.736

Asfia, U., Kamuni, V., Sutavani, S., Sheikh, A., Wagh, S., & Singh, N. M. (2019). A blockchain construct for energy trading against sybil attacks. *27th Mediterranean Conference on Control and Automation, MED 2019 - Proceedings*, 422–427. 10.1109/MED.2019.8798489

Autovista Group. (2020). *EV charging anxiety is the new range anxiety.* Autovista Group. https://autovistagroup.com/news-and-insights/ev-charging-anxiety-new-range-anxiety?fbclid=IwAR252W27xd4WfvupAdsGt0K2cjbIuewsMJ5BXuF0wptsvE9oM9JmIWsUgyE

Behrens, C. E., & Glover, C. (2012). U.S. energy: Overview and key statistics. In *Fossil Fuels*. Competition and Elasticities.

Ben-Sasson, E. (2014). Zerocash: Decentralized anonymous payments from bitcoin. *Proc. - IEEE Symp. Secur. Priv.*, 459–474. 10.1109/SP.2014.36

Beresford, C. (2020). GM Unveils Battery with Capacity Twice as Big as Tesla's. *Car and Driver*. https://www.caranddriver.com/news/a31226611/gm-ultium-electric-vehicle-battery-revealed/#:~:text=Rivian%2C

BitShares. (2014). https://bitshares.org/

Brandherm, Baus, & Frey. (2012). *Peer energy cloud-Civil marketplace for trading renewable energies.* . doi:10.1109/IE.2012.46

Brilliantova, V., & Thurner, T. W. (2019). Blockchain and the future of energy. *Technology in Society*, *57*, 38–45. Advance online publication. doi:10.1016/j.techsoc.2018.11.001

Buterin, V. (2014). *A next-generation smart contract and decentralized application platform*. Etherum.

Buterin, V. (2015). On Public and Private Blockchains. Ethereum Blog Crypto Renaiss. Salon, 1–24.

Casino, F., Dasaklis, T. K., & Patsakis, C. (2019). A systematic literature review of blockchain-based applications: Current status, classification and open issues. *Telematics and Informatics*, *36*, 55–81. Advance online publication. doi:10.1016/j.tele.2018.11.006

Celli, G., Ghiani, E., Pilo, F., Pisano, G., & Soma, G. G. (2012). *Particle Swarm Optimization for Minimizing the Burden of Electric Vehicles in Active Distribution Networks*. doi:10.1109/PESGM.2012.6345458

Centrica. (2018). *Centrica and LO3 Energy to deploy blockchain technology as part of Local Energy Market trial in Cornwall*. www.centrica.com/news/centrica-and-lo3-%0Aenergy-deploy-blockchain-technology-part-localenergy-%0Amarket-trial-cornwall

Chanthong, Ruangsakorn, & Glomglome. (2020). *Blockchain and Smart Contract Payment for Electric Vehicle Charging*. . doi:10.1109/ECTI-CON49241.2020.9158132

Chen, G., Xu, B., Lu, M., & Chen, N.-S. (2018). *Exploring blockchain technology and its potential applications for education*. Smart Learn. Environ. doi:10.118640561-017-0050-x

Chen, X., Zhang, T., Ye, W., Wang, Z., & Iu, H. H.-C. (2020). Blockchain-based Electric Vehicle Incentive System for Renewable Energy Consumption. *IEEE Transactions on Circuits and Wystems. II, Express Briefs*. Advance online publication. doi:10.1109/TCSII.2020.2996161

Chen, X., & Zhang, X. (2019). Secure Electricity Trading and Incentive Contract Model for Electric Vehicle Based on Energy Blockchain. *IEEE Access: Practical Innovations, Open Solutions*, *7*, 178763–178778. doi:10.1109/ACCESS.2019.2958122

Choubey, Behera, Patel, Mahidhar, & Misra. (2019). *EnergyTradingRank Algorithm for Truthful Auctions among EVs via Blockchain Analytics of Large Scale Transaction Graphs*. . doi:10.1109/COMSNETS.2019.8711249

Christidis, K., & Devetsikiotis, M. (2016). Blockchains and Smart Contracts for the Internet of Things. *IEEE Access: Practical Innovations, Open Solutions*, *4*, 2292–2303. Advance online publication. doi:10.1109/ACCESS.2016.2566339

Cong, L. W., & He, Z. (2019). Blockchain Disruption and Smart Contracts. *Review of Financial Studies*, *32*(5), 1754–1797. doi:10.1093/rfs/hhz007

Conoscenti, Vetro, & De Martin. (2016). *Blockchain for the Internet of Things: A systematic literature review*. . doi:10.1109/AICCSA.2016.7945805

Creyts, J., & Trbovich, A. (2018). *Can blockchain help us to address the world's energy issues?* World Economic Forum.

Crosby, M., & Nachiappan, P. (2016). *Blockchain Technology - Beyond Bitcoin*. Berkley Eng.

Das, H. S., Rahman, M. M., Li, S., & Tan, C. W. (2020). Electric vehicles standards, charging infrastructure, and impact on grid integration: A technological review. *Renewable & Sustainable Energy Reviews*, *120*, 109618. Advance online publication. doi:10.1016/j.rser.2019.109618

Deb, S., Tammi, K., Kalita, K., & Mahanta, P. (2018). Impact of electric vehicle charging station load on distribution network. *Energies*, *11*(1), 178. Advance online publication. doi:10.3390/en11010178

Denholm, P., Margolis, R., & Milford, J. (2009). Production cost modeling for high levels of photovoltaics penetration. *Renew. Energy Grid Interact. Bus. Photovoltaics*, 87–118.

Douceur, J. R. (2002). The sybil attack. Lecture Notes in Computer Science, 2429, 251–260. doi:10.1007/3-540-45748-8_24

Du, M., Ma, X., Zhang, Z., Wang, X., & Chen, Q. (2017). A review on consensus algorithm of blockchain. *2017 IEEE Int. Conf. Syst. Man, Cybern. SMC 2017*, 2567–2572. 10.1109/SMC.2017.8123011

Dwork & Roth. (2013). The algorithmic foundations of differential privacy. *Found. Trends Theor. Comput. Sci.* . doi:10.1561/0400000042

Electrify. (2020). *The Big Singapore's Energy Story*. https://electrify.asia/

Electron. (2020). *EmpoEnergy markets*. https://electron.net/

ElHusseini, H., Assi, C., Moussa, B., Attallah, R., & Ghrayeb, A. (2020). *Blockchain, AI and Smart Grids: The Three Musketeers to a Decentralized EV Charging Infrastructure. IEEE Internet Things Mag.* doi:10.1109/IOTM.0001.1900081

eMotorWerks. (2017). eMotorWerks and Share&Charge Deliver North America's First Peer to Peer Electric Vehicle Charging Network with Blockchain Payments. *PRNewswire*. https://emotorwerks.com/about/enewsso/press-releases/329-emotorwerks-and-share-charge-deliver-north-americas-first-peer-to-peer-electric-vehicle-charging-network-with-blockchain-payments

Energie, E. (2019). *A saving, generating and exchanging energy in Eemnes.* https://eemnesenergie.nl/

Energycentral. (2016). *Lights Out for Yeloha: Why We Shut Down the Solar Sharing Network.* Available: https://energycentral.com/c/um/lights-out-yeloha-why-we-shut-down-solar-sharing-network

Erdin, E., Cebe, M., Akkaya, K., Solak, S., Bulut, E., & Uluagac, S. (2018). Building a Private Bitcoin-Based Payment Network Among Electric Vehicles and Charging Stations. *2018 IEEE International Conference on Internet of Things (iThings) and IEEE Green Computing and Communications (GreenCom) and IEEE Cyber, Physical and Social Computing (CPSCom) and IEEE Smart Data (SmartData)*, 1609–1615. 10.1109/Cybermatics_2018.2018.00269

Ferrag, M. A., Derdour, M., Mukherjee, M., Derhab, A., Maglaras, L., & Janicke, H. (2019). Blockchain technologies for the internet of things: Research issues and challenges. *IEEE Internet of Things Journal*, *6*(2), 2188–2204. Advance online publication. doi:10.1109/JIOT.2018.2882794

Florea, B. C., & Taralunga, D. D. (2020). *Blockchain IoT for smart electric vehicles battery management.* Sustain. doi:10.3390u12103984

Fu, Z., Dong, P., & Ju, Y. (2020). An intelligent electric vehicle charging system for new energy companies based on consortium blockchain. *Journal of Cleaner Production*, *261*, 121219. Advance online publication. doi:10.1016/j.jclepro.2020.121219

Gai, K., Wu, Y., Zhu, L., Choo, K.-K. R., & Xiao, B. (2020). Blockchain-Enabled Trustworthy Group Communications in UAV Networks. *IEEE Transactions on Intelligent Transportation Systems*, 1–13. doi:10.1109/TITS.2020.3015862

Gentry. (2009). *Fully Homomorphic Encryption Using Ideal Lattices.* doi:10.1145/1536414.1536440

Goodin, D. (2009). *TJX suspect indicted in Heartland, Hannaford breaches - Networks pierced by garden-variety exploit.* http://www.theregister.co.uk/2009/08/17/heartland_payment_suspect/

Gorenflo, Golab, & Keshav. (2019). *Mitigating trust issues in electric vehicle charging using a blockchain.* . doi:10.1145/3307772.3328283

GrandView Research. (2019). *Global Blockchain Technology Market Size Report, 2019-2025.* GVR-1-68038-329-4.

Gridplus. (2017). *Grid+ Raises $29 Million as Blockchain Fever Grows.* https://gridplus.io/

Guarnieri, M. (2012). Looking back to electric cars. *3rd Reg. IEEE Hist. Electro - Technol. Conf. Orig. Electrotechnol. HISTELCON 2012 - Conf. Proc.* 10.1109/HISTELCON.2012.6487583

Guidehouse Insight. (2020). *Market Data: Blockchain-Enabled EV Charging and Integration.* Author.

Hanley, S. (2020). China Invests. In *EV Charging Infrastructure To Offset Coronavirus Economic Slump.* Clean Technica.

Hosseinian, Shahinzadeh, Gharehpetian, Azani, & Shaneh. (2020). Blockchain outlook for deployment of IoT in distribution networks and smart homes. *Int. J. Electr. Comput. Eng.* . doi:10.11591/ijece.v10i3.pp2787-2796

Huang, X., Xu, C., Wang, P., & Liu, H. (2018). LNSC: A Security Model for Electric Vehicle and Charging Pile Management Based on Blockchain Ecosystem. *IEEE Access: Practical Innovations, Open Solutions*, *6*, 13565–13574. Advance online publication. doi:10.1109/ACCESS.2018.2812176

Hunt, T. (2016). *Controlling vehicle features of Nissan LEAFs across the globe via vulnerable APIs*. https://www.troyhunt.com/controlling-vehicle-features-of-nissan/

Hyperledger project. (2015). Academic Press.

Iea. (2004). Electric vehicles in Europe. *Energy*, *3*, 33. Available: http://wrs.region-stuttgart.de/sixcms/media.php/923/ElectricMobilityPilotRegion_english.pdf%5Cn

Institute of Electrical and Electronics Engineers. (2011). *What are Standards? Why are They Important?* Institute of Electrical and Electronics Engineers. http://standardsinsight.com/ieee_company_detail/what-are-standards-why-are-they-important

International Energy Agency (IEA). (2020). Global EV Outlook 2020. Advance online publication. doi:10.1787/d394399e-en

IRENA. (2020). *Electricity trading arrangements*. Available: https://www.youtube.com/watch?v=UDv04q3U6e0

Islam, Shahjalal, Hasan, & Jang. (2020). *Blockchain-based Energy Transaction Model for Electric Vehicles in V2G Network*. . doi:10.1109/ICAIIC48513.2020.9065221

Javed & Javaid. (2019). *Scheduling charging of electric vehicles in a secured manner using blockchain technology*. . doi:10.1109/FIT47737.2019.00072

Jensen, O. (2018). *Disruption of the current utility business by utilizing EV charging infrastructure to unlock grid flexibility. Possibilities for Blockchain technology*. KTH Royal Institute of Technology.

Jeong, S., Dao, N.-N., Lee, Y., Lee, C., & Cho, S. (2018). Blockchain Based Billing System for Electric Vehicle and Charging Station. *2018 Tenth International Conference on Ubiquitous and Future Networks (ICUFN)*, 308–310. 10.1109/ICUFN.2018.8436987

Jouvray, C., Pellischek, G., & Tiguercha, M. (2013). Impact of a smart grid to the electric vehicle ecosystem from a privacy and security perspective. *World Electr. Veh. J.*, *6*(4), 1115–1124. doi:10.3390/wevj6041115

Kang, J., Yu, R., Huang, X., Maharjan, S., Zhang, Y., & Hossain, E. (2017, December). Enabling Localized Peer-to-Peer Electricity Trading among Plug-in Hybrid Electric Vehicles Using Consortium Blockchains. *IEEE Transactions on Industrial Informatics*, *13*(6), 3154–3164. doi:10.1109/TII.2017.2709784

Kerry, C. F. (2020). *Protecting Privacy in an AI-driven World*. Available: https://www.brookings.edu/research/protecting-privacy-in-an-ai-driven-world/

Kim, Kang, & Hong. (2017). *Mobile charger billing system using lightweight Blockchain*. . doi:10.1109/APNOMS.2017.8094151

Kim, M. (2019). *A secure charging system for electric vehicles based on blockchain*. Sensors. doi:10.339019133028

King & Nadal. (2012). *PPCoin: Peer-to-Peer Crypto-Currency with Proof-of-Stake*. Academic Press.

Kingdom. (2015). *Piclo peer-to-peer energy trading*. Academic Press.

Knirsch, Unterweger, & Engel. (2018). *Privacy-preserving blockchain-based electric vehicle charging with dynamic tariff decisions*. . doi:10.1007/s00450-017-0348-5

Kobashi, T. (2020). Smart city and ICT infrastructure with vehicle to X applications toward urban decarbonization. Urban Systems Design. doi:10.1016/B978-0-12-816055-8.00009-9

Kobashi, T., Yoshida, T., Yamagata, Y., Naito, K., Pfenninger, S., Say, K., Takeda, Y., Ahl, A., Yarime, M., & Hara, K. (2020). On the potential of 'Photovoltaics + Electric vehicles' for deep decarbonization of Kyoto's power systems: Techno-economic-social considerations. *Applied Energy*, *275*, 115419. Advance online publication. doi:10.1016/j.apenergy.2020.115419

Kohlbrenner, F., Nasirifard, P., Löbel, C., & Jacobsen, H.-A. (2019). A Blockchain-based Payment and Validity Check System for Vehicle Services. *Proceedings of the 20th International Middleware Conference Demos and Posters on ZZZ - Middleware '19*, 17–18. 10.1145/3366627.3368107

König, L., Korobeinikova, Y., Tjoa, S., & Kieseberg, P. (2020). *Comparing Blockchain Standards and Recommendations*. Futur. Internet. doi:10.3390/fi1212022

Kshetri, N. (2017). Blockchain's roles in strengthening cybersecurity and protecting privacy. *Telecommunications Policy*, *41*(10), 1027–1038. Advance online publication. doi:10.1016/j.telpol.2017.09.003

Kuo, T. T., Kim, H. E., & Ohno-Machado, L. (2017). Blockchain distributed ledger technologies for biomedical and health care applications. *Journal of the American Medical Informatics Association*, *24*(6), 1211–1220. Advance online publication. doi:10.1093/jamia/ocx068 PMID:29016974

Larminie, J., & Lowry, J. (2012). *Electric Vehicle Technology Explained* (2nd ed.). doi:10.1002/9781118361146

Lee & Clark. (2018). *Charging the Future: Challenges and Opportunities for Electric Vehicle Adoption*. Academic Press.

Lee, J.-H., & Kim, H. (2017, July). Security and Privacy Challenges in the Internet of Things [Security and Privacy Matters]. *IEEE Consum. Electron. Mag.*, *6*(3), 134–136. doi:10.1109/MCE.2017.2685019

Leou, R. C., Su, C. L., & Lu, C. N. (2014). Stochastic analyses of electric vehicle charging impacts on distribution network. *IEEE Transactions on Power Systems*, *29*(3), 1055–1063. Advance online publication. doi:10.1109/TPWRS.2013.2291556

Li, H., Lu, R., Misic, J., & Mahmoud, M. (2018, May). Security and Privacy of Connected Vehicular Cloud Computing. *IEEE Network*, *32*(3), 4–6. doi:10.1109/MNET.2018.8370870

Li, L., Liu, J., Cheng, L., Qiu, S., Wang, W., Zhang, X., & Zhang, Z. (2018). CreditCoin: A Privacy-Preserving Blockchain-Based Incentive Announcement Network for Communications of Smart Vehicles. *IEEE Transactions on Intelligent Transportation Systems*, *19*(7), 2204–2220. doi:10.1109/TITS.2017.2777990

Li, X., Jiang, P., Chen, T., Luo, X., & Wen, Q. (2020). A survey on the security of blockchain systems. *Future Generation Computer Systems*, *107*, 841–853. doi:10.1016/j.future.2017.08.020

LichtBlick. (2016). *German green utility goes international with 'sharing economy' concept for distributed energy*. https://www.lichtblick.de/privatkunden/schwarm-energie

Liu, Chai, Lau, & Chen. (2018). *Blockchain based energy trading model for electric vehicle charging schemes*. . doi:10.1007/978-3-319-94965-9_7

Liu, C., Chai, K. K., Zhang, X., Lau, E. T., & Chen, Y. (2018). Adaptive Blockchain-Based Electric Vehicle Participation Scheme in Smart Grid Platform. *IEEE Access: Practical Innovations, Open Solutions*, *6*, 25657–25665. doi:10.1109/ACCESS.2018.2835309

Lumenaza. (2020). *The Software for the energy revolution*. www.lumenaza.de/en

Machanavajjhala, Gehrke, Kifer, & Venkitasubramaniam. (2006). *ℓ-Diversity: Privacy beyond k-anonymity*. . doi:10.1109/ICDE.2006.1

Mannaro, K., Pinna, A., & Marchesi, M. (2017). Crypto-Trading: Blockchain-oriented energy market. *2017 AEIT Int. Annu. Conf. Infrastructures Energy ICT Oppor. Foster. Innov. AEIT 2017*, 1–5. 10.23919/AEIT.2017.8240547

Mao, T., Zhang, X., & Zhou, B. (2018). Modeling and solving method for supporting 'vehicle-to-anything' EV charging mode. *Applied Sciences (Basel, Switzerland)*, *8*(7), 1048. Advance online publication. doi:10.3390/app8071048

Matsuda, Y., & Tanaka, K. (2020). EV mobility charging service based on blockchain. *Proc. - 2020 IEEE Int. Conf. Environ. Electr. Eng. 2020 IEEE Ind. Commer. Power Syst. Eur. EEEIC / I CPS Eur.* 10.1109/EEEIC/ICPSEurope49358.2020.9160608

Mavrovouniotis, M., Ellinas, G., & Polycarpou, M. (2019). Electric Vehicle Charging Scheduling Using Ant Colony System. *2019 IEEE Congr. Evol. Comput. CEC 2019 - Proc.*, 2581–2588. 10.1109/CEC.2019.8789989

McCurry, J. (2016). 100 thieves steal $13m in three hours from cash machines across Japan. *The Guardian*. https://www.theguardian.com/world/2016/may/23/japan-cash-machine-100-thieves-steal-13m-dollars-three-hours

Mengelkamp, E., Gärttner, J., Rock, K., Kessler, S., Orsini, L., & Weinhardt, C. (2018). Designing microgrid energy markets: A case study: The Brooklyn Microgrid. *Applied Energy*, *210*, 870–880. doi:10.1016/j.apenergy.2017.06.054

Mobility Open Blockchain Initiative (MOBI). (2020). *Blockchains for Electric Vehicles & Grid Integration Review Contents*. Available: https://dlt.mobi/wp-content/uploads/2020/10/EVGI-Business-Review.pdf

Mukherjee, J. C., & Gupta, A. (2014). *A Review of Charge Scheduling of Electric Vehicles in Smart Grid*. doi:10.1109/JSYST.2014.2356559

Nagatsuka, Sano, & Yamaguchi. (2018). *Decentralized Transaction System of Surplus PV Output Using Blockchain*. Academic Press.

Nakamoto. (2008). *Bitcoin: A Peer-to-Peer Electronic Cash System*. Available: https://bitcoin.org/bitcoin.pdf

Narayanan, Bonneau, Felten, & Miller. (2016). Bitcoin and cryptocurrency technologies. Princeton University Press.

Newman, G. R., & Mcnally, M. M. (2005). Identity theft literature review. *U.S. Dep. Justice*, *210459*, 103. Available: https://www.ncjrs.gov/pdffiles1/nij/grants/210459.pdf

Ninghui, Tiancheng, & Venkatasubramanian. (2007). *t-Closeness: Privacy beyond k-anonymity and ℓ-diversity*. . doi:10.1109/ICDE.2007.367856

Ølnes, S., Ubacht, J., & Janssen, M. (2017). Blockchain in government: Benefits and implications of distributed ledger technology for information sharing. *Government Information Quarterly*, *34*(3), 355–364. Advance online publication. doi:10.1016/j.giq.2017.09.007

Olson, E. B. (2020). *Federal Investment in EV Charging Infrastructure for Economic Recovery*. Climate, and Public Health.

Open Utility. (2016). *A glimpse into the future of Britain's energy economy.* White Pap. *Open Utility Ltd.*

Orsini, L., Kessler, S., Wei, J., & Field, H. (2018). How the brooklyn microgrid and transactive grid are paving the way to next-gen energy markets. The Energy Internet: An Open Energy Platform to Transform Legacy Power Systems into Open Innovation and Global Economic Engines.

Pal. (2020). *Solving Duck Curve Problem Due to Solar Integration Using Blockchain Technology*. . doi:10.1007/978-981-32-9119-5_17

Patil, D., McDonough, M. K., Miller, J. M., Fahimi, B., & Balsara, P. T. (2017). *Wireless Power Transfer for Vehicular Applications: Overview and Challenges. IEEE Trans. Transp. Electrif.* doi:10.1109/TTE.2017.2780627

Pilkington, M. (2020). Blockchain technology: principles and applications. In *Research Handbook on Digital Transformations* (pp. 225–253). Edward Elgar Publishing.

Ping, J., Chen, S., Yan, Z., Wang, H., Yao, L., & Qian, M. (2019). EV charging coordination via blockchain-based charging power quota trading. 2019 IEEE PES Innov. Smart Grid Technol. Asia, ISGT 2019, 4362–4367. doi:10.1109/ISGT-Asia.2019.8881070

Poon & Dryja. (2016). *The Bitcoin Lightning Network: Scalable Off-Chain Instant Payments*. Academic Press.

PowerLedger. (2019). *Power Ledger completes blockchain-enabled P2P trial with KEPCO*. https://www.powerledger.io/

Praba Devi, G. S., & Miraclin Joyce Pamila, J. C. (2019). Accident Alert System Application Using a Privacy-Preserving Blockchain-Based Incentive Mechanism. *2019 5th Int. Conf. Adv. Comput. Commun. Syst. ICACCS 2019*, 390–394. 10.1109/ICACCS.2019.8728507

Pustišek, Kos, & Sedlar. (2018). *Blockchain based autonomous selection of electric vehicle charging station*. . doi:10.1109/IIKI.2016.60

PwC. (2018). *PwC's Global Blockchain Survey 2018.* PwC.

Qian, L. P., Wu, Y., Xu, X., Ji, B., Shi, Z., & Jia, W. (2020). Distributed Charging-Record Management for Electric Vehicle Networks via Blockchain. *IEEE Internet of Things Journal*, Advance online publication. doi:10.1109/JIOT.2020.3027482

Radi, Lasla, Bakiras, & Mahmoud. (2019). *Privacy-Preserving Electric Vehicle Charging for Peer-to-Peer Energy Trading Ecosystems*. doi:10.1109/ICC.2019.8761788

Rahman, S., & Shrestha, G. B. (1993). An investigation into the impact of electric vehicle load on the electric utility distribution system. *IEEE Transactions on Power Delivery*, *8*(2), 591–597. Advance online publication. doi:10.1109/61.216865

Rutherford & Yousefzadeh. (2011). *The impact of electric vehicle battery charging on distribution transformers*. . doi:10.1109/APEC.2011.5744627

Saberi, S., Kouhizadeh, M., Sarkis, J., & Shen, L. (2019). Blockchain technology and its relationships to sustainable supply chain management. *International Journal of Production Research*, *57*(7), 2117–2135. Advance online publication. doi:10.1080/00207543.2018.1533261

Saxena, N., Grijalva, S., Chukwuka, V., & Vasilakos, A. V. (2017, August). Network Security and Privacy Challenges in Smart Vehicle-to-Grid. *IEEE Wireless Communications*, *24*(4), 88–98. doi:10.1109/MWC.2016.1600039WC

Shahbaz, A. (2018). *The Rise of Digital Authoritarianism*. Available: https://freedomhouse.org/report/freedom-net/2018/rise-digital-authoritarianism

Share & Charge. (2020). Cross-provider smart charging based on OCPI Unlocking the innovation potential in EV charging. *Share & Charge*. https://shareandcharge.com/

Shaw, J. (2009). *Exposed: The erosion of privacy in the Internet era*. Harv. Mag. Available http://scholar.google.com/scholar?q=related:IK9kevLZKfkJ:scholar.google.com/&hl=en&num=30&as_sdt=0,5%5Cnfile:///Users/ajacinto/Dropbox/AppDocs/Papers2/Articles/Harvard Magazine/Shaw Shaw 2009.pdf%5Cnpapers2://publication/uuid/60B49E99-EAFA-44EB-833E-94B04

Shepard, J. (2017). Blockchain Payment Technology for EV Charging. *eepower*. https://eepower.com/news/blockchain-payment-technology-for-ev-charging/#

Silva, F. C., Ahmed, M. A., Martínez, J. M., & Kim, Y. C. (2019). Design and implementation of a blockchain-based energy trading platform for electric vehicles in smart campus parking lots. *Energies*. Advance online publication. doi:10.3390/en12244814

Sonnen. (2019). *What Is the Sonnen Community?* https://sonnengroup.com/sonnencommunity

Su, Z., Wang, Y., Xu, Q., Fei, M., Tian, Y. C., & Zhang, N. (2019). A secure charging scheme for electric vehicles with smart communities in energy blockchain. *IEEE Internet of Things Journal*, *6*(3), 4601–4613. doi:10.1109/JIOT.2018.2869297

Sun, Chen, Zhu, & Li. (2018). *Research on the Application of Blockchain Technology in Energy Internet*. . doi:10.1109/EI2.2018.8582599

Sun, G., Dai, M., Zhang, F., Yu, H., Du, X., & Guizani, M. (2020). Blockchain-Enhanced High-Confidence Energy Sharing in Internet of Electric Vehicles. *IEEE Internet of Things Journal*, *7*(9), 7868–7882. Advance online publication. doi:10.1109/JIOT.2020.2992994

Sun Contract. (2017). *An energy trading platform that utilises blockchain technology to create a new disruptive model for buying and selling electricity*. Available: https://suncontract.org/res/whitepaper.pdf

Sweeney, L. (2002, October). k-anonymity: A model for protecting privacy. *International Journal of Uncertainty, Fuzziness and Knowledge-based Systems*, *10*(5), 557–570. doi:10.1142/S0218488502001648

Tai, X., Sun, H., & Guo, Q. (2016, December). Electricity transactions and congestion management based on blockchain in energy internet. *Dianwang Jishu/Power Syst. Technol.*, *40*(12), 3630–3638. doi:10.13335/j.1000-3673.pst.2016.12.002

Tapscott, D., & Tapscott, A. (2016). *Blockchain Revolution*. Portfolio Penguin.

Thakur, Hayes, & Breslin. (2018). *A unified model of peer to peer energy trade and electric vehicle charging using blockchains*. . doi:10.1049/cp.2018.1909

Thakur, S., & Breslin, J. G. (2018). Electric Vehicle Charging Queue Management with Blockchain. Springer International Publishing. doi:10.1007/978-3-030-05081-8_18

The trust machine. (2020). *The Economist*.

Umoren, Jaffary, Shakir, Katzis, & Ahmadi. (2020). Blockchain-based energy trading in electric-vehicle-enabled microgrids. *IEEE Consum. Electron. Mag.* doi:10.1109/MCE.2020.2988904

Underwood, S. (2016). Blockchain beyond bitcoin. *Communications of the ACM*, *59*(11), 15–17. Advance online publication. doi:10.1145/2994581

UNFCCC. (2019). Me Solshare: Peer-To-Peer Smart Village Grids | Bangladesh. *United Nations Framework Convention on Climate Change*. http://unfccc.int/secretariat/momentum_for_change/items/9940.php

University of Waterloo. (2019). *Researchers use blockchain to drive electric-vehicle infrastructure.* https://uwaterloo.ca/math/research/research-stories/researchers-use-blockchain-drive-electric-vehicle

Unver, A. H. (2017). Digital Challenges to Democracy: Politics of Automation, Attention, and Engagement. *Journal of International Affairs*.

Vandebron. (2020). *Duurzame Energie Van Nederlandse Bodem*. https://vandebron.nl

Vaya, M. G., & Andersson, G. (2012). Centralized and decentralized approaches to smart charging of plug-in Vehicles. *IEEE Power Energy Soc. Gen. Meet.*, 1–8. 10.1109/PESGM.2012.6344902

Veldman, E., & Verzijlbergh, R. A. (2015). Distribution grid impacts of smart electric vehicle charging from different perspectives. *IEEE Transactions on Smart Grid*, *6*(1), 333–342. Advance online publication. doi:10.1109/TSG.2014.2355494

Wang, Su, & Zhang. (2019). BSIS-Blockchain based Secure Incentive Scheme for Energy Delivery. IEEE Trans. Ind. Informatics. doi:10.1109_TII.2019.2908497

Wang, Y., Su, Z., Xu, Q., & Zhang, N. (2018). Contract based energy blockchain for secure electric vehicles charging in smart community. *Proc. - IEEE 16th Int. Conf. Dependable, Auton. Secur. Comput. IEEE 16th Int. Conf. Pervasive Intell. Comput. IEEE 4th Int. Conf. Big Data Intell. Comput. IEEE, 3*, 328–333. 10.1109/DASC/PiCom/DataCom/CyberSciTec.2018.00062

Waseda & Tanaka. (2020). *Bidding Agent for Electric Vehicles in Peer-to-Peer Electricity Trading Market considering uncertainty*. . doi:10.1109/EEEIC/ICPSEurope49358.2020.9160846

Watanabe, H., Fujimura, S., Nakadaira, A., Miyazaki, Y., Akutsu, A., & Kishigami, J. J. (2016). Blockchain contract: A complete consensus using blockchain. *2015 IEEE 4th Glob. Conf. Consum. Electron. GCCE 2015*, 577–578. 10.1109/GCCE.2015.7398721

Westerström, E. (2020). *Barriers to blockchain adoption in the public electrical vehicle charging market*. Blekinge Institute of Technology.

Westin, A. F. (1966). Science, Privacy, and Freedom: Issues and Proposals for the 1970's. Part II: Balancing the Conflicting Demands of Privacy, Disclosure, and Surveillance. *Columbia Law Review, 66*(7), 1205. Advance online publication. doi:10.2307/1120983

Xia, S., Lin, F., Chen, Z., Tang, C., Ma, Y., & Yu, X. (2020). A Bayesian Game Based Vehicle-to-Vehicle Electricity Trading Scheme for Blockchain-Enabled Internet of Vehicles. *IEEE Transactions on Vehicular Technology, 69*(7), 6856–6868. Advance online publication. doi:10.1109/TVT.2020.2990443

Xu, Yan, Feng, & Zhao. (2019). Decentralized charging control strategy of the electric vehicle aggregator based on augmented Lagrangian method. *Int. J. Electr. Power Energy Syst., 104*, 673–679. . doi:10.1016/j.ijepes.2018.07.024

Xu, M., Chen, X., & Kou, G. (2019). A systematic review of blockchain. *Financ. Innov., 5*(1), 27. Advance online publication. doi:10.118640854-019-0147-z

Xue, Mao, Pan, Qin, Li, & Shi. (2020). *Design of blockchain-based trading mechanism under sharing mode of electric vehicle under smart grid*. doi:10.1109/ACPEE48638.2020.9136233

Zaheer, H. (2018). *Vehicle Charging: Is blockchain a driver of EV adoption? | Power Technology Research*. Power Technol. Res. Available https://powertechresearch.com/p2p-vehicle-charging-is-blockchain-a-driver-of-ev-adoption/

Zalnidzam, W. I. F. W., Mohamad, H., Salim, N. A., Mokhlis, H., & Yasin, Z. M. (2018). Optimal charging schedule coordination of electric vehicles in smart grid. *Indones. J. Electr. Eng. Comput. Sci., 11*(1), 82–89. doi:10.11591/ijeecs.v11.i1.pp82-89

ZF. (2017). *Mobile wallet: The electric car which pays for itself.* Blockchain Car eWallet.

Zhang, C., Wu, J., Long, C., & Cheng, M. (2017). Review of Existing Peer-to-Peer Energy Trading Projects. *Energy Procedia, 105*, 2563–2568. doi:10.1016/j.egypro.2017.03.737

Zheng, Z., Xie, S., Dai, H., Chen, X., & Wang, H. (2017). An Overview of Blockchain Technology: Architecture, Consensus, and Future Trends. *Proceedings - 2017 IEEE 6th International Congress on Big Data, BigData Congress 2017*, 557–564. 10.1109/BigDataCongress.2017.85

Zheng, Z., Xie, S., Dai, H. N., Chen, X., & Wang, H. (2018). Blockchain challenges and opportunities: A survey. *International Journal of Web and Grid Services*, *14*(4), 352. Advance online publication. doi:10.1504/IJWGS.2018.095647

Zhou, Tan, & Xu. (2018). *Blockchain and Edge Computing Based Vehicle-to-Grid Energy Trading in Energy Internet*. . doi:10.1109/EI2.2018.8582652

Zhou, G., Lv, L., Li, G., Huang, Y., Zhen, H., & Xiang, Z. (2019). Design of Energy Block Chain System Supporting Electric Vehicle Charge and Discharge Trading. *2019 IEEE 3rd International Electrical and Energy Conference (CIEEC)*, 526–533. 10.1109/CIEEC47146.2019.CIEEC-2019218

Zielinska, A., Skowron, M., & Bien, A. (2019). The concept of the blockchain technology model use to settle the charging process of an electric vehicle. *Appl. Electromagn. Mod. Eng. Med. PTZE, 2019*, 271–274. doi:10.23919/PTZE.2019.8781739

Zyskind, G., Nathan, O., & Pentland, A. S. (2015). Decentralizing privacy: Using blockchain to protect personal data. *Proc. - 2015 IEEE Secur. Priv. Work. SPW*, 180–184. doi:10.1109/SPW.2015.27

Chapter 4

Integrated Battery Chargers: Challenges and Opportunities in Design and Infrastructure

Syed Qaseem Ali
https://orcid.org/0000-0002-9212-2608
Quanta Technology, Canada

Geza Joos
McGill University, Canada

Chu Sun
Beihang University, Canada

ABSTRACT

Integrated battery chargers (IBCs) have been proposed as a low-weight, low-volume, and high-power solution to conventional conductive chargers. However, the design of such chargers is complicated, requiring special components or control techniques to solve inherent issues (such as galvanic isolation, torque generation, and system reconfiguration) associated with their design. Solutions vary based on charging power, drive topology, and motor technology. This chapter introduces designs for IBCs, including solutions as proposed in the literature. It also presents challenges in their industrial adoption. Finally, the chapter presents opportunities for fleet charging applications using IBCs.

INTRODUCTION

Transportation electrification can be challenging, particularly when needing to charge a battery in a minimal timeframe. Large batteries require high-power chargers for fast charging. One barrier to overcome is ensuring a longer travel distance on a single charge. Fast chargers are conventionally bulky and heavy; therefore, they cannot be placed on the vehicle. Portable chargers that can be placed on the vehicle must be designed to have a lower weight and volume, resulting in lower power. A compromise in the literature provides high-power charging without added bulk and heavy components on the vehicle. The

DOI: 10.4018/978-1-7998-6858-3.ch004

solution reuses existing components of the traction drive, reconfiguring components to implement an on-board integrated battery charger (IBC).

Background

Electric vehicle (EV) drivetrains usually consist of a battery, traction motor, and drive (see Figure 1). If a charger is needed, such as for plugin electric vehicles (PEVs), an additional power converter is required to connect to the direct current (DC) link created by the battery. The connected charger is classified as an on-board or off-board charger. On-board chargers must be on the vehicle and, therefore, are designed to have a small weight and size. This results in a low power charger. Most of the alternating current (AC) level 1 chargers are on-board chargers. Off-board chargers, on the other hand, are not placed on the vehicle. This gives them more flexibility in terms of size and weight; therefore, they can be designed for high-power charging. AC level 2 and DC chargers are often off-board and higher power as compared to AC level 1 chargers.

Figure 1. A typical EV drivetrain with battery charger

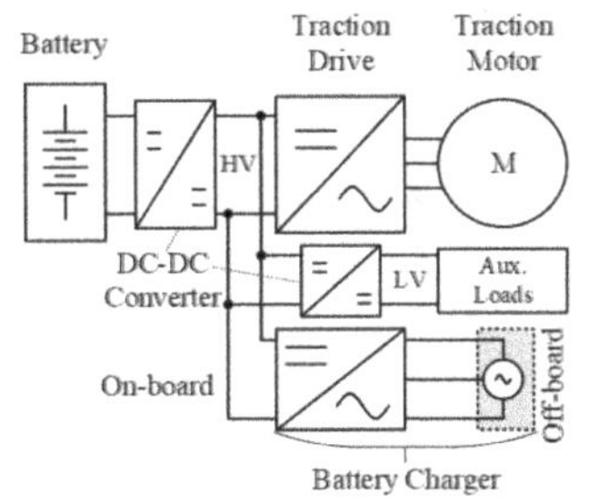

The main components of the charger include the power electronic converter and harmonic filter. An optional transformer can be added to provide galvanic isolation. However, if the personnel safety against electric shock is provided by another device, such as a residual current device (RCD), it can be omitted. On close observation, the traction drive has most, if not all, major components required by a charger.

Slicker (1985) and Thimmesch (1985) built on this to propose the IBC concept: reuse the existing components of the traction drive for charging. This is possible because of the mutual exclusivity of the two functions. When the EV traction drive is used for propulsion, its battery is not charged and when the EV battery is being charged, the vehicle is not required to move. In most cases, a topology change is required to switch between the propulsion mode and charging mode of the circuit. Therefore, most of the IBCs proposed require additional reconfiguration components (see Figure 2). The configuration of the battery charger in an IBC can be done in various ways and for different traction drive and motor topologies. The concept provides solutions from low-power level 1 chargers to high-power level 2 (and higher) chargers with an AC or DC supply interface.

Figure 2. Representation of an IBC that utilizes the motor windings as a filter

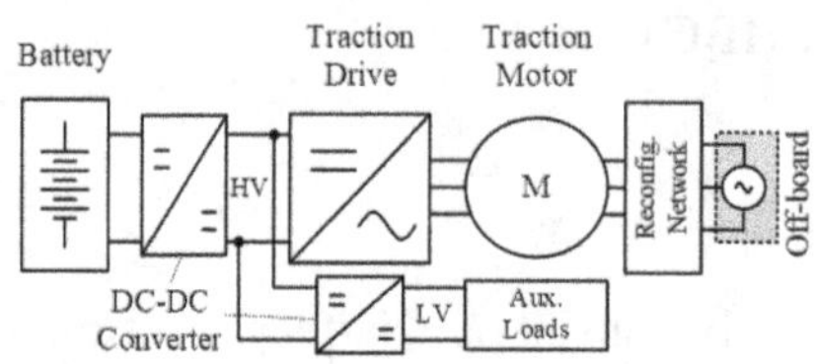

IBC TYPES

A detailed representation of the electrical architecture of an EV (as shown in Figure 1) shows three main components that can be repurposed into multifunction components either individually or in combination. Each component can be included or excluded in the active circuit based on the mode of operation via a reconfiguration network. A generalized representation of the EV drivetrains with the reconfiguration network is shown in Figure 3.

The IBCs proposed in the literature consists of three categories, depending on the traction drive component in the integration: (1) IBCs using theDC-DC converters; (2) IBCs using thetraction drive converter; and (3) IBCs using thetraction motor. The following sections explain each category and offer examples.

Figure 3. Detailed view of the EV electrical architecture

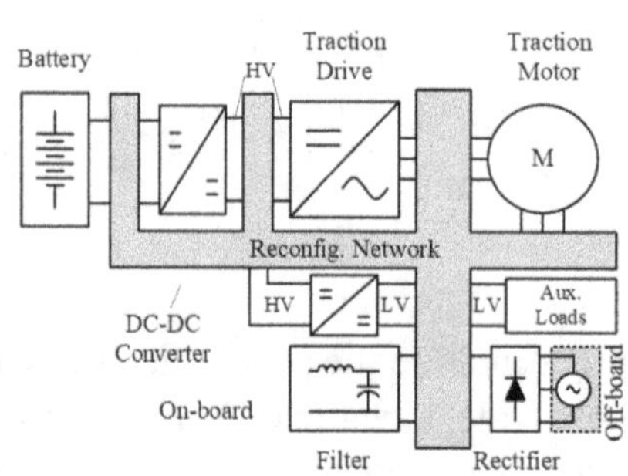

IBCs and DC-DC Converters

According to the architecture in Figure 3, there are two types of DC-DC converters on the EV electric drivetrain. The first's function is the creation of a low-voltage (LV) DC link for auxiliary loads from the high-voltage (HV) bus created for the traction inverter. The second's function is the creation of the HV bus from the available battery voltage, allowing the use of a battery with a wider voltage range. The second DC-DC converter may not be present due to the bulky high-current inductor required (due to increased size and weight of the drivetrain). Researchers have proposed various topologies that allow for the integration of the charging function into either of these converters. This category of IBCs only integrates the charger circuits in the power converter of the traction drive; therefore, they require additional components like filters, transformers, rectifiers, etc.

Lee, Khaligh, and Emadi (2009) proposed a topology that is capable of reusing a single inductor to achieve three mutually exclusive functions: (1) input power factor correction (PFC) while charging to provide power to the HV battery DC link; (2) step-up from the battery to the HV DC link for traction; and (3) step-down from the HV DC link to the LV battery for regenerative braking. Dusmez and Khaligh (2013, 2014) proposed a similar topology with a reduced switch count. Tang et al. (2018) proposed an isolated DC-DC converter with similar functionalities. Chinmaya and Singh (2018a, 2018b, 2019) proposed an IBC that integrates the charging function in the first stage of the traction drive DC-DC converter. Similarly, Anwar, Zhang, Wang, and Costinett (2016) proposed a converter that functions as an isolated interleaved boost DC-DC converter for the propulsion mode and a dual active bridge in the charging mode.

IBCs and Traction Drive Converter

The traction drive converter is specifically designed to feed the traction motor; therefore, the topology of the converter depends on the motor technology. Various charger configurations were proposed that reused the traction inverter by reconfiguring it into a charger suitable for three-phase or single-phase charging. Some required additional components like filters, isolation transformers, rectifiers, etc.

The IBC proposed by Slicker (1985) reconfigures a two-level, three-phase traction motor drive into a flyback DC-DC converter with an additional rectifier and a high-frequency isolation transformer. Rippel (1990) proposed adding a single-phase rectifier and inductance to form a boost converter with the traction inverter switches. Another proposed topology used one of the windings of a delta-connected motor to form the boost converter. The boost converter, in conjunction with the passive rectifier, can perform PFC. A bidirectional Cuk converter-based drive is proposed by Lacressonniere and Cassoret (2005) for a DC machine. The drive can disconnect the DC machine and connect a single-phase rectifier to form a battery charger with PFC. A reduced switch count converter with an integrated bidirectional boost converter is proposed in Hegazy et al. (2013) for a three-phase induction traction motor. Gui-Jia and Lixin (2011) proposed a current source converterbased traction drive in which the converter can be connected to a three-phase machine in propulsion mode and to a single- or three-phase supply voltage in charging mode. A traction drive inverter with an integrated isolated dual-active bridge (DAB) converter is proposed in Jinhao, Hui, Tengxin, and Yubin (2017). In propulsion mode, the converter is a regular three-phase, two-level converter. In charging mode, two of the legs from the traction inverter operate as the secondary side bridge of the DAB.

IBCs and the Traction Motor

The bulkiest part of the filter of most charger designs is the inductance, which is present as motor windings on the EV drivetrain. Therefore, the size and weight of the power converter can be reduced and optimized. Using reconfiguration networks, the motor windings can be configured to allow a connection to a single- or three-phase voltage supply. They can also be used as part of DC-DC converters, such as the energy buffer for a boost, buck, or a buck-boost converter. Depending on how the windings are configured, an AC or DC current flows in the motor windings during the charging mode of operation. This current may result in a torque on the machine shaft. This is an undesirable effect because the vehicle is not expected to move while parked during the charging session. The IBCs integrating the motor and the motor drive converter are organized into three categories: (1) single-phase chargers; (2) three-phase chargers with conventional three-phase machines; and (3) three-phase chargers based on special machines, including multiphase, switched reluctance, and flux switching machines.

Single-Phase Chargers

This subsection presents IBCs that utilize machine windings to create a single-phase charger. An IBC topology based on a three-phase, two-motor drivetrain is proposed in Rippel and Cocconi (1992). The topology is shown in Figure 4. The idea is to access the neutral of the machine windings and connect the single-phase voltage source to those points. A similar topology is proposed by Seung-Ki and Sang-Joon (1995) for a four-motor traction drive. Two drives are connected in the same configuration proposed by Rippel and Cocconi (1992); the other drives' machine neutrals are connected to the battery. This creates a six, two-stage battery charger (i.e., AC-DC-DC converter) connected in parallel. The topology, with an isolation stage and the possibility to charge the LV accessory battery in addition to the traction battery, is proposed in Gui-Jia and Lixin (2013). Similarly, Solero (2001) and Pellegrino, Armando, and Guglielmi (2010) proposed an IBC that allowed the connection of a rectified single-phase voltage source to the neutral of three-phase machine windings (see Figure 5). The connection forms three-phase interleaved boost converters to charge the battery and perform PFC. The current in the windings from the charging mode of operation does not produce a torque because it is common mode. Xiao, Liu, and Yu (2019) flipped the topology with reconfiguration contactors (see Figure 5). The battery is, therefore, connected to the machine windings neutral and the diode rectifier output on the DC link side. This maneuver creates a set of three parallel connected boost converters, controlling the charging current and operating interleaved to reduce the resulting charging current ripple in addition to performing PFC.

Shi, Tang, and Khaligh (2017) proposed an IBC in which a rectified voltage positive is connected to the first-phase terminal of a Y-connected machine winding. Additionally, the negative is connected to the negative of the DC link. The formed topology is an interleaved boost converter with PFC. It does not require a reconfiguration network. Wang, Zhang et al. (2019) proposed a charger front-end with a topology like Figure 5 (with an additional phase leg connected to the motor neutral in the traction mode). In the charging mode, the single-phase voltage supply is connected between the machine neutral and fourth-phase leg, allowing the converter to be operated as an interleaved PFC converter. The back end of the charger consists of an isolated DAB to control the charging current.

Figure 4. Two machine single-phase IBC

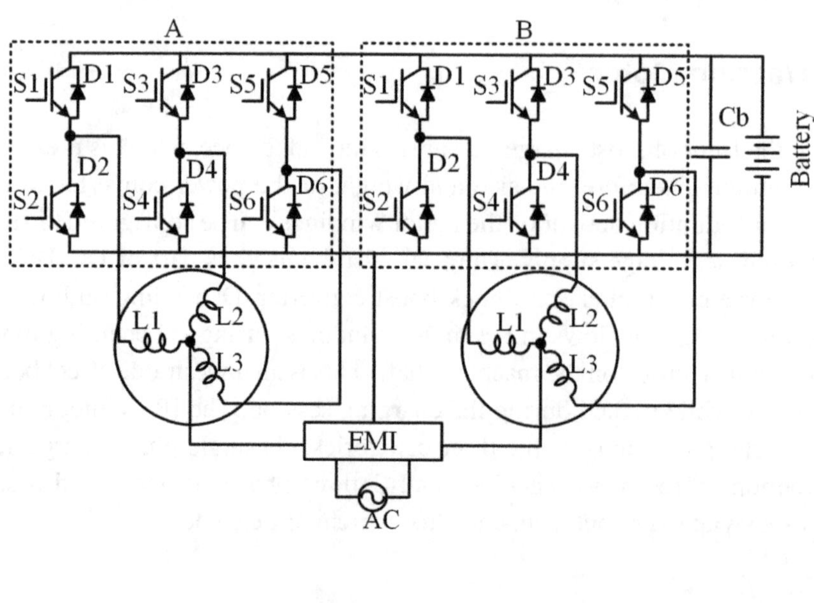

Figure 5. Topology with single-phase rectified voltage connected to machine neutral

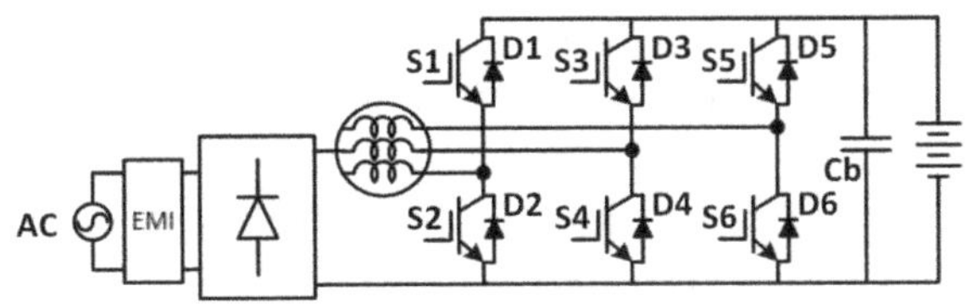

Cocconi (1994) proposed an IBC that allows for the connection of a single- or three-phase voltage source to the machine windings, creating a single- or three-phase boost rectifier that can be controlled to charge the battery. The reconfiguration network allows the motor windings to connect with a common neutral in propulsion mode and as individual branches in charging mode. Truntič et al. (2018) and Milanovic, Roskaric, and Auda (1999) proposed an IBC based on a two-level, three-phase traction motor drive, which can be reconfigured as a converter operating in both buck and boost modes to effectively perform the PFC function and control the battery charging power. The topology is shown in Figure 6.

A similar topology is proposed in Liu, Chen, Yi, and Chen (2015) for a conventional permanent magnetic synchronous motor-based (PMSM) propulsion drive that reconfigures the drive with a full-bridge diode rectifier into a single-phase boost or buck PFC charger based on battery voltage. The topology

Figure 6. IBC topology with converter reconfiguration

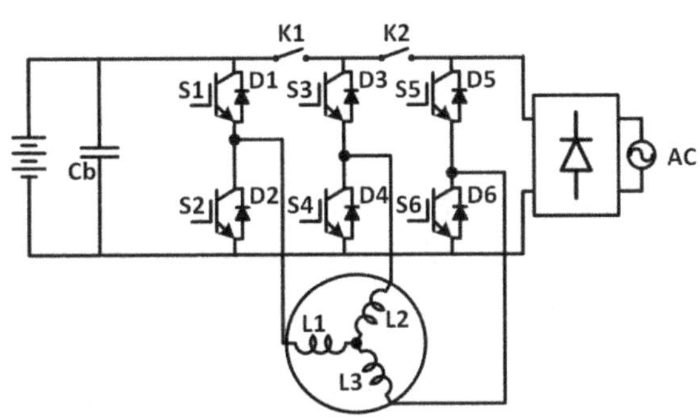

Figure 7. Single-phase charger based on a two-delta connection machine drive

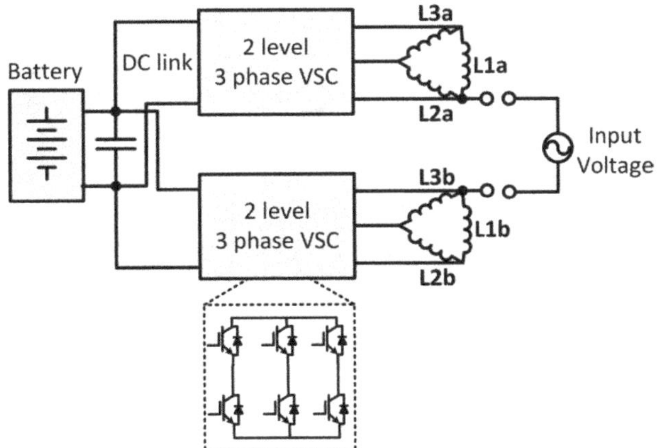

can operate by interleaving two legs to improve the current ripple. An IBC topology based on a two-machine drive with delta windings is proposed in Ali, Mascarella, and Joos (2014). It does not require an additional reconfiguration network; it allows direct connection of the AC supply to the machine terminals as shown in Figure 7.

Several topologies of single-phase IBCs are proposed in Subotic et al. (2016) for various induction machine types, including a nine-, six-, and five-phase machine. The machine windings are reconfigured so that the sinusoidal current in the windings result in exciting the nontorque-producing plane of the respective machines. The topology for the six-phase machine-based IBC is like Figure 4. Similarly, Meher, Banerjee, Vankayalapati, and Singh (2020) proposed a topology in which a conventional two-level converter with a three-phase brushless DC electric motor (BLDC) traction motor drive forms a full bridge boost rectifier for a single-phase input. An LLC resonant stage is added to the battery side interface, creating a charger with galvanic isolation.

Some single-phase IBC topologies are also proposed for specialized drives. An IBC topology for a three-phase machine with open-ended windings is proposed in Zhang et al. (2020). The voltage source is connected at the midpoints of two of the phase-windings. The two full-bridges connected to those phases are then disconnected from the third phase full-bridge, allowing it to be used for second harmonic ripple suppression on the DC link. IBCs based on SRM machines are proposed in Hung-Chun and Chang-Ming (2009, 2011) for traction drives based on four-phase SRM drives. The SRM winding inductances were used to form a buck-boost PFC charger for both topologies proposed. For the latter, an additional diode rectifier is required; in the former, the rectifier formed by the drive diodes is used.

Three-Phase Chargers with Conventional Three-Phase Machines

Various IBCs are proposed with a three-phase input supply and a motor drive with single or multiple three-phase machines. Patent (Kinoshita, 1996) proposed a reconfiguration network, which allows the connection of Y-connected or Δ-connected windings to a three-phase voltage source. Lacressonniere and Cassoret (2005) also proposed a three-phase IBC with a doubly fed induction motor (DFIM), allowing the rotor circuit to directly connect to the grid and the stator circuit to perform the charging operation. The rotor is locked; therefore, the DFIM is used as an isolation transformer. A topology based on an open-ended three-phase interior permanent magnet (IPM) machine drive is proposed in Bruye et al. (2010), De Sousa and Bouchez (2011), De Sousa, Silvestre, and Bouchez (2010), and Lacroix, Laboure, and Hilairet

Figure 8. IBC based on a traction drive with a three-phase open-ended winding machine

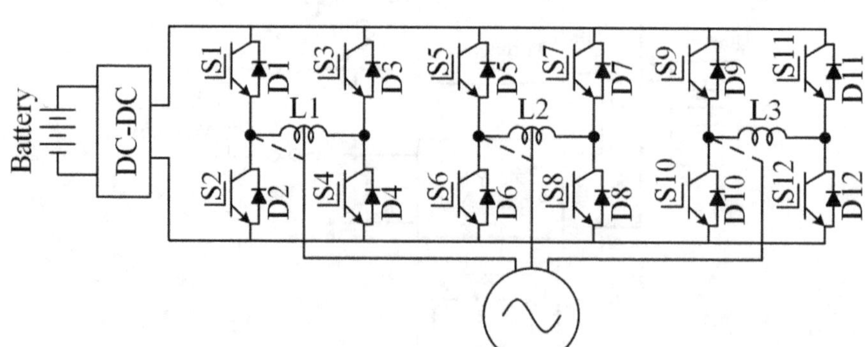

(2010). The authors proposed access to the midpoints of the machine to connect the three-phase voltage supply and control the charging current via the parallel-connected boost rectifiers thus formed. Another variant is to connect the three-phase voltage supply to either terminal of the windings, modulating the switches on the other end to form a three-phase boost rectifier. The topologies are shown in Figure 8.

Ali, Mascarella, Joos, and Tan (2017) proposed an IBC based on a drive with two traction motors on the same shaft. The topology reconfigures the machine windings such that the phase sequence seen by each machine is opposite. Thus, the torque generated by one machine is canceled on the shaft by the torque produced by the second machine. A similar topology is proposed in Sharma, Aware, and Bhowate (2020), in which the machine windings of a three-phase induction motor are split and reconfigured. Thus, the effective supply-phase sequence seen by the two sets of windings is opposite. The result of this reconfiguration is that the magnetic field inside the machine is pulsating rather than rotating. The effective charger topology for both cases is shown in Figure 9.

Figure 9. IBC proposed based on parallel-connected machine windings

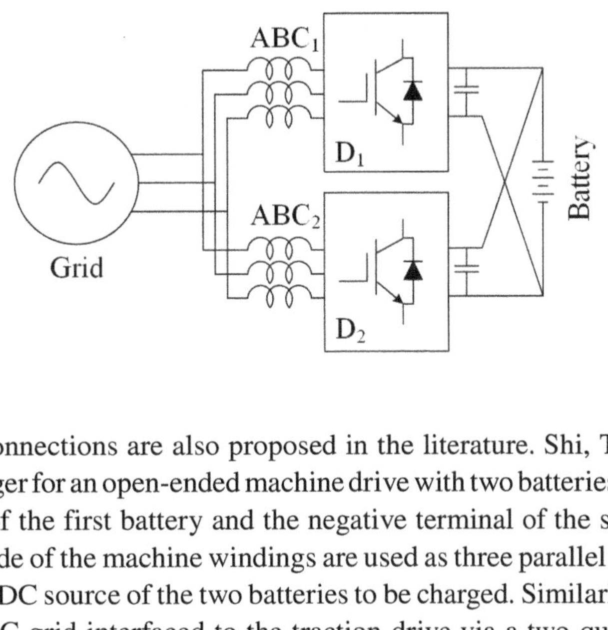

IBCs for direct DC connections are also proposed in the literature. Shi, Tang, and Khaligh (2017) proposed a DC input charger for an open-ended machine drive with two batteries. The DC source connects to the positive terminal of the first battery and the negative terminal of the second battery. The three-phase inverters on each side of the machine windings are used as three parallel full-bridges to control the current flow between the DC source of the two batteries to be charged. Similarly, Viana and Lehn (2019) proposed an IBC for a DC grid interfaced to the traction drive via a two-quadrant DC-DC converter. The output of the interface converter is directly connected to the machine neutral and, therefore, to the other phase legs via the machine windings. This creates synchronized buck-boost DC-DC converters.

Loudot, Briane, Ploix, and Villenueve (2014) proposed an IBC that requires an additional current source converter, the output of which connects to the machine winding neutral. The rest of the traction inverter operates as three parallel-connected boost converters (see Figure 10). Shi, Tang, and Khaligh (2018) proposed a similar charger; however, the current source converter is connected to the machine terminals (the dotted line in Figure 10). Access to the machine neutral is not required. A similar topology is proposed in Shi and Khaligh (2018), in which the current source inverter is a three-switch, three-phase buck rectifier. The topologies proposed are a two-stage battery charging circuit. The motor windings in all cases see a DC current; therefore, the torque can be minimized by locking the rotor at the zero-torque angle (i.e., $\frac{\pi}{2}$ rad).

Figure 10. Current source-based IBC topology

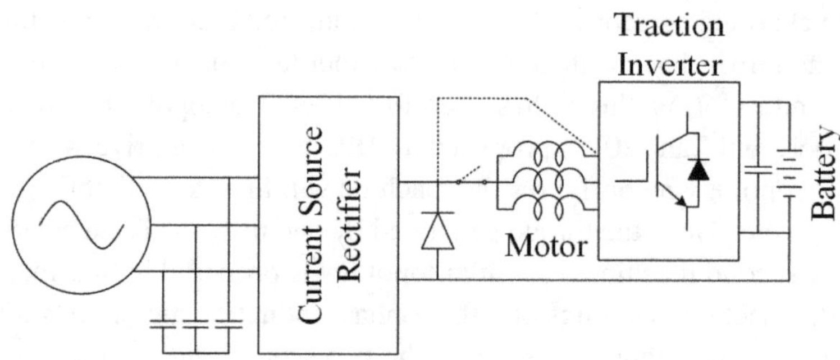

Three-Phase Chargers with Special Machines

IBCs with special traction drives with machines (i.e., multiphase machines, switched reluctance machines [SRMs], flux switching, and machines with reconfigurable windings) are also proposed in the literature. The motivation of such machines is that they provide additional degrees of freedom, which is usually used to address the problem of torque creation on a shaft for high-power charging solutions. Haghbin (2013) and Haghbin et al. (2013) proposed a reconfiguration scheme to split the three-phase windings of an IPM into two sets of galvanically isolated windings. During the charging operation, the first set of windings is used to bring the rotor to nominal speed (corresponding to the 50 Hz electrical frequency) in synchronism with the grid voltage. Once the rotor is in sync, the grid's three-phase voltage is connected and the machine starts to charge. To allow the rotor to rotate during the charging operation, the machine shaft must be decoupled from the mechanical drivetrain via a clutch. A similar concept is proposed in Abdel-Khalik, Massoud, and Ahmed (2018); however, it is for an IPM with fractional slot concentrated windings with either three or six phases.

IBCs based on multiphase machines like five-phase (Subotic et al., 2016), six-phase (Ali, Mascarella, Joos, & Tan, 2018; Subotic, Bodo, Levi, et al., 2016; Subotic, Bodo, & Levi, 2017; Wang, Liu et al., 2019) or nine-phase machines (Bodo et al., 2017; Subotic, Bodo, Levi, & Jones, 2015) are proposed for high-power charging for an isolated or nonisolated case. However, in each case, the high-power charging results in a significant current flowing through the machine windings. This could potentially move the rotor or add a mechanical vibration. The extra degree of freedom afforded by the nature of multiphase machines is exploited in each case to direct the current through the machine windings in a way where the torque-producing plane is not excited, resulting in a pulsating magnetic field in the machine.

IBC COMPARISON

The topologies presented in the previous sections are compared and summarized in terms of their performance and features. Table summarizes the topologies, which are based on multifunctional DC-DC converters. At this time, the IBC topologies proposed for DC-DC converters with multiple function integration require additional components, such as a rectifier. They are proposed for single-phase level 1 charging. The power levels associated with level 1 (typically 1.92 kW) charging do not require bulky, heavy components when compared to high-power charging. Therefore, the benefits provided by integrating multiple functions into a single charger are limited.

Table 2 summarizes the topologies that integrate the battery charging function with the traction drive converter. These topologies are mainly dependent on the traction converter designed for the propulsion mode of operation. This is usually rated at a much higher power than the charging power proposed, meaning the converter switches are over-designed for the charging operation. Therefore, the charging efficiency is compromised (as evidenced by the efficiency numbers in the table). The exception, as proposed by Gui-Jia and Lixin (2011), uses a current-source inverter as the traction drive converter. This design requires a bulky DC link inductor, which adds losses compared to a DC link capacitor. In addition, a voltage-to-current (V/I) converter is required on the battery to interface it to a current source converter, resulting in additional losses.

The topologies proposed that integrate the battery charging function using both the traction drive and traction motor are presented in Table 3 and Table 4 (single-phase and three-phase, respectively). Despite the use of the traction motor windings, the chargers proposed for single-phase use are rated for low power, except for the IBCs proposed by of Ali et al. (2014) and Shi, Tang, and Khaligh (2017). These chargers do not have the same issue as the other single-phase IBCs (for example, the charging power is comparable to the power rating of the electric power train, which limits efficiency compromise). Also, single-phase chargers use machine windings. Therefore, there is neither a rotating magnetic field produced in the machine nor a risk of torque generation by design. However, the same is not true for three-phase IBCs integrating the traction drive converter and traction motor. The three-phase currents in the machine windings create a torque, which must be minimized by either topology reconfiguration, control, or a combination of both. These IBCs are often proposed for higher powers, comparable to the traction drive power. Therefore, the efficiency loss due to mismatched power ratings is not as significant. There is limited winding inductance provided by the machines; therefore, it may not be enough to filter the switching harmonics injected into the grid while charging. If such is the case, additional filters are needed, which may overshadow the benefit provided by integrating the charging function.

For multiphase machines, significant reconfiguration is needed. This requires contactors. Adding contactors in the current path of the propulsion mode compromises the efficiency of the propulsion mode of operation. Therefore, a cost benefit analysis is required before deploying an integrated charger design for EVs. Some topologies also provide the function of injecting power back to the grid (V2G), which may benefit the grid if the charger power is high and the infrastructure supports such functions. The efficiencies reported for the laboratory prototypes of such chargers are not as high as with dedicated charger designs. However, this could be improved by unified optimal design of the electric power train for propulsion and charging mode of operation.

Proposed Features

This chapter reviews IBCs with different topologies and configuration options. It finds that the proposed chargers have the following features: (1) PFC; (2) charging torque elimination; (3) zero-sequence and harmonic cancellation; and (4) V2G. The following section discusses the feasibility of each feature.

PFC

For the on-board charger, especially single phase and low-power level, a diode rectifier is usually used on the grid side. This may result in a grid current distortion and reduced power factor. The integrated charger can reutilize the components to reconstruct a PFC circuit. There are generally two ways of re-configuration:

1. **Reuse Motor Winding as Inductor of PFC Circuit:** This is usually achieved by reconfiguring the windings and the charging/drive converter as an interleaved buck or boost converter. The positive end of the diode rectifier can be connected to different positions, such as the positive end of the DC bus (Shi et al., 2017), one-phase terminal of the charging/drive converter (Xiao et al., 2019), or the neutral point of the motor (Pellegrino et al., 2010).
2. **Reuse Inductor of Intermediate DC-DC Converter Interfacing the Battery:** This may be impossible for a single-stage battery charger (Chinmaya & G. K. Singh, 2019, 2018a, 2018b; Dusmez & Khaligh, 2013, 2014; Tang et al., 2018), which may be impossible for single stage battery charger.

Charging Torque Mitigation

There may be torque generation on the shaft of the motor in the charging mode when the windings of the traction motor are used as a filter inductor, transformer interfacing AC converter, or energy buffer in DC-DC converter. For a three-phase synchronous generator, if the rotor is assumed standstill and a three-phase balanced current flows through the windings, there will be pulsating magneto-motive force (MMF) as the magnetic field of the rotor remains static. Consequently, the average torque of the motor will be zero. There will be a pulsating torque with a fundamental component due to the rotor magnetic field. Additionally, there will be a second and higher-order component due to the saliency of the machine structure (Zhuolin et al., 2014). To derive the torque during the charging operation of a multiphase machine, a generic Park transformation is usually adopted to express the current and voltage in a flux/torque producing plane ($\alpha - \beta$) and nontorque producing plane ($x - y$). A review of the torque generation principle is given in Subotic et al. (2016).

Apart from using mechanical brakes to stop the motor from rotating, there are generally three ways to manage the charging torque. Typical work on torque cancellation is listed in Table 3 and Table 4. Several practical aspects relating to torque cancellation are as follows:

1. Active torque cancellation using winding connections, current control, and a combination of the two:
    a. The use of center-tapped windings cause the current of each phase to split in two opposite directions equally. The MMF of the whole stator will be canceled by the connection. This can be found in the single-phase IBC in Zhang et al. (2020) or three-phase IBC (L. De Sousa et al., 2010).
    b. The current in the machine windings are controlled, making the average torque on the shaft zero. A general law in He et al. (2020) achieves zero torque of IBC. For the single-phase AC input, zero output torque can be achieved by ensuring that the current is equal through all the phases. For three-phase AC input, torque elimination can be achieved by making the phase sequence of the multiple winding set opposite. The effectiveness of the general law is also

reflected in the topologies and operation of the IBC in Bodo et al. (2017), Subotic, Bodo, Levi, Jones, and Levi (2016), Ali et al. (2017, 2018), and Wang, Liu, et al. (2019).

2. The rotor position dependence of the torque cancellation strategy can be an issue for overall charging operation, as demonstrated by Ali et al. (2018). The inclusion of the rotor position in the current reference can completely cancel the torque with the aid of a high-resolution rotor position sensor. Therefore, a circulating reactive power must be tolerated between the converters.
3. Although the rotor is controlled stationary by the above approaches, there will be demagnetization due to the stationary rotor (Abdel-Khalik et al., 2018; Ali, 2018). To address this, a control approach is proposed in Haghbin et al. (2013), which makes the rotor rotate at a synchronous speed. This reduces the demagnetization effect.

Harmonic and Zero-Sequence Current Suppression

When windings of traction motors are used as an AC filter, the impedance of each phase may be unequal, depending on the rotor position with respect to the stator windings. There will, hence, be negative-sequence current and harmonic (especially second order) power flow through the IBC. Although the harmonic torque can be cancelled, the power and current should also be controlled to eliminate the low-frequency harmonics and avoid a DC voltage ripple. For this, a vector resonant PI control in a dual or multiple sequence coordinate can be adopted to suppress the negative-sequence and harmonic current (Bodo et al., 2017; Subotic, Bodo, & Levi, 2016).

For IBC using single-phase PWM rectifier, there will be a second harmonic power and current on the DC side. This passive filter, such as a larger capacitor or LC filter, should be designed to reduce the second harmonics on the DC side. In Zhang et al. (2020), the active filter based on bidirectional H bridge is restructured from dual paralleled three-phase converters.

Regarding paralleled converters with common DC bus in the IBC, a common-mode voltage and resulting zero-sequence circulating power will be found between them. This may cause grid current distortion, extra losses, and magnetic saturation. The issue can be mitigated by special space vector pulse width modulation (SVPWM) methods with carefully designed zero vector placement (Somasekhar, Srinivas, & Kumar, 2008), such as Near State PWM (NSPWM) (Un & Hava, 2007), alternate sub-hexagonal center PWM switching scheme (ASHCSS) (Somasekhar et al., 2008), and interleaved PWM with angle flip at the transition boundary of voltage sectors (Ali, Mascarella, Joos, & Moschopoulos, 2015). This can also be avoided via hardware components like adding a common-mode choke or an isolating transformer before the grid interface (Kinoshita, 1996; Lacressonniere & Cassoret, 2005). The isolating transformer can also block the flow of the triplen and its multiple harmonic components into the grid side. However, additional hardware may result in increased weight and volume of the charging system. Moreover, additional current balancing control may be added to achieve the phase current equalization at different winding sets (Ali et al., 2018; Sharma et al., 2020).

Vehicle to Grid (V2G)

When a bidirectional power electronic converter is used as the grid interface, IBC has V2G capability, which can send active power back to the grid and provide reactive power (Abdel-Khalik et al., 2018; Hegazy et al., 2013; Tang et al., 2018; Wang, Zhang et al., 2019). The potential V2G functions include frequency regulation, power smoothing, load/generation following, load shifting, spinning reserve, and voltage and reactive power supportc. However, the constraints of the transferrable power and the impact on battery storage and EV customers should be considered when providing grid services. Given that the power level of IBC is low, the V2G capability of a single IBC may be limited. Therefore, a proper market scheme is needed to aggregate, schedule, and dispatch EVs to harness the required power and energy for grid support. Given the complexity of the market mechanism and need for an optimal design of the charging system of different modes, the research on V2G based on IBC is relatively rare. However, many topologies have V2G capabilities (see Tables 1 through 4).

Table 1. Comparison of single-phase IBCs integrated in the DC-DC converter of the EV electric power train

Reference	Voltage (VLL)	Max Ch. Power (kW)	Reconf.	Add. Comp.	V2G	Isolation	Eff. (%)
(Lee et al., 2009)	120	1.44	Y	RectifierSwitches	N	N	97
(Dusmez & Khaligh, 2013)	120/240	1.8/8.4	Y	RectifierSwitches	N	N	91.6
(Dusmez & Khaligh, 2014)	120/240	1.8/6.6	N	None	N	N	92.1
(Tang et al., 2018)	120/240	3.3	N	PFC Rectifier	Y	Y	92.75
(Chinmaya & Singh, 2019, 2018a, 2018b)	120/240	1	N	Rectifier	N	N	NA
(Anwar et al., 2016)	120/240	5	Y	PFC Rectifier	N	Y	98.5
(Lacressonniere & Cassoret, 2005)	220	1.2	Y	RectifierContactors	N	N	90

Table 2. Comparison of single-phase IBCs integrated in the traction drive converter of the EV electric power train

Reference	Voltage (VLL)	Max Ch. Power (kW)	Reconf.	Add. Comp.	V2G	Isolation	Eff. (%)
(Slicker, 1985)	120	NA	Y	Rectifier FilterContactors	N	N	NA
(Thimmesch, 1985)	120	3.6	Y	Contactors	N	Y	86
(Rippel, 1990)	120/240	NA	N	RectifierFilter	N	N	NA
(Hegazy et al., 2013)	220	NA	N	Switches	Y	N	91.5
(Gui-Jia & Lixin, 2011)	120/240	22.5	Y	Contactors	Y	N	NA

Table 3. Comparison of single-phase IBCs integrated in the traction drive converter and traction motor of the EV electric power train

Reference	Voltage (VLL)	Max Ch. Power (kW)	Reconf.	Add. Comp.	Torque Canc.	V2G	Isolation	Motor	Eff. (%)
(Rippel & Cocconi, 1992)	120/240	NA	N	None	N	N	N	IM+PMSM	NA
(Seung-Ki & Sang-Joon, 1995)	120/240	1.4	Y	Contactors	N	N	N	IM	NA
(Gui-Jia & Lixin, 2013)	120/240	5	N	Filter	N	N	N	IM+PMSM	92
(Solero, 2001)	120	0.36	N	Rectifier Filter	N	N	N	PMSM	NA
(Pellegrino et al., 2010)	220	1.2	N	Rectifier Filter	N	N	N	IPM	87.5
(Xiao et al., 2019)	230	1.8	Y	RectifierContactors	N	N	N	IPM	94
(Shi, Tang et al., 2017)	120/240	3	N	Rectifier	N	N	N	IPM	93.1
(Wang, Zhang et al., 2019)	120/240	1.2	Y	Contactor	N	Y	N	IM	92
(Truntič et al., 2018)	120	0.4	Y	RectifierContactors	N	N	N	NR	94
(Milanovic et al., 1999)	220	1.2	Y	RectifierContactors	N	N	N	NR	NA
(Liu et al., 2015)	120	0.55	Y	RectifierContactors	N	N	N	NR	97
(Ali et al., 2014)	240	19.2	N	None	N	Y	N	2 x PMSM	NA
(Subotic et al., 2016)	240	0.4	N	None	Y	Y	N	9ph, 6ph 5ph IM	86
(Meher et al., 2020)	123	0.87	Y	Contactors	N	N	Y	BLDC	97.6
(Zhang et al., 2020)	100	0.4	N	None	Y	N	N	IPM	NA
(Hung-Chun & Chang-Ming, 2009)	100	0.47	Y	Contactors	N	N	N	SRM	82.8
(Hung-Chun & Chang-Ming, 2011)	105	0.44	N	None	N	N	N	SRM	NA
(Shi et al., 2017)	380	50	N	None	NA	N	N	PMSM	NA
(Viana & Lehn, 2019)	260	9	N	DC Converter	NA	N	N	SM	NA

Table 4. Comparison of three-phase IBCs integrated in the traction drive converter and traction motor of the EV electric power train

Reference	Voltage (VLL)	Max Ch. Power (kW)	Reconf.	Add. Comp.	Torque Canc.	V2G	Isolation	Motor	Eff. (%)
(Cocconi, 1994)	120/240	NA	Y	Contactor	N	N	N	PMSM	NA
(De Sousa et al., 2010; De Sousa & Bouchez, 2011; Lacroix et al., 2010)	240	3.3	N	None	Y	N	N	IPM	NA
(Ali et al., 2017)	208	28.8	Y	Contactors	Y	Y	N	2 x PMSM	NA
(Sharma et al., 2020)	100	1.2	Y	Contactors	Y	N	N	IM	86
(Loudot et al., 2014)	400	NA	N	CS Rectifier	NA	N	N	any	NA
(Shi et al., 2018)	208	3.3	N	CS Rectifier	NA	N	N	any	92.6
(Shi & Khaligh, 2018)	190	1.5	N	CS Rectifier	NA	N	N	PM	95.5
(Haghbin, 2013; Haghbin et al., 2013)	197	5	Y	Contactors	N	Y	Y	IPM	79
(Abdel-Khalik et al., 2018)	14	10	Y	Contactors	N	Y	Y	IPM	79
(Ali et al., 2018)	208	28.8	Y	Contactors	Y	Y	N	6ph PMSM	NA
(Subotic et al., 2016)	400	1	Y	ContactorsTransformer	Y	Y	Y	6ph IM	NA
(Subotic et al., 2017)	415	0.8	Y	Contactors	Y	Y	N	6ph IM	NA
(Wang, Liu et al., 2019)	208	5	Y	Contactors	Y	Y	Y	6ph PMSM	90
(Bodo et al., 2017; I. Subotic et al., 2015)	415	2	N	None	Y	Y	N	9ph IM9ph PMSM	9190

IBC OPPORTUNITIES

IBCs provide an opportunity to reduce the size, weight, volume, and cost of a charger. The case for cost savings can be convincingly made when considering the system level (i.e., the utility grid point of connection to the EV battery). IBCs reuse the existing components of the traction drive; therefore, such integration has the potential to provide high-power charging to the user without the user investing in the infrastructure. Such chargers have value for services requiring a fleet of EVs. To provide dedicated fast chargers for vehicles, significant investment in infrastructure is required, including a power connection to the utility, a transformer, and a fast charger (the costliest item). With such costs, only a limited number of chargers can be bought; therefore, the vehicles must be optimally scheduled to ensure complete recharge before the next trip. With an IBC, the charger electronics are on all fleet vehicles. The investment is only on the grid connection and required transformers, allowing for more vehicles being charged at a given time.

IBC CHALLENGES IN ADOPTION

Various IBC solutions have been proposed, including functional validation results. However, the technology is yet to be widely adopted by the industry. Several original equipment manufacturers (OEMs) have patented the design of IBC technology, including Eaton Corporation (Slicker, 1985) and General Motors (Rippel & Cocconi, 1992). However, this product is not commercially available to EV manufacturers and operators. Some OEMs proposed products by AC Propulsion (Cocconi, 1994) for a 150 kW drive with an 18 kW charger, systems from Valeo (De Sousa & Bouchez, 2011) for a 22 kW drive and charger, and Renault (Briane & Loudot, 2011; Loudot et al., 2014) for systems up to 43 kW, the Chameleon charger integrated in an EV called Zoe. At this time, AC Propulsion and Valeo have discontinued their IBCs. Renault has reduced the integrated charging power for their Chameleon charger in recent Zoe models (since 2020).

Two issues, reliability and grid code/standards compliance and component design, are identified as remining challenges. Each issue is discussed in detail.

Reliability

IBCs reuse the existing components of the drivetrain for an additional function. Therefore, the reliability of the drive is not the same as if it were solely used for traction. This implies that the lifecycle of the drivetrain decreases. IBCs that require reconfiguration networks need electromechanical contactors for implementation. Such contactors can be bulky if designed for high currents. Additionally, they introduce losses due to contact resistance for both traction and charging modes of operation.

Many IBCs that use motor windings as filters have currents flowing through them that may result in a torque on the shaft. For most IBCs, locking the rotor is enough because the torque produced is not significant. However, for high-power IBCs, the solution is to allow the motor shaft to rotate (by having a clutch) or have a control strategy that prevents a resultant torque on the shaft by converting the rotating magnetic field inside the machine to a pulsating one. For motor drives with permanent magnets, the effect of pulsating fields for torque cancellation strategies and rotating fields in locked rotor strategies on the permanent magnets is not completely understood. However, some attempts have been made for certain types of IBCs and certain machine topologies (Ali, 2018; Li et al., 2019).

Grid/Standards Compliance and Component Design

A battery charger design is expected to meet specific standards related to efficiency, harmonic distortion, and electromagnetic interference and compatibility (EMI/EMC). Most of the proposed IBCs adopt a strategy of repurposing a drive that was designed and optimized for the traction mode of operation that are suboptimal for chargers. For example, for efficient traction operation, motors are designed to have minimum inductance. However, if machine windings are used for the charging application, they must be designed to ensure a minimum current ripple on the AC side. Similarly, the DC link capacitance is designed to filter the switching harmonics of the converter and minimize the effect of sudden load changes. Regarding the case of single-phase chargers, the DC link capacitor must also be designed to sustain the second harmonic components. Another area for exploration is the component rating. The traction drive DC link is designed to provide the voltage required by the motor to achieve its rated speed. Regarding the charger, the DC link voltage is designed for the input supply voltage (it may change based on region). So far, a unified IBC design strategy that designs a system that meets all traction and charger design requirements in an optimal fashion does not exist. Mukundan et al. (2018) attempted to design the traction motor. However, the concept needs to be extended to the system level design of the traction drive and battery charger as a unit.

CASE STUDIES

The two case studies presented in this section provide implementation details of IBCs. The first case study is an IBC proposed by the authors in the literature (Ali et al., 2018). It is based on a six-phase permanent magnet synchronous machine (PMSM) for light- and medium-duty EV drivetrains. The second case study was commercialized for Renault Zoe in Briane and Loudot (2011) and Loudot et al. (2014).

Three-Phase IBC with Six-Phase PMSM

A three-phase IBC based on a six-phase PMSM was proposed in Ali (2018) and Ali et al. (2018). The proposed IBC features bidirectionality and torque cancellation on the machine shaft. The charger was proposed for light/medium duty vehicles for up to 28.8 kW charging.

Topology and Components

The topology of the proposed charger, as shown in Figure 11, requires a reconfiguration of the machine winding connections using contactors. The topology is proposed for a three-phase 208 V_{LL}/60 Hz. The battery voltage supported is between 400 V and 250 V. A higher voltage can be supported if the DC link is chosen to be higher and the proposed topology is 400 V_{DC}. The motor used in the proposed charger is a 100 kW six-phase PMSM. It has a torque cancellation strategy, so the charging does not produce a torque on the shaft of the machine.

Figure 11. IBC proposed based on a six-phase PMSM traction drive

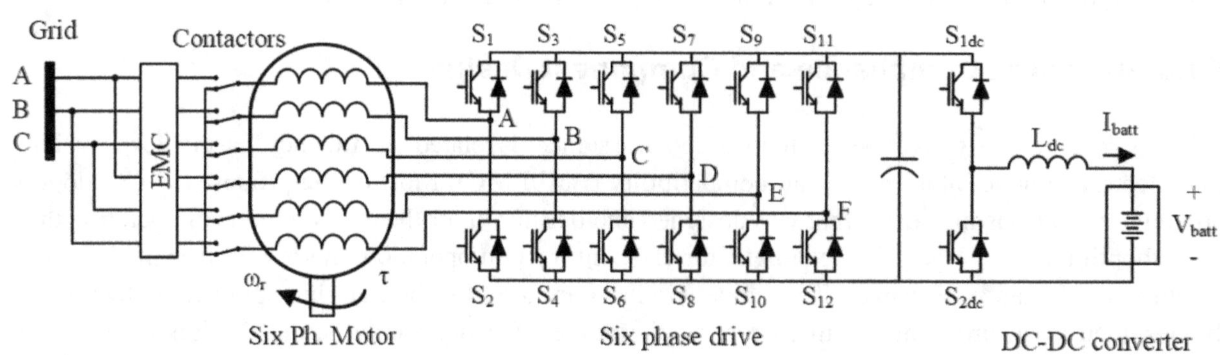

Operation and Control

In the traction mode of operation, the contactors remain in the position as shown in Figure 11. All six phases of the machine connect with a common neutral. The topology can also work if the two sets of three-phase windings, ABC and DEF, have their respective isolated neutrals. The machine is operated as a six-phase motor. The six-phase, two-level drive formed by switches S_1-S_{12} control the phase currents of the motor, which control the torque and speed of the motor.

In the charging mode of operation, the contactors disconnect the motor windings from the neutrals and connect them to the three-phase grid supply. In the reconfiguration, the phase sequence of the first three-phase set (ABC) is opposite from the second three-phase set (DEF). In other words, phase ABC of the motor is connected to the ABC phase of the grid; the DEF phases of the motor are connected to the ACB phase of the grid. This controls the current in the windings so the total resulting torque on the motor shaft is eliminated.

Torque Cancellation

By transforming the motor currents on a rotating frame aligned with the magnetic axis of the rotor, as shown in Figure 12, the total torque produced on the rotor, τ, can be written as Eq. (1). Angle α in the figure is the angular displacement between the two three-phase sets of the motor windings. The common values for α are 0°, 30°, and 60°.

Figure 12. Phase relationships between stator winding magnetic axes and rotor magnetic axis on the rotor frame of reference

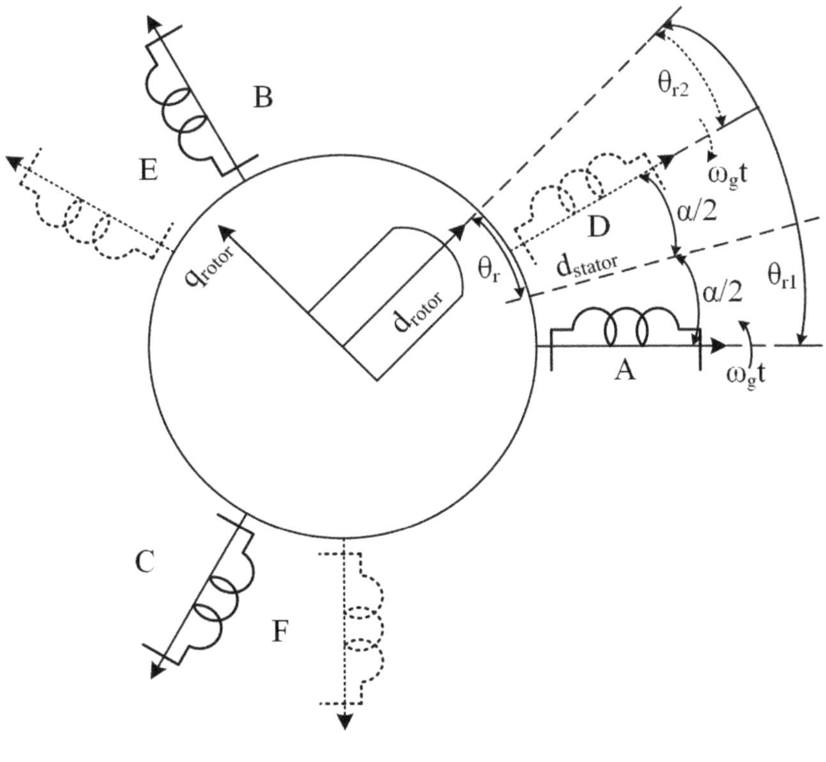

$$\tau = \frac{3}{2} p\left[\lambda_{pm}\left(I_{qs1} + I_{qs1}\right) - \left(L_q - L_d\right)\left(I_{ds1}I_{qs1} + I_{ds2}I_{qs2}\right) + M_d\left(I_{ds2}I_{qs1} + I_{ds1}I_{qs2}\right) - M_q\left(I_{ds1}I_{qs2} + I_{ds2}I_{qs1}\right)\right] \tag{1}$$

- p is the number of poles of the machine.
- λp_m is the flux of the permanent magnets on the rotor d-axis.
- L_d and L_q are the direct and quadrature axis inductances, respectively.
- M_d is the mutual inductance between the stator *d*-axes of set 1 and 2.
- M_q is the mutual inductance between the stator *q*-axes of the two sets.
- $I_{qs1(2)}$ is the stator current component of set 1 (or 2) aligned to the rotor q-axis.
- $I_{ds1(2)}$ is the stator current component of set 1 (or 2) aligned to the rotor d-axis.

To cancel the torque on the shaft, the stator current references should be as given by Eq. (2).

$$\begin{bmatrix} I_{qs1} \\ I_{qs2} \\ I_{ds1} \\ I_{ds2} \end{bmatrix} = \begin{bmatrix} I_{pk}\sin\left(\omega_g t - \theta_r\right) \\ I_{pk}\sin\left(-\omega_g t - \theta_r\right) \\ I_{pk}\cos\left(\omega_g t - \theta_r\right) \\ I_{pk}\cos\left(-\omega_g t - \theta_r\right) \end{bmatrix} \tag{2}$$

- I_{pk} is the peak magnitude of the stator current.
- ωg is the grid frequency in rad/s.
- t is the time in seconds.
- θr is the rotor angle with respect to the middle of the stator magnetic d-axes of the stator.
- The value of I_{pk} is calculated based on the charging power to be absorbed from the grid supply.

The total torque produced on the motor shaft is cancelled when the values in (2) are substituted in (1). The implication of torque cancelation is that the current references of the phase sets must be displaced from each other by an angle equal to the rotor angle. If the value of $\theta r_=$ 0, the phase sets produce a current that is in phase. However, if the value of θr $_i$s non-zero, some amount of circulating current flows between the phase sets. The total current drawn from the grid is in phase with the grid voltage.

Results for the case of 7.2 kW charging are shown in Figure 13 and Figure 14. Figure 13 shows the stator currents in the stator d and q axes following the references by Eq. (2). Figure 14 shows the rotor position, which is measured by a displacement sensor. It is both with and without the torque cancellation scheme. Figure 14 shows that the vibrations on the rotor motor shaft are effectively eliminated.

Figure 13. Stator d and q axes currents for 7.2 kW charging with torque cancellation

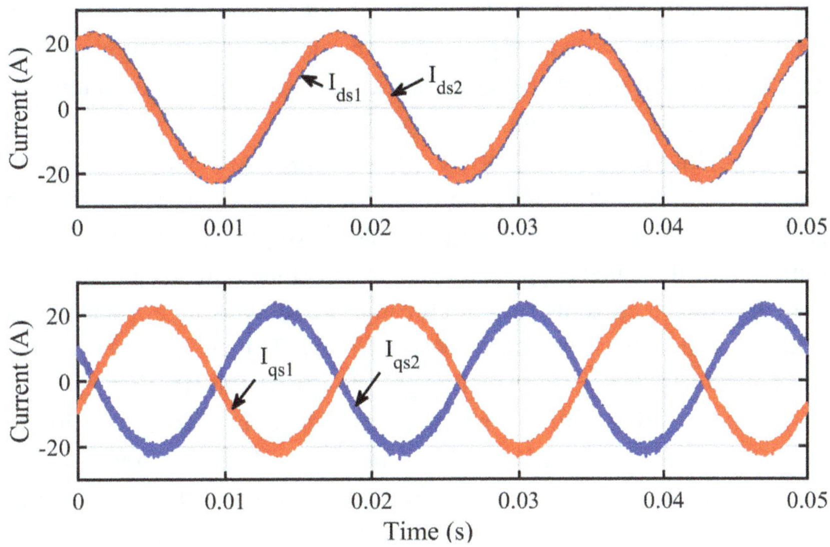

Figure 14. Rotor position measured using a displacement sensor for with and without control at 7.2 kW charging power

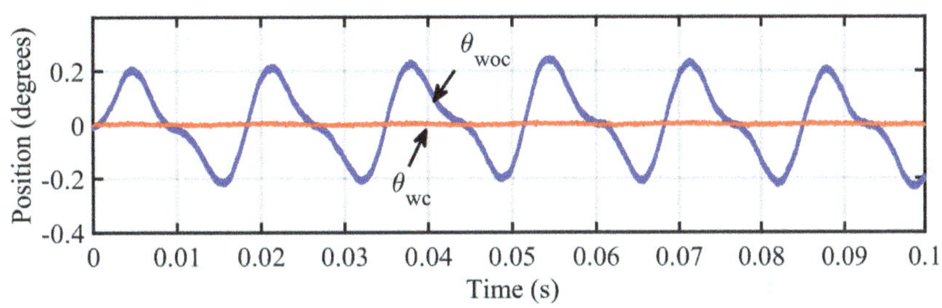

Discussion

The IBC in this section has practical aspects that must be considered when implementing the charger.

1. **Machine Winding Inductance:** This is conventionally designed to be small, which minimizes the voltage drops at rated operations. However, this directly conflicts with the charger design requirement of an inductance that is large enough to filter the harmonics of the current injected into the grid. Particularly, in the charging mode of operation, the winding inductances must address the following:
    a. **Total Harmonic Distortion:** The charger must limit the harmonics injected into the grid; therefore, the machine inductances for charging operation would be designed to meet a certain value. If this design objective does not satisfy other design objectives, such as torque and speed maximization, it can be relaxed by including a capacitive filter. If it still does not meet criteria, an inductance can be added in series with machine windings. However, this results in added weight and volume to the charger.

   b. **Circulating Current:** Parallel connected converters as proposed in this IBC have a common mode circulating current path between them. Therefore, the machine windings will either provide a common mode impedance to reduce the common mode current between the converters or be added after the contactors so that it is not in the conduction path when the system operates in the traction mode of operation.
2. **Increased Cycling, Machine Losses, and Thermal Implications:** When the machine windings and inverter are also used for charging, the usage of the drive increases. This may have a detrimental impact on the lifetime of the drive motor and drive power electronics. To address this, the component must be overdesigned to ensure durability. Furthermore, despite the torque cancellation, the machine windings have currents flowing through them, causing copper losses in the coils and eddy current losses in the stator iron and rotor magnets. This leads to a temperature increase of both the stator and rotor. Extraction of heat is possible from the stator if there is a liquid cooling loop. However, heat extraction from the rotor is a challenge, especially if stationary.
3. **Rotor Saliency and Stator Saturation:** The torque cancellation strategy is proposed for torque components up to the second order component (reluctance torque). However, if the machine design is such that higher order harmonic components of the torque exist, the proposed strategy does not guarantee cancellation. Similarly, if a lower ratings motor is used, the charging operation may saturate the stator core of the motor. This will also result in a failure of the strategy to effectively cancel the torque as the assumption that the d-axis inductance for both phase sets is equal is no longer true.

Chameleon Charger for Renault Zoe

A bidirectional IBC proposed by Briane and Loudot (2011) utilized the windings of the machine as part of the charger topology. The topology and controls of the charger are discussed in this subsection.

Topology and Components

Figure 15 shows the topology of the drivetrain and the proposed charger. The figure shows on-board components; only the connection to the grid is offboard. The topology can be used for connection to a three- or single-phase AC voltage network or DC network. For the case of a three-phase network, the points of connection (L_1, L_2, L_3, and N) can be used to connect a three-phase, four-wire connection (the neutral connection can be omitted if the network is a three-phase, three-wire connection). Single-phase AC or DC power supply networks can be connected between points L_1 and L_2. The charger supports the universal input supply (i.e., it can operate with a 230V/50Hz, 100V/50Hz, 100V/60Hz, and 120V/60Hz electric supply network). The voltage of the battery supported by the charger can vary from 250V to 400V, with a maximum charging current of up to 200A for when the battery is discharged (at 250V).

The main feature of the charger is that it allows for the reuse of motor windings as a filter for the DC-DC converters. If designed for a 400V battery, the 63A charging current with a three-phase AC interface can be as heavy as 30 kg and as large as 20 liters. Another feature of the charger is that it does not require an additional contactor or circuit reconfiguration. An additional design feature of the charger is that it supports a single (or double for a three-phase four-wire power supply) free-wheeling diode to avoid using an arm of the current source converter to short circuit the stator winding current. This feature allows for a reduction in conduction and switching losses.

Figure 15. Topology of the IBC proposed by Renault

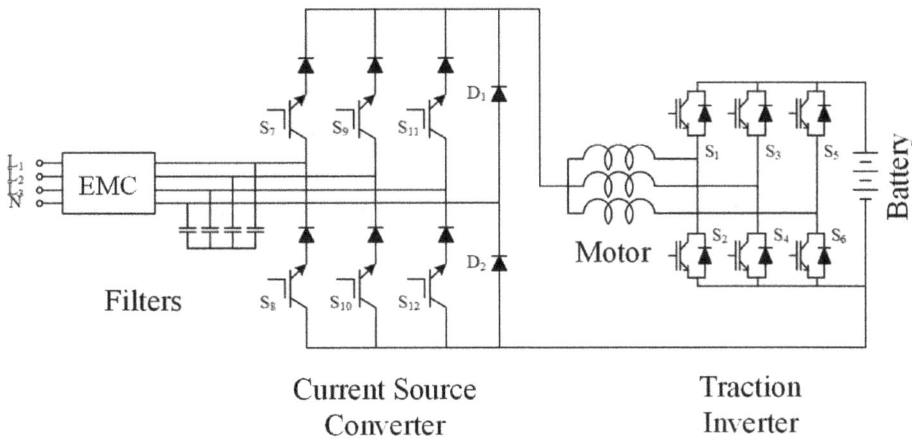

Operation and Control

In the traction mode of operation, the power supply is not connected; the current source converter (S_7-S_{12}) and diode D_1 (and D_2) are not operated. The motor windings connect in a star with a floating neutral. The traction inverter is a conventional two-level voltage source converter (S_1-S_6), with the traction battery connected on the DC side. In this mode of operation, the traction inverter is pulsed to control the motor current, which, in turn, controls the motor torque and speed.

In the charging mode of operation, a power supply is connected to the charger input terminals. There are three possible power supply connections for the charger: single-phase or three-phase (with or without neutral) AC and DC. For each case, the operation and control of the traction inverter portion of the charger remains unchanged. Therefore, the control and operation of the second stage is explained first.

The second stage of the charger consists of the battery, traction inverter, and motor windings. Each arm of the traction inverter with motor windings forms a buck converter with the IGBT of the top switch and the diode of the second switch. In total, the traction inverter converts into three buck converters, which can be interleaved to produce a charging current for the battery that has lesser harmonics when compared to the case of noninterleaved converters. The current of each arm can be individually controlled or controlled together to produce a charging current for the battery. The charging current reference depends on the charging strategy. This is limited by the maximum charging power available at the AC interface.

The first stage of the charger consists of the interface to the power supply. This interface has a three-phase, two-level current source converter, using unidirectional IGBTs. The current source converter uses three or two legs, depending on the supply nature. For a three-phase AC system (three or four wire), all three legs (switches S_7-S_{12}) are used. For a single-phase AC or DC power supply, the first two legs (S_7-S_{10}) are used. In all cases, the current drawn by the charger is controlled to be in synchronism with the supply voltage. The output current provided by this stage cannot be more than 10% higher than the battery charging current.

Discussion

The charger presented in this section does not require a contactor for reconfiguration. However, it requires six additional switches and seven diodes. Furthermore, the control of the output stage of the charger requires current sensors for each phase to control the individual buck converter current. Most three-phase motor drives with motor windings connected in star use two current sensors. The third-phase current is derived from the two measured currents. This does not apply in the charging case. Moreover, points raised about the increased cycling of the drive components due to an added charging function affects the overall life of the components.

CONCLUSION

This chapter provides a comprehensive introduction and review of the state-of-the-art characteristics of IBCs. Various topologies, as reported in the literature, showcase solutions that either exist in the market or are presented as proof-of-concepts in the literature. Opportunities and challenges faced by the technological adoption of IBCs are presented. At this time, there has been one case of an IBC featured in an EV. IBCs must overcome many obstacles before they gain wider industrial acceptance.

FUTURE RESEARCH

Future trends in IBC research and development include:

1. Quantifying the impact of IBCs on the reliability of the traction drive and electric drivetrain
2. Defining a unified optimal design strategy for the electric drivetrain, meeting both traction drive and charger requirements
3. Developing a strategy to unlock charging opportunities offered by high-power IBCs by increasing the availability of required charging outlets, including three-phase systems, and formalizing the transactional nature and benefits of this charging option

ACKNOWLEDGMENT

This research was partially supported by the Natural Science and Engineering Research Council of Canada (NSERC) and the industrial partners TM4 Electrodynamic Systems, Linamar, and Mentor Infolytica for their support under the Automotive Partnership Canada (APC) project.

REFERENCES

Abdel-Khalik, A. S., Massoud, A., & Ahmed, S. (2018). Interior permanent magnet motor-based isolated on-board integrated battery charger for electric vehicles. *IET Electric Power Applications*, *12*(1), 124–134. doi:10.1049/iet-epa.2017.0319

Ali, Mascarella, Joos, & Moschopoulos. (2015). Circulating current minimization for dual three phase motor integrated battery charger. *2015 IEEE Energy Conversion Congress and Exposition (ECCE)*, 3651–3658. doi:10.1109/ECCE.2015.7310176

Ali, S. Q. (2018). *Integrated Battery Chargers for Electric Vehicle Traction Drives with Synchronous Machines*. McGill University Libraries.

Ali, S. Q., Mascarella, D., & Joos, G. (2014). Integrated battery charger for delta connected machines in plug-in hybrid electric vehicles. *2014 IEEE 27th Canadian Conference on Electrical and Computer Engineering (CCECE)*, 1–6. 10.1109/CCECE.2014.6901072

Ali, S. Q., Mascarella, D., Joos, G., & Tan, L. (2017). Torque elimination for integrated battery charger based on two permanent magnet synchronous motor drives for electric vehicles. *IET Electric Power Applications*, *11*(9), 1627–1635. doi:10.1049/iet-epa.2017.0302

Ali, S. Q., Mascarella, D., Joos, G., & Tan, L. (2018). Torque Cancelation of Integrated Battery Charger Based on Six-Phase Permanent Magnet Synchronous Motor Drives for Electric Vehicles. *IEEE Transactions on Transportation Electrification*, *4*(2), 344–354. doi:10.1109/TTE.2017.2788190

Anwar, S., Zhang, W., Wang, F., & Costinett, D. J. (2016). Integrated DC-DC converter design for Electric Vehicle powertrains. *2016 IEEE Applied Power Electronics Conference and Exposition (APEC)*, 424–431. 10.1109/APEC.2016.7467907

Bodo, N., Levi, E., Subotic, I., Espina, J., Empringham, L., & Johnson, C. M. (2017). Efficiency Evaluation of Fully Integrated On-Board EV Battery Chargers With Nine-Phase Machines. *IEEE Transactions on Energy Conversion*, *32*(1), 257–266. doi:10.1109/TEC.2016.2606657

Briane, B., & Loudot, S. (2011). Rapid reversible charging device for an electric vehicle. *Patent* No. US20110254494, A1.

Bruye, X., Re, A., De Sousa, L., Bouchez, B., Sandulescu, P., Kestelyn, X., & Semail, E. (2010). *A multiphase traction/fast-battery-charger drive for electric or plug-in hybrid vehicles: Solutions for control in traction mode*. doi:10.1109/VPPC.2010.5729220

Chinmaya, K. A., & Singh, G. K. (2018a). A Multifunctional Integrated Onboard Battery Charger for Plug-in Electric Vehicles (PEVs). *2018 IEEE 18th International Power Electronics and Motion Control Conference (PEMC)*, 28–33. 10.1109/EPEPEMC.2018.8521990

Chinmaya, K. A., & Singh, G. K. (2018b). A Single-stage Integrated Charger for Electric Vehicles (EVs) and Plug—In Electric Vehicles (PEVs) Incorporating Induction Motor Drive. *IECON 2018 - 44th Annual Conference of the IEEE Industrial Electronics Society*, 954–959. 10.1109/IECON.2018.8591669

Chinmaya, K. A., & Singh, G. K. (2019). Integrated onboard single-stage battery charger for PEVs incorporating asymmetrical six-phase induction machine. *IET Electrical Systems in Transportation*, *9*(1), 8–15. doi:10.1049/iet-est.2018.5015

Cocconi, A. G. (1994). *Combined Motor Drive and Battery Recharge System*. US Patent No. US 5341075.

De Sousa, L., & Bouchez, B. (2011). *Combined Electric Device for Powering and Charging*. Patent No. 0221363 A1.

De Sousa, L., Silvestre, B., & Bouchez, B. (2010). *A combined multiphase electric drive and fast battery charger for Electric Vehicles*. doi:10.1109/VPPC.2010.5729057

Duan, Feng, Wei, Yongxing, Dong, & Xuhui. (2014). Analysis of charging torque in a three-phase integrated charger of electrical vehicle. *2014 IEEE Conference and Expo Transportation Electrification Asia-Pacific (ITEC Asia-Pacific)*, 1–5. 10.1109/ITEC-AP.2014.6940908

Dusmez, S., & Khaligh, A. (2013). A Compact and Integrated Multifunctional Power Electronic Interface for Plug-in Electric Vehicles. *IEEE Transactions on Power Electronics*, *28*(12), 5690–5701. doi:10.1109/TPEL.2012.2233763

Dusmez, S., & Khaligh, A. (2014). A Charge-Nonlinear-Carrier-Controlled Reduced-Part Single-Stage Integrated Power Electronics Interface for Automotive Applications. *IEEE Transactions on Vehicular Technology*, *63*(3), 1091–1103. doi:10.1109/TVT.2013.2284592

Gui-Jia, S., & Lixin, T. (2011). *Current source inverter based traction drive for EV battery charging applications*. doi:10.1109/VPPC.2011.6043143

Gui-Jia, S., & Lixin, T. (2013). *An integrated onboard charger and accessory power converter for plug-in electric vehicles*. doi:10.1109/ECCE.2013.6646895

Haghbin, S. (2013). *Integrated Motor Drives and Battery Chargers for Electric or Plug-in Hybrid Electric Vehicles*. Chalmers University of Technology. https://publications.lib.chalmers.se/records/fulltext/173471/173471.pdf

Haghbin, S., Khan, K., Zhao, S., Alakula, M., Lundmark, S., & Carlson, O. (2013). An Integrated 20-kW Motor Drive and Isolated Battery Charger for Plug-in Vehicles. *IEEE Transactions on Power Electronics*, *28*(8), 4013–4029. doi:10.1109/TPEL.2012.2230274

He, S., Xu, Z., Chen, M., Yang, H., & Li, W. (2020). General Derivation Law with Torque-Free Achieving of Integral On-Board Charger on Compact Powertrains. *IEEE Transactions on Industrial Electronics*, 1–1. doi:10.1109/TIE.2020.3005063

Hegazy, O., Barrero, R., Van Mierlo, J., Lataire, P., Omar, N., & Coosemans, T. (2013). An advanced power electronics interface for electric vehicles applications. *IEEE Transactions on Power Electronics*, *28*(12), 5508–5521. doi:10.1109/TPEL.2013.2256469

Hung-Chun, C., & Chang-Ming, L. (2009). Development of a Compact Switched-Reluctance Motor Drive for EV Propulsion With Voltage-Boosting and PFC Charging Capabilities. *IEEE Transactions on Vehicular Technology*, *58*(7), 3198–3215. doi:10.1109/TVT.2009.2017546

Hung-Chun, C., & Chang-Ming, L. (2011). An Integrated Driving/Charging Switched Reluctance Motor Drive Using Three-Phase Power Module. *IEEE Transactions on Industrial Electronics*, *58*(5), 1763–1775. doi:10.1109/TIE.2010.2051938

Jinhao, G., Hui, W., Tengxin, W., & Yubin, W. (2017). An integrated topology for on-board charger and driven of electric vehicle. *2017 IEEE International Conference on Industrial Technology (ICIT)*, 504–508. 10.1109/ICIT.2017.7913283

Kinoshita, S. (1996). *Electric System for Electrical Vehicle—Has Power Converter which Rectifies AC Power to DC Power so that DC Power is Regenerative to the Secondary Battery*. Patent No. US 5504414.

Lacressonniere, F., & Cassoret, B. (2005). *Converter Used as a Battery Charger and a Motor Speed Controller in an Industrial Truck*. doi:10.1109/EPE.2005.219286

Lacroix, S., Laboure, E., & Hilairet, M. (2010). *An Integrated Fast Battery Charger for Electric Vehicle*. doi:10.1109/VPPC.2010.5729063

Lee, Y.-J., Khaligh, A., & Emadi, A. (2009). Advanced integrated bidirectional AC/DC and DC/DC converter for plug-in hybrid electric vehicles. *IEEE Transactions on Vehicular Technology*, *58*(8), 3970–3980. doi:10.1109/TVT.2009.2028070

Li, W., Feng, G., Lai, C., Li, Z., Tian, J., & Kar, N. C. (2019). Demagnetization Analysis of Interior Permanent Magnet Machines Under Integrated Charging Operation. *IEEE Transactions on Industry Applications*, *55*(5), 5204–5213. doi:10.1109/TIA.2019.2922209

Liu, T. H., Chen, Y., Yi, P. H., & Chen, J. L. (2015). Integrated battery charger with power factor correction for electric-propulsion systems. *IET Electric Power Applications*, *9*(3), 229–238. doi:10.1049/iet-epa.2014.0168

Loudot, S., Briane, B., Ploix, O., & Villeneuve, A. (2014). *Fast Charging Device For An Electric Vehicle*. US Patent No. US 8847555 B2.

Meher, S. R., Banerjee, S., Vankayalapati, B., & Singh, R. K. (2020). A Reconfigurable On-board Power Converter for Electric Vehicle with Reduced Switch Count. *IEEE Transactions on Vehicular Technology*, *69*(4), 1–1. doi:10.1109/TVT.2020.2973316

Milanovic, M., Roskaric, A., & Auda, M. (1999). Battery Charger Based on Double-Buck and Boost Converter. doi:10.1109/ISIE.1999.798706

Mukundan, S., Dhulipati, H., Lai, C., Mukherjee, K., Tjong, J., & Kar, N. C. (2018). Design and Optimization of Traction IPMSM With Asymmetrical Damper Bars for Integrated Charging Capability Using Evolutionary Algorithm. *IEEE Transactions on Energy Conversion*, *33*(4), 2060–2069. doi:10.1109/TEC.2018.2877104

Pellegrino, G., Armando, E., & Guglielmi, P. (2010). An Integral Battery Charger With Power Factor Correction for Electric Scooter. *IEEE Transactions on Power Electronics*, *25*(3), 751–759. doi:10.1109/TPEL.2009.2033187

Rippel, W. E. (1990). *Integrated Traction Inverter and Battery Charger Apparatus*. Patent No. US 4920475.

Rippel, W. E., & Cocconi, A. G. (1992). *Integrated Motor Drive and Recharge System*. Patent, US Patent No. US 5099186.

Seung-Ki, S., & Sang-Joon, L. (1995). An Integral Battery Charger for Four-Wheel Drive Electric Vehicle. *IEEE Transactions on Industry Applications*, *31*(5), 1096–1099. doi:10.1109/28.464524

Sharma, Aware, & Bhowate. (2020). Integrated Battery Charger for EV by using Three-Phase Induction Motor Stator Windings as Filter. *IEEE Transactions on Transportation Electrification*, 1–1. doi:10.1109/TTE.2020.2972765

Shi, C., & Khaligh, A. (2018). A Two-Stage Three-Phase Integrated Charger for Electric Vehicles With Dual Cascaded Control Strategy. *IEEE Journal of Emerging and Selected Topics in Power Electronics*, *6*(2), 898–909. doi:10.1109/JESTPE.2018.2797913

Shi, C., Tang, Y., & Khaligh, A. (2017). A Single-Phase Integrated Onboard Battery Charger Using Propulsion System for Plug-in Electric Vehicles. *IEEE Transactions on Vehicular Technology*, *66*(12), 10899–10910. doi:10.1109/TVT.2017.2729345

Shi, C., Tang, Y., & Khaligh, A. (2018). A Three-Phase Integrated Onboard Charger for Plug-In Electric Vehicles. *IEEE Transactions on Power Electronics*, *33*(6), 4716–4725. doi:10.1109/TPEL.2017.2727398

Shi, R., Semsar, S., & Lehn, P. W. (2017). Constant Current Fast Charging of Electric Vehicles via a DC Grid Using a Dual-Inverter Drive. *IEEE Transactions on Industrial Electronics*, *64*(9), 6940–6949. doi:10.1109/TIE.2017.2686362

Slicker, J. M. (1985). *Pulse Width Modulation Inverter With Battery Charger*. Patent No. US 4491768. https://www.google.com/patents/US4491768

Solero, L. (2001). Nonconventional On-board Charger for Electric Vehicle Propulsion Batteries. *IEEE Transactions on Vehicular Technology*, *50*(1), 144–149. doi:10.1109/25.917904

Somasekhar, V. T., Srinivas, S., & Kumar, K. K. (2008). Effect of Zero-Vector Placement in a Dual-Inverter Fed Open-End Winding Induction-Motor Drive With a Decoupled Space-Vector PWM Strategy. *IEEE Transactions on Industrial Electronics*, *55*(6), 2497–2505. doi:10.1109/TIE.2008.918644

Subotic, I., Bodo, N., & Levi, E. (2016). Single-Phase On-Board Integrated Battery Chargers for EVs Based on Multiphase Machines. *IEEE Transactions on Power Electronics*, *31*(9), 6511–6523. doi:10.1109/TPEL.2015.2504400

Subotic, I., Bodo, N., & Levi, E. (2017). Integration of Six-Phase EV Drivetrains into Battery Charging Process with Direct Grid Connection. *IEEE Transactions on Energy Conversion, 32*(99), 1–1. doi:10.1109/TEC.2017.2679280

Subotic, I., Bodo, N., Levi, E., Dumnic, B., Milicevic, D., & Katic, V. (2016). Overview of fast on-board integrated battery chargers for electric vehicles based on multiphase machines and power electronics. *IET Electric Power Applications*, *10*(3), 217–229. doi:10.1049/iet-epa.2015.0292

Subotic, I., Bodo, N., Levi, E., & Jones, M. (2015). On-board Integrated Battery Charger for EVs Using an Asymmetrical Nine-Phase Machine. *IEEE Transactions on Industrial Electronics, 62*(99), 1–1. doi:10.1109/TIE.2014.2345341

Subotic, I., Bodo, N., Levi, E., Jones, M., & Levi, V. (2016). Isolated Chargers for EVs Incorporating Six-Phase Machines. *IEEE Transactions on Industrial Electronics*, *63*(1), 653–664. doi:10.1109/TIE.2015.2412516

Tang, Y., Lu, J., Wu, B., Zou, S., Ding, W., & Khaligh, A. (2018). An Integrated Dual-Output Isolated Converter for Plug-in Electric Vehicles. *IEEE Transactions on Vehicular Technology*, *67*(2), 966–976. doi:10.1109/TVT.2017.2750076

Thimmesch, D. (1985). An SCR Inverter with an Integral Battery Charger for Electric Vehicles. *IEEE Transactions on Industry Applications*, *IA-21*(4), 1023–1029. doi:10.1109/TIA.1985.349573

Truntič, M., Konjedic, T., Milanovič, M., Šlibar, P., & Rodič, M. (2018). Control of integrated single-phase PFC charger for EVs. *IET Power Electronics*, *11*(11), 1804–1812. doi:10.1049/iet-pel.2017.0663

Un, E., & Hava, A. M. (2007). A Near State PWM Method With Reduced Switching Frequency And Reduced Common Mode Voltage For Three-Phase Voltage Source Inverters. *2007 IEEE International Electric Machines & Drives Conference*, 235–240. doi:10.1109/IEMDC.2007.383583

Viana, C., & Lehn, P. W. (2019). A Drivetrain Integrated DC Fast Charger With Buck and Boost Functionality and Simultaneous Drive/Charge Capability. *IEEE Transactions on Transportation Electrification*, *5*(4), 903–911. doi:10.1109/TTE.2019.2925211

Wang, Z., Liu, B., Guan, L., Zhang, Y., Cheng, M., Zhang, B., & Xu, L. (2019). A Dual-Channel Magnetically Integrated EV Chargers Based on Double-Stator-Winding Permanent-Magnet Synchronous Machines. *IEEE Transactions on Industry Applications*, *55*(2), 1941–1953. doi:10.1109/TIA.2018.2879869

Wang, Z., Zhang, Y., You, S., Xiao, H., & Cheng, M. (2019). An Integrated Power Conversion System for Electric Traction and V2G Operation in Electric Vehicles With A Small Film Capacitor. *IEEE Transactions on Power Electronics*, 1–1. doi:10.1109/TPEL.2019.2944276

Xiao, Y., Liu, C., & Yu, F. (2019). An Integrated On-Board EV Charger with Safe Charging Operation for Three-Phase IPM Motor. *IEEE Transactions on Industrial Electronics*, *66*(10), 7551–7560. doi:10.1109/TIE.2018.2880712

Zhang, Y., Fang, J., Gao, F., Song, T., Gao, S., & Rogers, D. J. (2020). Second-Harmonic Ripple Voltage Suppression of Integrated Single-Phase Pulsewidth Modulation Rectifier Charging System for EVs. *IEEE Transactions on Power Electronics*, *35*(4), 3616–3626. doi:10.1109/TPEL.2019.2937745

KEY TERMS AND DEFINITIONS

Auxiliary Load: On-board components in an electric vehicle that consume power and are not part of the main traction drive (e.g., heating, ventilation, lights, etc.).

Integrated Battery Charger (IBC): A system that reuses existing on-board components of an electric vehicle for charging purposes.

Special Machines: Machines other than conventional three-phase machines (e.g., six-phase or five-phase machines).

Torque Cancelation: The mitigation or complete cancelation of the unwanted torque produced on the traction motor shaft because of the current flowing through the motor windings during the charging operation.

Traction Drive: The components that constitute the on-board system in an electric vehicle responsible for controlling the electric vehicle traction. This includes the traction power converter and traction motor.

Vehicle-to-Grid (V2G): The ability of an electric vehicle charger to feed power to the grid. In this chapter, this term encompasses all vehicle-to-X technologies like Vehicle-to-Home, Vehicle-to-Load, etc.

Chapter 5
Impact of Charging on Battery Life and Battery Degradation in Electric Vehicles

Mamdouh Ahmed Ezzeldin
https://orcid.org/0000-0002-0033-0275
Nile University, Egypt

Ahmed Mohsen Ali
Nile University, Egypt

Jomana Ashraf Mahmoud
Nile University, Egypt

Sohaila Ahmed Rabie
Nile University, Egypt

Hossam Hassan Ammar
Nile University, Egypt

ABSTRACT

Electrical vehicles are the future of the world; hence, there is a necessity to pave the way for the upcoming technology and to ensure its contribution to the society fairly. Nevertheless, if the EVs completely replaced the fuel-based cars, more EV charging stations would be needed which might develop overconsumption of the main grid power causing remarkable instability. Consequently, the micro grids become the solution to this problem, in which they are defined as relatively small networks of energy sources and loads at the distribution level that aim to provide electricity to remote locations where the charging stations are located. In this chapter, the EV is considered as a load to the micro grid indirectly through the EV charging stations. Thus, micro grid loads will be retrieved from experimental data of an actual prototype electric vehicle to reflect on the battery degradation in a micro-grid connected system.

DOI: 10.4018/978-1-7998-6858-3.ch005

INTRODUCTION

Recalling the environmental conditions prevailing the whole world recently, there is a tendency to replace energy sources that are harmful to the environment dramatically and reduce their usage as possible. One of these initiatives are related with vehicles. Since they are one of the most useful objects in our daily life as it contributes to the least tasks of our day and they have sundry benefits and applications mainly to humans, it is necessary to focus on how they function, the type of energy they use and how they impact our surrounding environment. Due to the numerous advantages of the electrical energy, the concept of electric vehicle becomes very promising.

Furthermore, those electric vehicles will require recharging through an electrical energy supply that could be provided through charging stations; however, those EVs will act as an extra load to the main grid. Thus, a compromised solution has been obtained lately, which is developing micro grids. The micro grids act as a network of distributed generations (DGs), which are mostly based on extracting energy from renewable energy sources, in which those sources are found to be intermittent and suffers from high fluctuations. These key drawbacks of renewable sources may lead to low power quality and transmission limitations through micro grids and large power systems.

Hence, Energy Storage Systems (ESS) is considered as one of the solutions for overcoming the surging need of storing energy and changing the form of energy from one type to another. It is compulsory to review and select the optimum methods to save and store energy efficiently for later usage, in other words, picking the most efficient ESS based on the desired application and some other chief factors such as the storage capacity, the scalability, the cost, the transmission method, the amount of energy to be transmitted, the rate of transferring, the response of the system, the supply and demand of energy at some specific region, etc. The five main forms of categorizing ESS are dragged under the classes of mechanical, electrical, electro-chemical, chemical, thermal energy storage.

For illustration, in the scope of interest of EVs, batteries are found to be the main storage element based on the scalability of the required energy to be stored in EVs. Amongst several researches, Lithium-ion batteries are agreed on as the optimum and most promising battery, especially in the application of electric vehicles as they run for both relatively high cycling capability and elevated energy density; however, their main disadvantage is the high initial cost. So, it is necessary to take into consideration the main factors that affect the life of the battery.

Meanwhile, a case study for a proposed urban EV model was carried out to know the load that the batteries will be exposed to. The load identification was done based on some aspects such as providing a full description of the dynamical model, the forces exerted on the vehicle, inspecting its kinematics at certain cases; in addition to, assessing the dynamic responses of the system and obtaining the power losses of the EV at variable loads with respect to the velocity. Also, battery and motor sizing were carried out based on the battery and motor specifications provided by the manufacturer. All the previous points contribute to determining the total load on the micro grid.

LITERATURE REVIEW

In general, batteries work on the concept of electro-chemical potential which is the tendency of the metal to lose electrons. As shown by the electro-chemical series, Lithium in specific has high tendency to lose electrons as they have only one electron in their outer shell and always wants to lose it, so they are preferred to be used as battery cells. Pure lithium is a reactive metal but when it is combined with metal oxide, it becomes stable and when the lithium atom is separated from the metal oxide, it splits into an electron and a lithium ion (Van Schalkwijk & Scrosati 2002). If a path is provided for both the electron and lithium ion to return to the metal oxide forming a lithium atom, electricity is produced from this electron flow through that path and that is how electricity is generated in a battery cell in a simple form (Angulakshmi et al. 2014). The battery cell contains multiple layers of chemical compounds such as the copper layer which has graphite coated on them that act as storage area. Graphite is well known of its loosely bounded structure layers (Angulakshmi et al. 2014). Secondly the electrolyte layer, which is found between the copper and the aluminum layers, acts as a filter which only allows lithium ion to flow through. On the electrolyte, the separator is found. The main task of the separator appears when the temperature in the battery cell increases due to failure or something so the electrolyte dries out and can result in a short circuit and explosion (Van Schalkwijk & Scrosati 2002). The separator is found to prevent such incidents by preventing the flow of electrons thought them. Finally, the aluminum layer, which the metal oxide is coated on it, acts as a current collector. When a power source is connected to the battery cell, the lithium atom in the metal oxide will lose its electron to the positive side of the power source leaving lithium ion alone. As mentioned above, the electrons cannot pass through the electrolyte, so the external source is the only way to flow through to reach the graphite layers (Angulakshmi et al. 2014). On the other hand, the lithium ion will be attracted to the negative side of the power source passing through the electrolyte reaching the graphite layers. Once both lithium ions and the electrons reach the graphite layers the cell will be fully charged. And as the power source is replaced with a load, the lithium ions want to return to their stable state in the metal oxide, so the lithium ions flow back though the electrolyte while the electrons flow through the load generating electric current in it (Gong et al. 2014).

The Life Cycle of the Battery

The battery has a lifespan in the car and after a certain period, it must be replaced by a new battery after fully wearing out. Studies showed that for 26.6 kWh battery cell, the efficiency of the battery under the normal use is between 95% to 96% and can perform in an efficient way till 80% of its initial capacity (Ellingsen et al. 2014). Other studies showed that after 80% of the initial capacity of the battery, the battery can be used in the car but will drain quickly so it is used in other applications such as houses till 65% of its initial capacity, a feature known as the battery life cycle. The life cycle of the batteries are sequential stages the battery passes through them, targeting to reflect the impact of the batteries on the environment. Starting from mining for the material, manufacturing the battery cells, usage in the car, reusing it in different applications such as houses, and till its life ends or becomes drained. The life cycle of the battery differs from one type of lithium ion battery to the others such as Lithium manganese oxide (LMO), lithium iron phosphate (LFP), and lithium manganese cobalt oxide (NMC). These different types of batteries are due to different material types of cathode, while using the same graphite for anode. The life cycle for these batteries can be more than 5,000 cycles based on its type, where the cycles represent the number of times the battery is fully charged and discharged. A fully charged battery can range up to 500 km and its life cycle reaches more than 400,000 km (Ellingsen et al. 2014). The lifespan of the battery can decrease due to various reasons such as the surrounding operating temperature, Depth of Discharge (DOD), and charging and discharging current (Mehmood et al. 2017). If these factors were not taken into consideration, batteries would drain quickly, resulting in additional costs.

Parameters That Affect Battery Life Cycle

This section lists the parameters that affect the life cycle of the battery such as the battery temperature, the depth of discharge, charging and discharging current and the Degradation.

Battery Temperature

The temperature is a major factor affecting the life cycle of the battery. The Arrhenius equation is used to generate a relation between the temperature increase and the rate of chemical reaction in the batteries, as shown in equation (5.1) (Qian et al. 2010),

$$\Phi = Ae^{-\frac{E_a}{RT}} \tag{5.1}$$

φ is the rate constant of the chemical reaction, A is the pre-exponential factor, E_a is the activation energy, and the R is the gas constant $8.3144JK^{-1}mol^{-1}$. It was noticed by many experiments that every 10 Celsius degree the rate of the chemical reaction becomes the double. A study was made on lithium ion battery by keeping the 80% DOD constant, to show the effect of temperature on the batteries and how it will operate (Qian et al. 2010). It was found that lithium ion battery at $20°C$ can still operate till 2,000 cycle, while at $55°C$ the battery can stand for only 800 cycle. The decrease in the life cycle is due to the temperature and it affects the react rate as shown in the Arrhenius equation. Although high temperature can result in better performance, but unfortunately, the chemical reaction also increases resulting in battery life loss (Garche & Jossen 2000). Unfortunately, the study is not totally realistic as it is different for the automotive application as the temperature changes in the same day, month, year, and across different countries (Qian et al. 2010). The effect of the temperature on the battery can be determined by the thickness of the electrode of the battery (Huang et al. 2018). The cell oxidation caused by temperature result in growing a layer on the electrode which increases the resistance and causes failure by time (Huang et al. 2018). After several experiments, it was found that the resistance (r_f) is proportional to the thickness (Qian et al. 2010). Based on equation (5.1), the change in resistance (Δr_f) can said to be proportional to the change in the capacity (ΔC) delivered at a given time and as a result an equation is made which is temperature dependent, as shown in equation (5.2).

$$r_f\left(n_c,T\right) = n_c.e^{-\frac{E_a}{R}\left(\frac{1}{T_{ref}}-\frac{1}{T}\right)} \tag{5.2}$$

n_c is the previous number of cycles, and T_{ref} is the reference temperature which mainly equal to $20°C$. Several researches have been done showing that at a certain temperature and after several cycles, a capacity fading phenomenon occurs which reduces the total capacity of the battery (Erdinc et al. 2009). Other research showed that both low and high temperature affect the life cycle. At low temperature, the internal resistance increases as the low temperature affect the material making it almost inactive. At high temperature, the electrode degrades as the internal chemical reaction rate of the battery increases (Mehmood et al. 2017). A study is done to validate these researches and plot it at several high and low temperatures such as $60°C$, $50°C$, $40°C$, $10°C$, $0°C$ and $-10°C$ to determine the change of the life cycle.

As the temperature increased, the number of cycles decreased. When the temperature 60°*C* was reached, which is very far from the normal conditions, the number of cycles decreased rapidly to be only 20 cycles to reach 80.63% of the discharge capacity percentage (Liu et al. 2012). The number of cycles for both 0°*C* and 10°*C* is almost the same with average discharge capacity percentage 97.77%, while at −10°*C*, the battery cycle decreases rapidly with average discharge capacity percentage 85%. After experimenting the battery, a graph is plotted (Liu et al. 2012), as shown in figure 1, it is observed from the graph that the temperature affects the battery cycle negatively more than the low temperature. In addition, the study also showed the effect of temperature on the internal resistance of the battery with the same temperature (Liu et al. 2012), as shown in figure 2. It is observed that extremely high and low temperature affects the internal resistance of the battery heavily making it fail to operate. So, it is recommended to maintain the normal temperature for the batteries to achieve the maximum number of cycles. An experiment was done to show the optimal temperature for the battery to operate. The experiment showed range of temperature between 15°*C* and 50°*C* which is the ideal working temperature range. Above and Below the ranged temperature the life cycle of the battery decreases rapidly (Qian et al. 2010).

Figure 1. The capacity of discharge at different temperature

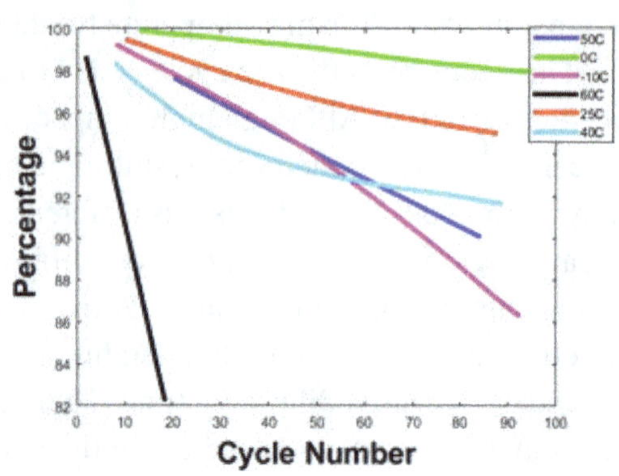

Figure 2. The internal resistance using different temperature

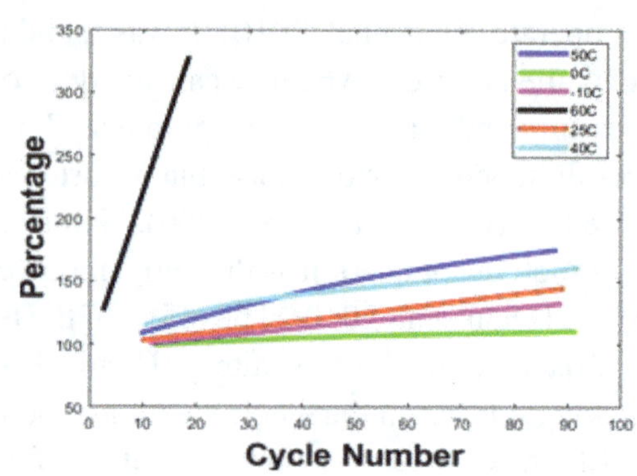

Depth of Discharge (DOD)

Depth of Discharge (DOD) is another main factor affecting the life cycle of the battery negatively. The life cycle decreases as the battery is discharged at high DOD level. Moreover, the depth of discharge is inversely proportional to the battery's state of charge (SOC). An experiment on several lithium ion batteries was executed showing the effect of DOD on the number of life cycles (Qian et al. 2010). For example, at 100% DOD, 3,000 life cycles were observed, and at 3% DOD, 1,000,000 cycle were observed. Another study was made to show the effect on the capacity fade by using different 50% and 100% DOD cycling (Guena & Leblanc 2006). It was observed that as the DOD decreases the capacity fade decreases and vice versa. The results showed that capacity of the battery lost 10% after 370 cycles for a 50% DOD. On the other hand, 10% of the capacity was lost after only 90 cycles for 100% DOD.

Charging and Discharging Current

Charging and discharging current is the third factor affecting the life cycle of the battery negatively. Several studies showed that both drawing or supplying high current to the battery for recharging or discharging would decrease the life cycle of the batteries (Garche & Jossen 2000). After many researches, it was observed that the internal resistance increases as large current capacity (4 to 5 C) is supplied to the battery, causing capacity fading (Mehmood et al. 2017).

Degradation

Degradation occurs in case of the electrons in the battery reach the electrolyte. According to research, a proposed solution to minimize the lithium ion batteries from degradation is to use solid electrolyte interface (SEI), which is a layer to protect the electrolyte from degrading. It is formed when the battery is charged for the first time, as the lithium ions flow through the electrolyte, it passes with a solvent molecule covering it. The lithium ion with the solvent flows to the graphite layers and react with it forming the SEI layer, which prevent any contact between the electrons and the electrolyte, preventing degradation (Xu et al. 2016).

SOLUTIONS TO MAINTAIN THE LIFECYCLE OF THE BATTERY

This section provides some solutions to maintain the life cycle of the battery such as the cooling system and BMS.

Cooling System

Battery cells generates lots of heat while operating; Hence, a cooling system must be supplied to the battery. Battery management system (BMS) is used to manage the temperature of the cells by using the glycol coolant, to provide and maintain the suitable temperature for the batteries (Ellingsen et al. 2014). In addition, Aluminum Radiator is added which act as the main component for cooling the cells.

BMS

While the battery is being charged, the higher capacity cell will be charged more than other cells. To solve this problem, the BMS uses cell balancing, which makes all the batteries equally charged and discharged at the same time and protect them from the hazard of over and under voltage (Garche & Jossen 2000)

FUTURE TECHNOLOGIES TO INCREASE THE LIFECYCLE

Enhancing the efficiency of the lithium ion battery either by trying to increase its life cycle or reducing the times of changing the batteries have been the target of sundry researches. Also, it has been proposed by researchers to replace Graphite with Silicon, which is expected to increase the energy density of the cell by five more times (Romare & Dahllo¨f 2017). Improving the battery's storage capacity is one of the ways of to improve the batteries and its life cycle (Mehmood et al. 2017).

EV CHARGING STATIONS

An electric vehicle charging station is a component in an infrastructure that supplies electric energy to recharge the electric vehicles that includes electric cars, neighborhood electric vehicles and plug-in hybrid cars by plugging them into an energy source. The charging station components is composed of a fixture that is connected to an electrical plug or distribution panel that has one or more charging cables. These cables are supplied by a connector that is like the gas nozzle. Some electric vehicles have converters on board, which will plug into a typical electrical plug or a high-capacity appliance outlet when it is needed to be charged at home or work.

Types of Electric Vehicle Charging

This section categorizes the types of electric vehicle charging into Level 1, Level 2, and DC Fast Charging.

Level 1 (120 V)

This type is used by all types of electric vehicles; It uses the 120 volts that is supplied from any household outlet (Falvo et al. 2014). It is considered as a slow charging type that can charge the car in eight hours to be used in 50 Km (as shown in table 1).

Table 1. Charging time based on distance driven (Falvo et al. 2014)

Type of Station	Distance Traveled (Km)	Approximated Energy Consumption (KWh)	Charging Station Power (KW)	Approximated Charging Time (hr.)
120-V and 12 A Charging Cable	25	5.2	1.4	4
	50	10.4		8
	100	20.7		15

Level 2 (240 V)

These stations depend on the 240 V that is same as the operating volts for the oven or the dryer in the house (Falvo et al. 2014). It is commonly spread in public places that any EV can fit with its plug except some models such as Tesla which will need an adapter. With 30-amp circuit, the EV can be charged in 8 hr. to take a distance of 290 Km as explained in table 2.

Table 2. Charging time based on distance driven (Falvo et al. 2014)

Type of Station	Distance Traveled (Km)	Approximated Energy Consumption (KWh)	Charging Station Power (KW)	Approximated Charging Time (hr.)
240-V outlet and 15 A	25	5.2	3.6	1.5
	50	10.4		3
	20.7	6		6
240-V outlet and 30 A	25	5.2	7.2	0.75
	50	10.4		1.5
	20.7	6		3

DC Fast Charging (480 V)

It converts the AC power to DC power that can be directly stored in the EV batteries which charge 80% of the battery within 30 to 60 minutes. As in level 2, all these plugs are applicable for all EVs types except for Tesla that requires an adapter to recharge (Falvo et al. 2014) as in table 3. This type is useful for long trips and the cars that are in use most of the day. Table 4 explains the difference between the three types of charging stations.

Table 3. DC Fast Charging time based on distance traveled (Falvo et al. 2014)

Type of Station	Distance Traveled (Km)	Approximated Energy Consumption (KWh)	Charging Station Power (KW)	Approximated Charging Time (min.)
Dc Fast Charging	25	5.2	40	8
	50	10.4		16
	100	20.7		25

Table 4. Comparison between three types of stations (Falvo et al. 2014)

Point of Comparison	Level 1	Level 2	DC Fast Charge
Voltage	120 V	240 V	480 V
Current Type	AC	AC	DC
Power	1.4 KW	7.2 KW	50 KW
Max. Output	1.9 KW	19.2 KW	150 KW
Charging Time	12 hr.	3 hr.	20 min.

METHODOLGY

In this section, the degradation of battery in microgrid system will be explained and discussed along with the relevant equations. Based on the determined loads from proposed EV car in the previous section, the number of cycles needed and battery discharging rate can also be approximated.

Microgrids Degradation

The micro-grid degradation depends on the degradation of the battery energy storage system; This degradation will be optimized under the capacity change of the battery. This capacity is known as state of health (SOH) of the battery and is usually compared to its optimal conditions. The units of SOH are in percentage where 100% represent the battery at its ideal condition at the time of manufacture and it will diminish after some time and use. The SOH can be analyzed from equation (5.3):

$$SOH = \frac{C_{curr}}{C_{nomi}} \times 100\% \tag{5.3}$$

Where C_{curr} is the capacity of battery in Ah at current time, C_{nomi} is the nominal capacity of battery in Ah.

The capacity fading ratio is also another essential term to be introduced which is a phenomenon in rechargeable battery utilization in which the sum of charge a battery can provide at the evaluated voltage. This charge decreases with the use and it can be expressed by equation (5.4):

$$y = 1 - SOH = \frac{C_{nomi} - C_{curr}}{C_{nomi}} \times 100\% \tag{5.4}$$

Where y is the capacity fading ratio.

The mechanism of degradation of battery is affected by two condition; The first one is the wastage of usable lithium ions which is caused by the developing of the strong electrolyte interface film (SEI). The second condition is the low material effectiveness of the anode and cathode. In addition, Arrhenius law also shows that temperature has a vital role in this condition in chemical reactions like the formation of SEI film as expressed in equation (5.5).

$$y_k = Ae^{\left(\frac{-E_a}{RT}\right)K^z} \tag{5.5}$$

Where k is the cycle number from the battery's initial state, A is constant, Ea is the activation energy in J/mol, T is temperature in Kelvin, R is the gas constant in J/(mol•K), Z is power law factor.

If the capacity fade ratio will be decided by the increase in the SEI; so, the power law factor (z) will be equal to ½.

The current rate of the discharging process has also an influence on the capacity degradation beside the cycle number as in each cycle number, the current discharging affect the degradation. Equation (5.6) determine the relation between the capacity fade ratio (y) and the current degradation at each cycle.

$$\gamma_{I,K} = \frac{y_{I,K}}{y_{1C,K}} \tag{5.6}$$

Where $\gamma_{I,K}$ is relative change between the capacity fade ratio and the current degradation at each cycle, $y_{I,K}$ is the actual capacity fade ratio at cycle (k) at the current discharge (I), $y_{1C,K}$ is the capacity fade ratio at cycle (k) at current discharge 1C.

After analyzing equation (5.6) at different cycle numbers, we get that the relation between (γ) and (I) is exponential function despite the cycle number as shown in figure (3).

Figure 3.

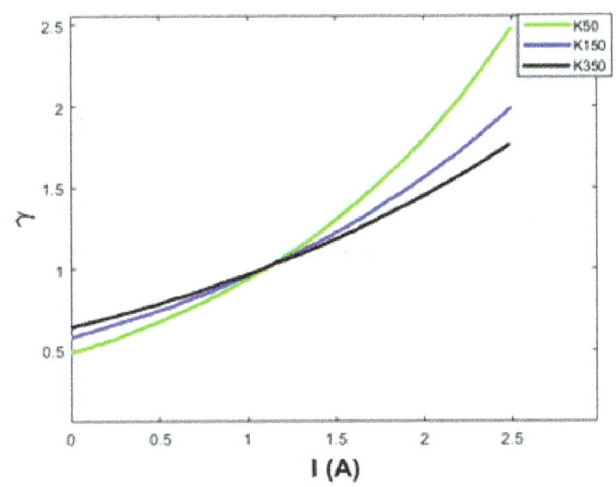

So, the capacity degradation model can be given by the current rate which can be calculated from equation (5.5) if the current fade ratio was under current discharging of 1C; but if the current fade ratio was not under 1C so the equation (5.7) will be used:

$$y_{I,K} = \gamma_{I,K} y_{1C,K} \tag{5.7}$$

CASE STUDY

In order to study the battery degradation issues and their results on a micro grid, micro grid loads must be identified first to know the characteristics and specification of this micro grid. Based on the case study of an urban electric vehicle, the needed loads will be extracted and, thus, a model of a DC micro grid can be built. Figure 4 shows the sequence used to model the micro grid in this section.

Figure 4. Steps to identify battery degradation

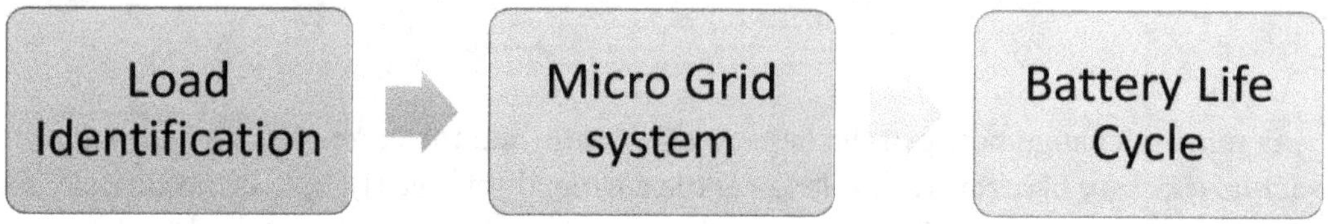

Load Identification (Proposed EV)

The electric vehicle has an urban style, manufactured to carry two persons, and is designed for domestic use and moderate distance. This car is aimed to have low production cost, be able to mass production, and optimize battery consumption. Based on these requirements, a delicate car design and components sizing were made for the new car.

Dynamic Model

The power of the car must overcome the forces exerted on it. Equation (5.8) stands for friction force, equation (5.9) for drag force, equation (5.10) for gradient force as follow:

$$F_d = \frac{\rho}{2} c_D A_f v^2 \quad (5.8)$$

$$F_f = fw \quad (5.9)$$

$$F_g = mg \sin(\alpha) \quad (5.10)$$

Where ρ is the air density, c_D is the drag coefficient, A_f is the frontal area of the car, v is the velocity of the car relative to the wind, f is the friction coefficient, w is the vehicle weight, and α is the incline angle of the road.

Through iterative design approaches, we were able to find the most optimized design that can achieve this model's targets and at the same time has minimum power consumption. Figure 5, 6, 7 show the car CAD design and ANSYS Analysis respectively, which were used to calculate the drag coefficients and other car parameters.

The Ansys analysis was conducted at 60km/hr. and the resulted parameters were assembled in table 5 below.

To ensure the effectiveness of the proposed design, a kinematics model was also implemented to test the braking system of the car. Using MATLAB model (figure 8), the car can decelerate from 60km/hr. to 0km/hr. in almost 1.2 s with stopping distance of 13.02m/s^2.

Figure 5. CAD Design of the proposed EV Car

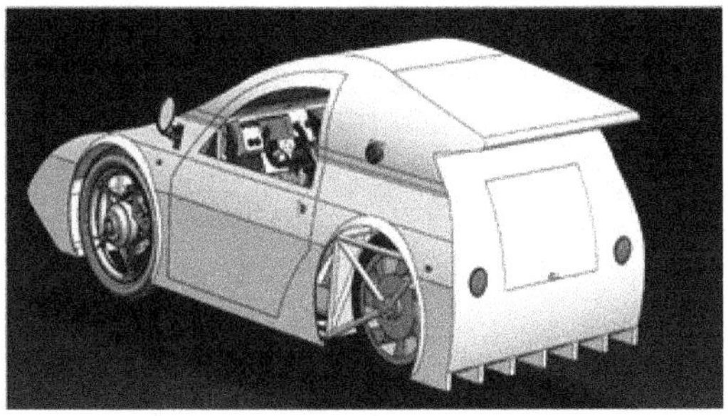

Figure 6. Ansys Analysis for the aerodynamic force on the car (side view)

Figure 7. Ansys Analysis for the aerodynamic force on the car (3D view)

Table 5. Proposed EV Parameters

Drag Force (at 60km/hr.)	Drag Coefficient	Lift Force	Frontal Area	Car mass
95 N	0.3	40 N	1.53m^2	360 kg

Figure 8. MATLAB Model for the braking system

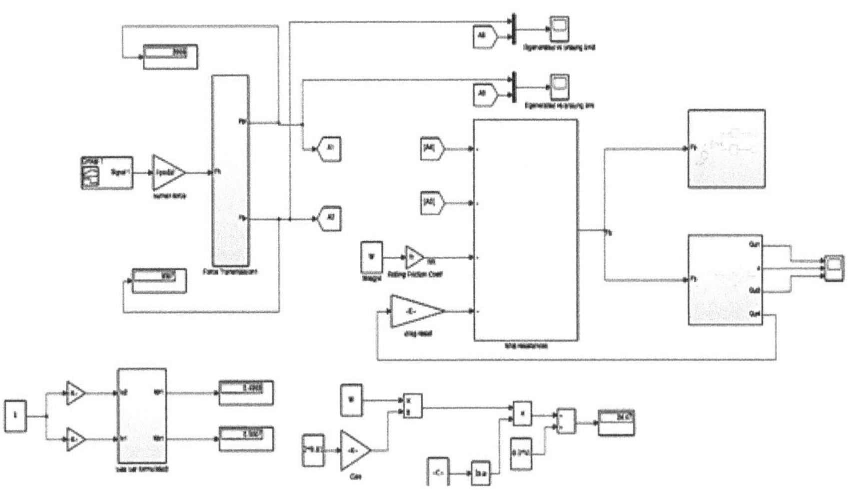

Dynamic Response and Power Consumption

By using the previous equations and models to test the response of the car under the dynamic model, more information about the velocity and current of the proposed car in a certain track simulation in SIMULINK were obtained. The results are shown in figure 9 below.

Figure 9. Dynamic Response of the proposed EV

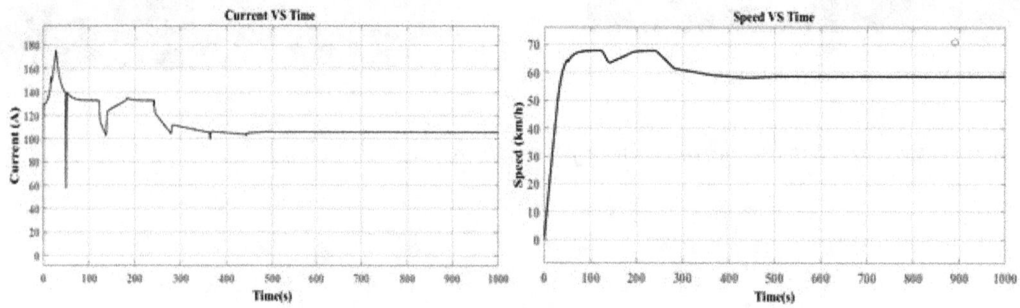

In addition, Since the dynamic forces applied on the car vary with the vehicle velocity, a graph was conducted relating the total losses for the car with the velocity in order to know the required power needed to reach the vehicle top speed which is 95 km/hr. as shown in figure 10.

Figure 10. Power losses with velocity

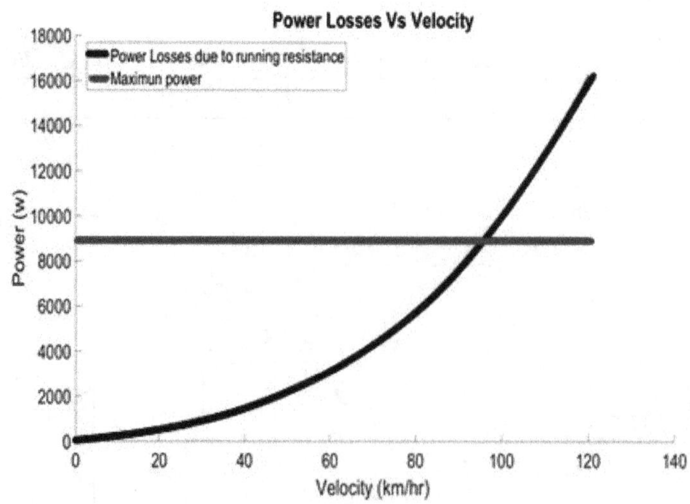

Battery and Motor Sizing

The propulsion system of the car was calculated to meet the top speed of the car. Therefore, the car needed two motors with 4500 watt each (figure 11). To fulfil such performance, two Lithium Ion batteries were used. According to the manufacturer, each battery has a nominal capacity of 80Ah and nominal voltage of 25.6 V (figure 12).

Figure 11. Hub Motor used in the car

Figure 12. LIFEBO4 Battery used in the car

Power Train and Wiring

As mentioned before, the power train of the proposed car consists mainly of the two lithium batteries, two controllers, and two hub motors. In addition, there is also an auxiliary system responsible for the lighting systems, horn, and other light current accessories. The performance of this power train was tested using Simulink (figure 13) and the results were quite satisfactory as shown in the section of (Dynamic Response & Power Consumption).

Also, to calculate all the factors of power consumption of the proposed model, Total losses in the car wires were calculated according equation (5.11) as sated below.

$$V_{drop} = \frac{length * current * resistivity}{Cross\ sectional\ area} \quad (5.11)$$

The wiring for the whole power train can be seen in figure 14.

Total Load on the Microgrid

To know the micro grid loads that the proposed EV vehicles will impose, charging specifications are shown in table 6 below.

Figure 13. Model of the car on Simulink

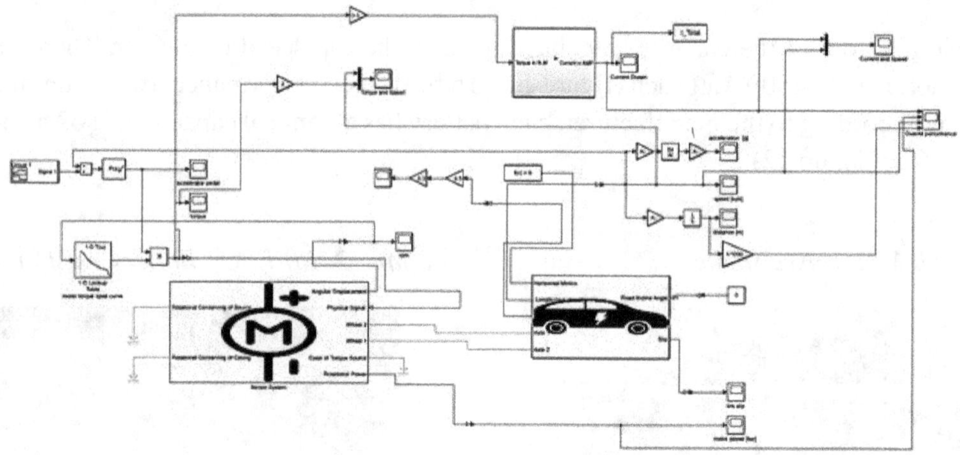

Figure 14. Power Train Wiring.

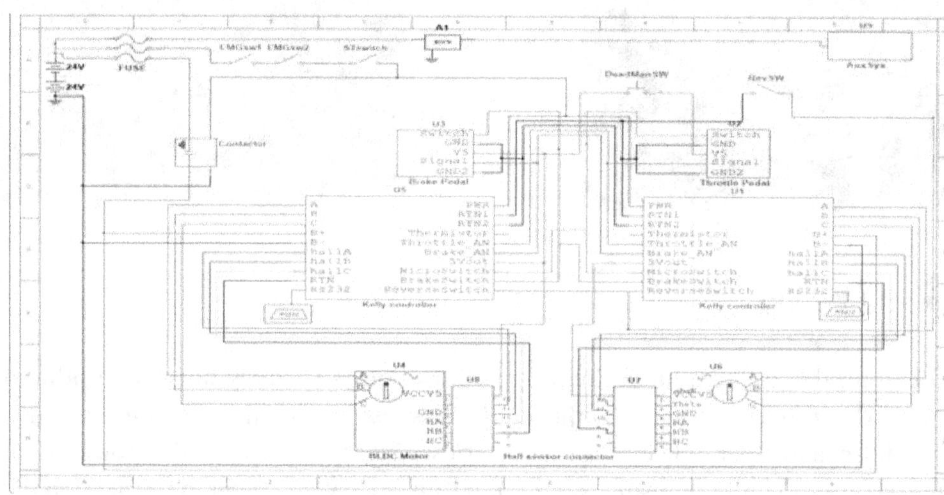

To develop a model of the micro grid for relatively small town, efficient information about the total loads exposed to the micro grid must be known. Total number of 50 EVs will be operating within the range of the proposed micro grid. The small working area of these EVs means they will be travelling relatively small distance (ex. town or rural areas). Thus, the following data can be obtained (given the assumptions made in this section).

The micro grid in the model is assumed to provide 60 KW for the whole charging time which is 40 min. In addition, EVs introduces a new challenge. They may need to charge in certain times and works normally for the rest of the day, presenting a critical peak time to the micro grid. The charging stations in the proposed micro grid need to meet these requirements for the total number of cars operating within the micro grid. The usage of renewable energy sources is recommended but the intermittency of such approach can lead to some power outages, so an energy source sizing and efficient ways of energy storage are required to withstand the demand needed in table 7.

Table 6. Charging Specification of the EV

Items Needed	Specifications
Nominal Charging Rate	0.2 C
Maximum Charging Rate	1.5 C
Minimum Time for Charging	40 Minutes
Total Charging Power/Car	6000 watts

Table 7. The expected total loads for the micro grid

Points of the expected load	Assumed Data
Total Number of Cars	50
Number of Charges/Week (Per Car)	2
Number of Required Charging Stations	10
Maximum Power Demand (at any time)	60 kw

CONCLUSION

To sum up, the various factors that impact the life cycle of the battery and causes its degradation such as the DOD, the charging and discharging current, etc. were discussed thoroughly in this chapter. In particular, the theory behind the lithium ion batteries was explained since it is the most used either in electric vehicles or in BESS for a grid-connected system. Several solutions and futuristic technologies for maintaining and increasing the battery life cycle were also considered. Given that the constant charging and discharging is one of the crucial factors that certainly affect the life cycle of the battery; therefore, the EV charging stations and their types should be included. Furthermore, the microgrids degradation and the state of health of the battery was studied in terms of mathematical equations. Eventually, we constructed our own case study of an urban electric vehicle to calculate and identify the microgrid load, which indeed affects the battery life cycle.

ACKNOWLEDGMENT

We would like to thank Prof. Dr. Ahmed Radwan (Nile university Vice-president of research) for his helpful efforts and continued supervision throughout this chapter. Also, our sincere gratitude to Nile University for providing us with the all the facility and equipment used in the case study.

We thank you.

REFERENCES

Angulakshmi, N., Kumar, R. S., Kulandainathan, M. A., & Stephan, A. M. (2014). Composite polymer electrolytes encompassing metal organic frame works: A new strategy for all-solid-state lithium batteries. *The Journal of Physical Chemistry C, 118*(42), 24240–24247. doi:10.1021/jp506464v

Ellingsen, L. A.-W., Majeau-Bettez, G., Singh, B., Srivastava, A. K., Valøen, L. O., & Strømman, A. H. (2014). Life cycle assessment of a lithium-ion battery vehicle pack. *Journal of Industrial Ecology, 18*(1), 113–124. doi:10.1111/jiec.12072

Erdinc, O., Vural, B., & Uzunoglu, M. (2009). A dynamic lithium-ion battery model considering the effects of temperature and capacity fading. *2009 International Conference on Clean Electrical Power*, 383–386. 10.1109/ICCEP.2009.5212025

Falvo, M., Sbordone, D., Bayram, I., & Devetsikiotis, M. (2014). EV charging stations and modes: International standards. Academic Press.

Gamez, M. R., Perez, A. V., Sera, A. S., & Ronquillo, Z. M. (2017). Renewable energy sources and local development. *International Journal of Social Science and Humanity*, *1*(2), 10–19.

Garche, J., & Jossen, A. (2000). Battery management systems (bms) for increasing battery lifetime. *TELESCON 2000. Third International Telecommunications Energy Special Conference*, 81–88.

Gong, X., Xiong, R., & Mi, C. C. (2014). Study of the characteristics of battery packs in electric vehicles with parallel-connected lithium-ion battery cells. *IEEE Transactions on Industry Applications*, *51*(2), 1872–1879. doi:10.1109/TIA.2014.2345951

Guena, T., & Leblanc, P. (2006). How depth of discharge affects the cycle life of lithium-metal-polymer batteries. *INTELEC 06-Twenty-Eighth International Telecommunications Energy Conference*, 1–8. 10.1109/INTLEC.2006.251641

Huang, X., Ke, S., Lv, H., & Liu, Y. (2018). A dynamic capacity fading model with thermal evolution considering variable electrode thickness for lithium-ion batteries. *Ionics*, *24*(11), 3439–3450. doi:10.100711581-018-2476-8

Liu, J., Gao, D., & Cao, J. (2012). *Study on the effects of temperature on lifepo 4 battery life. In 2012 IEEE Vehicle Power and Propulsion Conference*. IEEE.

Mehmood, K. K., Khan, S. U., Lee, S.-J., Haider, Z. M., Rafique, M. K., & Kim, C.-H. (2017). Optimal sizing and allocation of battery energy storage systems with wind and solar power dgs in a distribution network for voltage regulation considering the lifespan of batteries. *IET Renewable Power Generation*, *11*(10), 1305–1315. doi:10.1049/iet-rpg.2016.0938

Qian, K., Zhou, C., Yuan, Y., & Allan, M. (2010). Temperature effect on electric vehicle battery cycle life in vehicle-to-grid applications. In *CICED 2010 Proceedings* (pp. 1–6). IEEE.

Romare, M. & Dahll¨of, L. (2017). The life cycle energy consumption and greenhouse gas emissions from lithium-ion batteries. *Stockholm. Zugriff am, 23*.

Van Schalkwijk, W., & Scrosati, B. (2002). Advances in lithium ion batteries introduction. In *Advances in Lithium-Ion Batteries* (pp. 1–5). Springer. doi:10.1007/0-306-47508-1_1

Xu, B., Oudalov, A., Ulbig, A., Andersson, G., & Kirschen, D. S. (2016). Modeling of lithium-ion battery degradation for cell life assessment. *IEEE Transactions on Smart Grid*, *9*(2), 1131–1140. doi:10.1109/TSG.2016.2578950

KEY TERMS AND DEFINITIONS

Battery Degradation: It is the reduction that happens to the battery in its efficiency, power, and energy capacity and it is a dangerous economic problem.

Battery Life: The amount of time that the battery can charge and discharge to reach its maximum number of charges then it will not be usable.

Capacity Fading: A phenomenon in rechargeable battery utilization in which the quantity of charge a battery is determined at a rated voltage when it is in use.

Electric Vehicle Station: A component in an infrastructure that supplies electric energy to recharge the electric vehicles by plugging them into an energy source.

Energy Sources: The materials that valuable energy can be extracted or recouped straightforward or by implies some transformations; the energy sources can be renewable or non-renewable.

Energy Storage System: The equipment that is used to store the various energy types safely to be used in other times as the lithium-ion batteries.

Micro Grid: Small networks of energy sources and loads at the distribution level that aim to provide electricity to remote locations.

Chapter 6
Impact of Electric Vehicle Charging Infrastructure on the Distribution Network Power Quality:
Challenges and Opportunities

Gagandeep Sharma
https://orcid.org/0000-0002-3155-3586
Ontario Tech University, Oshawa, Canada

Vijay K. Sood
Ontario Tech University, Oshawa, Canada

ABSTRACT

This chapter discusses the available charging systems for electric vehicles (EV) which include battery electric vehicles (BEV) and plugged hybrid electric vehicles (PHEV). These architectures are categorized as common DC bus charging (CDCB) station and common AC bus charging (CACB) station. CACB charging stations are generally used as slow chargers or semi-fast chargers (on-board chargers). CDCB charging stations are used as fast chargers (off-board chargers). These chargers are vital to popularize the electric vehicles (EVs) as a green alternative to the internal combustion engine (ICE) vehicles. Further, this chapter covers the power quality problems related to the grid-connected fast charging stations (FCS), AC-DC converter, control strategies for converters, proposed system of architectures, methodology, system results with comparisons, and finally, a conclusion.

DOI: 10.4018/978-1-7998-6858-3.ch006

1 INTRODUCTION

Batteries are the most critical component in the xEVs (BEV and PHEV). Batteries need to have high energy density, be cost-effective, safe, and efficient for a wider market penetration. Lithium-Ion (Li-Ion) batteries are presently the most popular electrochemical battery structure used in EVs (El Kharbachi et al., 2020)[2]. The car battery is typically about 25-50% of the cost of the total vehicle cost. *Bloomberg News Energy Finance* predicts that there would be a decrease in the Li-ion battery price by one-fourth to today's price, and the market is predicted to touch USD 95.3 billion by 2030 (BNEF, n.d.).

Bloomberg NEF also speculates that 1 in 10 vehicles will be an EV by 2025, and there will be a requirement for 12 million public FCS by 2040. These number of FCS need an estimated $ 400 billion in investments on infrastructure. The EV market is projected to rise from USD 146,902.20 Million in 2019 to USD 359,854.56 Million by the end of year 2025. The Compound Annual Growth Rate (CAGR) is expected to be 16.10% in the EV market. Fast charging stations components will also be a major part of this unprecedented development. CHAdeMO 4 charger is predicted to be the highest gainer in this period (Globe Newswire, n.d.).

The number of EVs are growing worldwide. They were expected to increase from 5 hundred thousand in 2012 to 2.2 million globally by year 2019. This growth rate is expected to continue in future years, and therefore FCS development is also required. Table 1 shows the current number of EV charging stations in a few leading countries:

Table 1. Number of Charging Stations (leading EV user countries) as per 2019

Country	Number of Charging Stations
Norway	7,065 (approx. 400 FCS and semi FCS)
Canada	5,004 (approx. 772 FCS)
USA	78,500 (approx. 11,000 FCS)
China	4,96,000 (approx. 2,00,000 FCS)
United Kingdom	9,200 (approx. 1,600 FCS)

It has been predicted that AC supply is the source for most of the power required to charge the EV batteries. There are chances that power demand will increase to 300 Twh to charge the expected 140 million EVs by year 2030 (Expansion, challenges, and opportunities in the EV market, n.d.). FCS will emerge as a new load to utility companies This higher amount of power demand will have a negative impact on the power quality indices of the grid.

Battery chargers can be categorized as (a) on-board, and (b) off-board chargers. On-board chargers are having a lower kW rating and weight, while off-board chargers have higher kW rating and are mostly grid-connected. xEV charging needs reliable, fast, and safe charging infrastructure to cope up with the range anxiety of consumers and to meet the charging needs. Fast Chargers can play an important role in making EVs as an alternative for ICE vehicles (Moradewicz, 2019).

On-board chargers are located inside the vehicle and need very little infrastructure outside the vehicle expect power outlet for Level 1 chargers. The power rating of these chargers is low to medium level. **Level 1** (single-phase, 120 V_{AC} (North America), 230 V_{AC} (British System)) and **Level 2** (single-phase or three-phase supply, 230 V_{AC} (North America), 400 V_{AC} (British System)) charging technology is mostly used in on-board chargers.

DC fast chargers or **Level 3** chargers are connected to the grid through a 3-phase supply link. There is requirement of hardware infrastructure for the deployment of these chargers. It includes of 3-phase AC supply or DC supply, 208-600 $V_{AC,}$ 208-600 V_{DC} and time taken by PHEVs is 0.2 to 0.5 hours and 0.4 to 1 hours for EVs to fully charge. Even so, DC FCS are much needed to encourage people to adopt xEVs without worrying about charging and range anxiety (Shaukat, 2018). Since charging stations are a vital part of the EV infrastructure; a study of different architectures and associated components is required.

2 RELATED LITERATURE TO FAST CHARGING INFRASTRUCTURE

In fast-charging infrastructure, CACB and CDCB charging are popular charging architectures. Fast Charging Stations (FCS) are still not widespread. There is a need for more FCS to increase the EV adoption amongst the masses. Ref. (Tran et al., 2017) reviewed the existing state of EV charging arrangements and concluded that FCS is a viable solution to lessen the impacts of EV charging on the electric grid. Ref. (Tan et al., 2016) described two basic designs for fast EV charging, i.e. CACB and CDCB charging. In the CACB architectures, all the chargers have separate AC-DC converters; and in CDCB one main AC-DC converter is required. The vehicles getting charged via household charging stations in a congested neighbourhood can be considered as operating off a common AC bus. Ref. (Naveen et al., 2014) explained the working of the control system for AC-DC converter which includes a cascaded control in the dq-frame. This strategy produces reference signals for the PWM generator to send the gating pulses to converter.

It is evident that AC-DC converters (Erb et al., 2010; Monteiro et al., 2016; Rajashekara, 2013) play a vital role in the EV charging infrastructure, irrespective of topology employed. In these converters, the Voltage Source Converter (VSC) is a key component in the exchange of apparent power with the grid. The maximum amount of apparent power exchanged between the charger and grid is governed by the equation #.1 (Mojdehi & Ghosh, 2016):

$$P_k^2 + Q_k^2 \leq \overline{S}^2 \tag{1}$$

Where $\overline{S}$ is the maximum apparent power of the inverter. P_k is the real power, and Q_k is the reactive power.

2.1 Common AC Bus Architecture (CACB)

In the common AC bus architecture, a shared AC bus is used. MV-LV transformer is a vital equipment to supply the AC bus. Afterward, there is a separate AC-DC rectifier and DC-DC converter for each EV. The charging station can be a combination of Level 2 (medium charging speed) and Level 3 chargers (fast charging speed) in common AC bus architecture (Ben Jar & Watson, 2016). In CACB, power electronic equipment can be used as on-board and the solution will be inexpensive as compared to the common DC bus architecture.

The literature shows that there are additional conversion losses in case of AC bus architecture as compared to DC bus architecture. There are about 22% more conversion losses in AC bus architecture (Khan et al., 2019). AC bus architecture can be used in public places, homes or at commercial entities. AC dedicated charging comes under Level 3 charging, and there is a provision by many car manufacturing companies that Level-2 accessories can be used in Level-3 charging accessories too. A 3-phase system is generally available in European electrical distribution systems, and they can use it for fast charging or can form an AC common bus charging system (Van den Bossche et al., 2016).

2.2 Common DC Bus (CDCB)Architecture

Common DC bus charging architecture is a widely popular for xEV charging. It can be used for both conductive and inductive charging. The major advantage of the CDCB charging station is the use of one AC-DC conversion step as compared to multiple converters in common AC bus charging or on-board chargers. CDCB architecture mainly consists of the main AC-DC converter and further DC-DC converters for regulation. Their aim is to preserve the power quality indices of the system during the charging of multiple EVs (Gray & Morsi, 2015).

It is evident from the literature that common DC bus FCS is beneficial in many aspects if properly designed. Many operations like G2V, V2G and V2V etc. are possible with the proper control and coordination of DC FCS.

Many car companies are now shifting production towards EVs. Charging stations are an essential infrastructure to escalate the sale of EVs. Therefore, many companies are entering into this business such as; Charge Point, ABB, BP, Shell, Hyundai, Webasto, RWE, Daimler Mercedes-Benz, Siemens, EVgo etc. (Francis, n.d.).

Mostly, both CDCB and CACB FCS architectures are grid-connected; therefore, there is an impact of these on the power quality at the PCC.

3 POWER QUALITY PROBLEMS ASSOCIATED WITH EV CHARGING

It has been mentioned in many studies that if a FCS is supplied only by an electric-grid without using any electrical storage system or renewable energy sources, then there is direct impact of charging/discharging of EV on the power quality of grid. Power quality (PQ) is a broader term and is now widely used and is defined by IEEE (IEEE 1159:2009, IEEE 1100:2005) as "The concept of powering and grounding sensitive electronic equipment in a manner that is suitable to the operation of that equipment and compatible with the supply system and other connected load."

Power quality problems can be categorized as (i) Generation of Harmonics including; characteristic and non-characteristics harmonics, sub-harmonics, and inter-harmonics (ii) Voltage dips (iii) Voltage imbalance (iv) Voltage variations (sags and swells) (v) Frequency variations (vi) DC components in AC networks (vii) power system transients.

These events are undesirable in the power-system, and can cause short or long-term problems. These are explained as follows:

(a) Short-Duration Variations

These kind of problems may be instantaneous, momentary and/or temporary in the form of interruption, sag and swell. The cause of these kind of problems may be a fault in the system, large load energization and loose connections. As per IEEE-1250,

- Instantaneous interruption - between 0.5 and 30 cycles
- Momentary interruption - between 30 cycles and 2 seconds
- Temporary interruption – between 2 seconds and 2 minutes.

Similarly, sags and swells are the short duration undesired events. Sag is the temporary reduction in the rms voltage between 0.1 and 0.9 pu, and swell is the increase in voltage magnitude between 1.1 and 1.8 which can last between 0.5 cycles and 1 pu.

(b) Long-Duration Variations

These problems can persist for longer than 1 minute as per IEEE-1159 and ANSI-C84.1 standard. The reason for these variations are switching operations and load variations. These disturbances are categorized as sustained interruptions, causing over-voltages, and under-voltages.

- Sustained interruptions – as per IEEE standards, duration is more than 1 minute. These are the most serious kind of power quality problems often caused by fault incidences, incorrect relay operation, and scheduled maintenance.
- Over-voltages - increase in rms voltage from 1.1 to 1.2 pu for more than 1 minute. These over-voltages can be further classified as a lightning overvoltage, switching overvoltage, and over-voltages caused by insulation failure, ferro-resonance, tap changer transformer etc.
- Under-voltages - duration is more than 1 minute, and increase in rms voltage from 1.1 to 1.2 pu.

The other major power quality problems are voltage imbalance and waveform distortions.

1. Voltage imbalance occurs in 3-phase systems. In it, 3-phases are not the same and the phase difference in each phase is not identical i.e. 120°. The reasons for this are unbalances in 1-phase loading, single-phasing, blown-out fuses in capacitor banks etc.
2. Waveform distortions are termed as a steady-state nonconformity from the sinusoidal wave-shape. It may be caused by harmonics, inter-harmonics, DC offset, notching and electric noise.
    a. Harmonics- Harmonics are a major source of distortions. Mostly, harmonics are sinusoidal in shape, but the frequency is not same as power frequency (50 or 60 Hz). These are integer multiples of the power fundamental frequency. The reason for harmonics is non-linear load, communication network frequency interference, faults, malfunctioning of control devices, resonance frequencies due to cable capacitance and PFC capacitor. Most predominant harmonics are triplen harmonics, sub-harmonics, and inter-harmonics.

i. Triplen harmonics are the odd multiples of the 3rd order harmonics (3,9,15,21,27, ...) which results into higher distortions. A higher proportion of triplen harmonics is harmful to grounded-star connected system as excessive currents will flow through the neutral wire.
ii. Sub-harmonics are lower order of frequencies i.e. below the fundamental frequency. These subharmonics are produced in the input side and may be due to resonance between the harmonics currents/voltages and system capacitance and inductance. These subharmonics are generated in highly inductive or capacitive systems.
iii. Inter-harmonics appears as discrete or wide band frequencies, not as integer multiple of fundamental frequency. These harmonics occur in between the fundamental and integer multiples of frequency. In it, multiple is a non-integer number. The cause of these frequencies is modulation of current and voltage for control purposes and transient phenomenon in the system.

The other prominent harmonics that can occur are characteristic and uncharacteristic harmonics, positive, negative and zero sequence harmonics, spatial harmonics.

EV charging can have an adverse impact on the PQ of the conventional grid. In this chapter, PQ problems due to EV charging are studied in context of system harmonics (IEEE Std. 1100-2005, 2006) (Fuchs & Masoum, 2015).

In Table 2 and 3 (Duffey & Stratford, 1989), current and voltage harmonic limits are listed. These limits are very important for the safe and secure operation of the FCS.

Table 2. IEEE Standard 519-1992: Current harmonic limits for general distribution systems (129-69000 V)

Maximum harmonic currents (in percent of I_L)						
Individual harmonic order (odd harmonics)						
I_{sc}/I_L	h<11	11≤h<17	17≤h<23	23≤h<35	35£h	TDD(%)
<20*	4.0	2.0	1.5	0.6	0.3	5.0
20 to <50	7.0	3.5	2.5	1.0	0.5	8.0
50 to <100	10.0	4.5	4.0	1.5	0.7	12.0
100 to <1000	12.0	5.5	5.0	2.0	1.0	15.0
>1000	15.0	7.0	6.0	2.5	1.4	20.0

• Even harmonics are limited to 25% of the odd harmonics above.
• Current distortions that result in a DC offset, for example, half wave converters, are not allowed.
• I_{SC}=maximum short-circuit current at PCC and I_L= maximum demand load current (fundamental frequency component) at PCC.

Table 3. IEEE Standard 519-1992: Voltage harmonic limits

Bus Voltage at PCC	Individual voltage distortion (%)	Total Voltage distortion (%)
69 kV and below	3.0	5.0
69.001-161 kV	1.5	2.5
161.0001kV and above	1.0	1.5

Standards are framed to regulate the extent of system harmonics such as IEEE 519-1992, IEC 61000-3-12/2-4 and EN 50160:2000 (Duffey & Stratford, 1989) (Watson & Watson, 2017).

EVs are a new and dynamic load which may cause power quality-related problems having detrimental impacts on the electric grid. These effects include harmonic distortion, voltage unbalance, voltage drop, equipment overloading, phase unbalance and problems related to power system stability etc.

4 VOLTAGE SOURCE CONVERTER (AC-DC CONVERTER)

Converters are the core of EV charging infrastructure to convert infeed AC from the grid to DC via. Various converter arrangements (Iyer et al., 2018) such as (i) Diode rectifiers (ii) 2-level VSC (iii) 3-level Neutral Point Clamped (NPC) VSC, and (iv) Vienna Rectifier etc. to connect DC FCS with either a 1-phase or 3-phase grid supply. Diode rectifiers are a simple and economical technology as compared to other expensive alternatives, but a 2-level VSC using Insulated-Gate Bipolar Transistors (IGBTs) is proposed in this study (Sharma et al., 2020) since they offer higher efficiency, low harmonic generation, ability to have bi-directional operation with flexible gate control, snubber-less operation, higher voltage regulation and possibility for energy recuperation.

4.1 Two-Level VSC

Figure 1. Two-level, 3-phase VSC (Sharma et al., 2020)

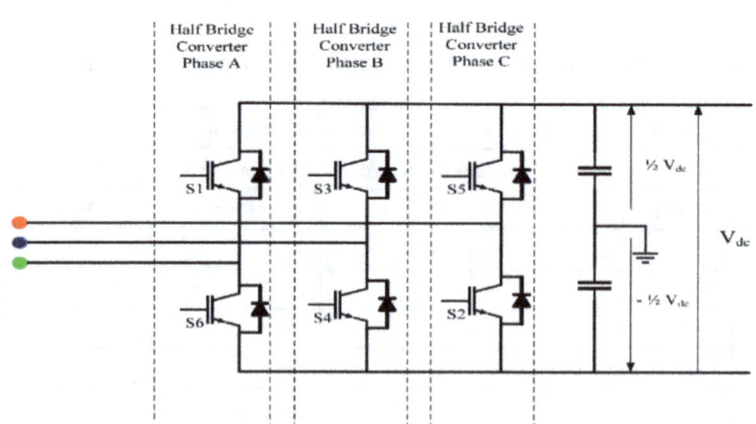

A 2-level VSC converter is shown in Fig. 1 2(Expansion, challenges, and opportunities in the EV market, n.d.). This VSC can be operated, with suitable control algorithms, in either uni-polar or bi-polar modes. AC-side terminal voltages can be deduced for each phase as

$$V_{TR}^{'}(t) = m_R(t)\frac{V_{DC}}{2} - r_{on}i_R(t) \tag{2}$$

$$V_{TB}^{'}(t) = m_B(t)\frac{V_{DC}}{2} - r_{on}i_B(t) \quad (3)$$

$$V_{TY}^{'}(t) = m_Y(t)\frac{V_{DC}}{2} - r_{on}i_Y(t) \quad (4)$$

Where $V_{TR}^{'}, V_{TB}^{'}, V_{TY}^{'}$ are terminal voltages of phase R, B and Y respectively, $m_R(t)$, $m_B(t)$, and $m_Y(t)$ are modulating signals for the respective phases, similarly $r_{on}i_R(t)$, $r_{on}i_B(t)$, and $r_{on}i_Y(t)$ are ohmic voltage drops of the respective phases.

5 CONTROL SCHEMES FOR VSC

The aim of control scheme is to estimate the reference values of currents from the sensed signals at the PCC with the help of feedback signals. The reference currents are then used to derive gating signals for the converter switches. Pulse Width Modulation (PWM) or Hysteresis Current Controller (HCC) can be used for generating gate pulses. In this study, Unit Template Control (UTC) strategy is implemented for generating the reference currents.

5.1 UTC Strategy

Unit Template Control (UTC) is a simple method for producing reference source currents to generate gating pulses of VSC. UTC is an efficient control strategy based on SRFT and supports grid-connected VSC. It is used as switching governor of the converter and reactive power control of the grid (Sharma et al., 2020). These applications could be understood with the help of Block diagram #.2.

Figure 2. Block diagram of UTC strategy (Bangarraju et al., 2014)

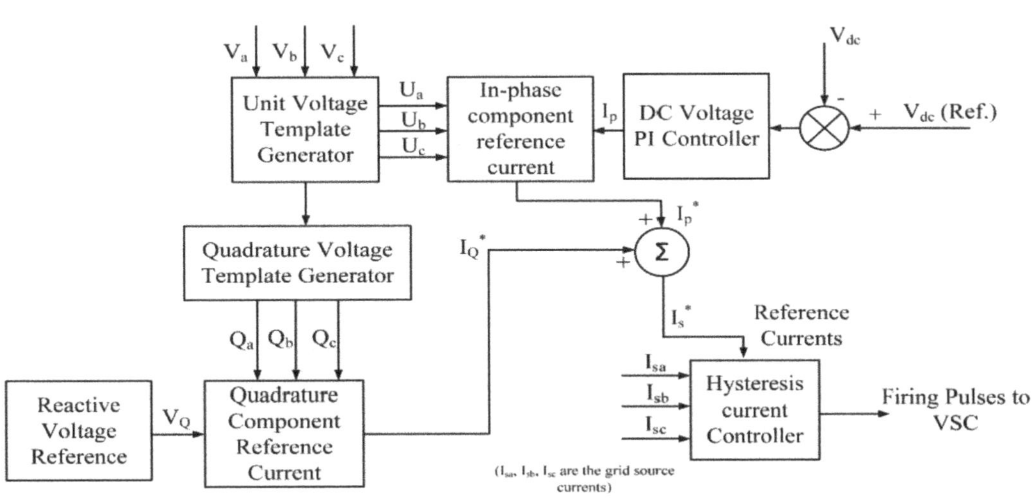

An indirect current control method is used to get reference signals (I_s*) for HCC switching of the VSC, as shown in Fig. 2 The reactive power is controlled by obtaining quadrature unit templates and by comparing it with the desired value. Similarly, active power is controlled through unit templates and addition of these serve as reference signals to obtain switching pulses.

6 PROPOSED SYSTEM OF ARCHITECTURES

Two architectures for the FCS for EVs are compared in this study. The two FCS architectures are assumed to be fed from the grid and feed into an assumed charging load on the basis of 10 bays having a typical EV (e.g. Nissan Leaf S Plus 2020 with a 62 kWh battery, 214-hp, 160 kW motor) in each bay. The vehicles are to attain up to 80% of their charge within 25 minutes. The criteria used to calculate the rating of the station is P = E/t where P = charging power in watts, E = Battery Energy in Wh, t = charging time in hours, a charging station with a rating of 118 *10 kW, i.e. 1.18 MW is required. Assuming some contingency and allowances for different types of vehicles to be charged, the FCS power rating is, therefore, estimated to be about 2 MVA. It has been considered that FCS is grid-connected. Therefore, there are certain strategies required to maintain the seamless operation of the charger and grid (Sharma et al., 2020). Following conditions as shown in Table #.3 are required at charger and grid to achieve efficient operation of the system:

Table 3. Required conditions on Grid and charger (Sharma et al., 2020)

Grid Side	Charger Side
UPF operation is required for the grid connected operation. Appropriate control algorithm must be used to achieve this.	Controller should be capable of maintaining Sufficient Voltage level ($V \pm \Delta V$) at DC link in common DC bus to charge the EVs with lesser ripples. (In this model battery voltage is 360 V and ΔV = 10%). In common AC bus, individual rectifier stage is required for each bay and afterwards voltage level is maintained.
As per IEEE 519-1992 (revised in 2014), 5% voltage harmonic distortion limit and 8% THD limit should be maintained at PCC. In case of current harmonics, IEEE and IEC standards should also be maintained and it depends upon the ratio of I_{sc} (Short-circuit current) and I_L (Load current). If I_{sc}/I_L is 20 to <50, then allowed TDD is less than 8%. THD is the ratio of sum of powers of all harmonic components to the power of fundamental frequency. TDD is calculated harmonic current distortion against the full demand level of the electrical system. At full load, THD(I) = TDD(I).	Charger side converter and controls are responsible to maintain the voltage and current harmonic limits as per International standards (IEEE 519-1992 and IEC 61000-3-12/2-4). CC-CV method is used at the charger end as DC-DC converter for regulation. It saves the battery from overcharging and increase the life of battery.

In this study, we focused on charging operation only and the impact of it on the power quality. The required important calculations for designing are made. Demand estimation of charging slots, location and area of the charging station, rate of allowable charging power for each vehicle, nominal voltage level and permissible power levels at the point of common coupling are the important aspects that should be considered while designing a charging station (Sharma et al., 2020).

The two FCS architectures are (a) DC bus architecture and (b) AC bus architecture

6.1 The DC Bus Architecture

The DC bus architecture (Fig. 3) is a well-established architecture. There are two power conversion stages in this architecture. First is the common (central) AC-DC rectifier stage, which is further connected to the DC-DC converter for regulation purpose. The proposed architecture is consisting of 10 bays.

The grid connection is at 12 kV, 2 MVA, 60 Hz, which feeds into the main FCS star-delta (or other configuration) 2MVA, 1200/600V, 60Hz transformer T1. The secondary of the DT has an AC filter (60kVA, 5Ω, 20μF) and 4 mH smoothing reactor for filtering purposes and feeds into the 2-level, 3-phase, active AC-DC, 6 pulse, VSC (IGBT/diode). The DC-link capacitor is 3300 μF at the common DC bus voltage of 400 V_{DC}. Ten EV bays (EV_1 to EV_{10}) are connected to the CDCB. Each EV bay consists of a DC-DC converter with a DC capacitor filter of 700 μF. The EVs can then be connected to this stage. In between each bay, an inductive link with 0.3 μH is assumed (this is assumed to be part of the leakage/ parasitic inductance between bays). The battery is 360V, 1000 Ah Li-ion battery. The system becomes more efficient as less number of conversion stages are required. Energy storage devices and super capacitors can also be used in this system, so, if there is any problem on the utility side, then also least impact on the DC bus system.

There are certain drawbacks to the system too. The rating of the central converter is higher, so there are more stern requirements for THD as per grid code.

Figure 3. Common DC Bus FCS

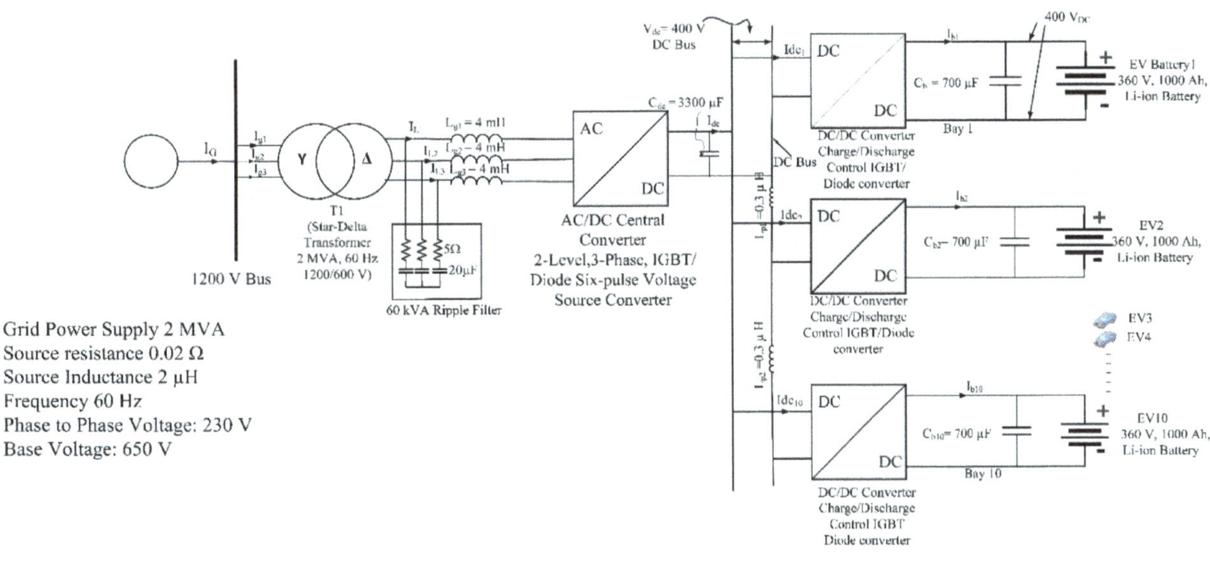

6.2 The AC Bus Architecture

AC bus architecture is a common architecture, which can be used to connect the charging points in FCS to the electric grid. There are also two conversion stages in this structure. First is AC to DC Conversion for each bay (decentralized) and then DC to DC conversion for regulation. It needs a separate firing, control and filtering system for each stage. Hence, it makes the system complex and expensive.

Three-phase AC bus is used to tap the connections for each and every charging point, and every point can work independently. But, due to the use of several rectifiers and different stages, there are chances of the production of the unwanted harmonic at the utility grid. It is an expensive operation as there is a need for several rectifiers with as many filters and sensors.

Even both of these fast chargers have a deleterious impact on the power quality of the grid. Foremost problems comprise of harmonics, voltage deviations, DC offset, stray fluxes and phantom loading. Power electronic converters are non-linear by nature, and due to it inject harmonics in the line current drawn by them. These harmonics added up the I^2R losses in the winding of DT and detrimental for its age.

The control strategy for the VSC is responsible for the smooth and efficient operation of FCS. UTC algorithm is proposed in this study for producing stable and sinusoidal grid currents. Generally, the PLL technique is used to connect the AC and DC architectures with the grid, but it is a complex and time-consuming method. UTC method is found comparatively sturdy and simple with both the architecture, which is discussed earlier in section 5.1.

7 SYSTEM METHODOLOGY

Both the AC and DC bus architectures are simulated in MATLAB/Simulink® environment to study the impact on the power quality due to the EV charging Infrastructure. A Distribution Transformer (DT) is used as an interface among the electric grid and power electronic converters, which are further connected to the EV batteries. DT is important and expensive equipment of these two architectures and different transformer configurations are studied to propose a suitable configuration. The magnitude of the third harmonic component depends upon the transformer configuration.

The following test cases are studied by using both systems:

1. Fully Loaded system (with 10 EVs)
2. Comparative PQ analysis (Fully (10 EVs) and Lightly Loaded system (with 5 EVs))
3. Analysis during a fault on the bus (dynamic-state)

In these cases, the behavior of FCS during steady-state and dynamic state is examined. The voltage and current waveforms are studied to evaluate the power quality. In the CDCB architecture, the DC bus link voltage is an important factor during charging.

8 SYSTEM RESULTS AND COMPARISONS

In section 7, methodology and cases to study the FCS were discussed. The results are compiled according to different cases, as follows:

8.1 Case I (Fully Loaded FCS)

In the following result waveforms, comparison of State of Charge (SOC) of battery during charging, PCC voltages, PCC currents are shown for CDCB and CACB architectures.

Figure 4. (a) SOC of EV Battery (CDCB Arch.); (b) SOC of EV Battery (CACB Arch.)

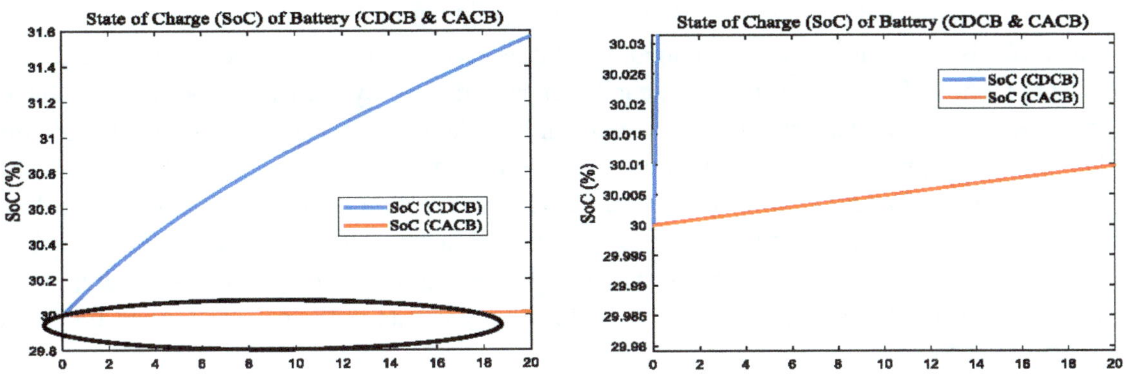

(a) (b)

Figure 5. (a) Input Voltage (Phase-to-Phase) CDCB; (b) Input Voltage (Phase-to-Phase) CACB

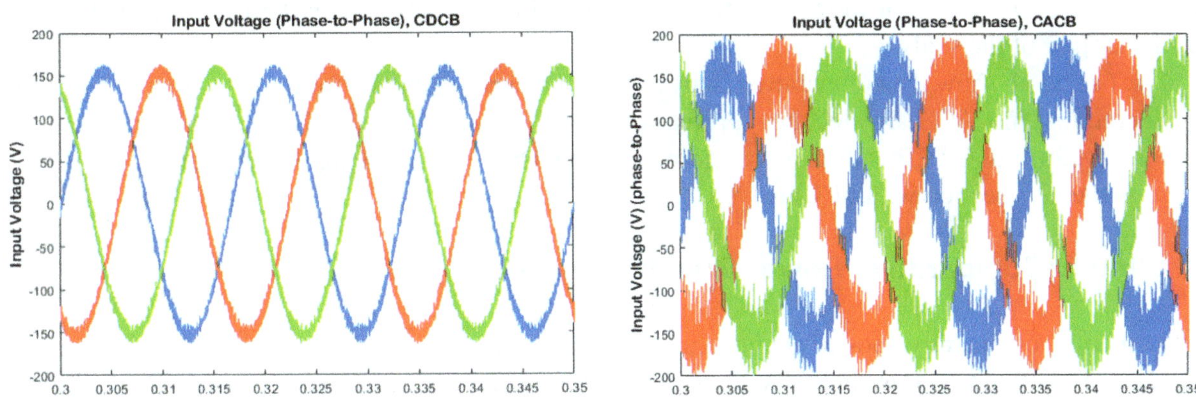

(a) (b)

Figure 6. (a) Input Current (CDCB); (b) Input Current (CACB)

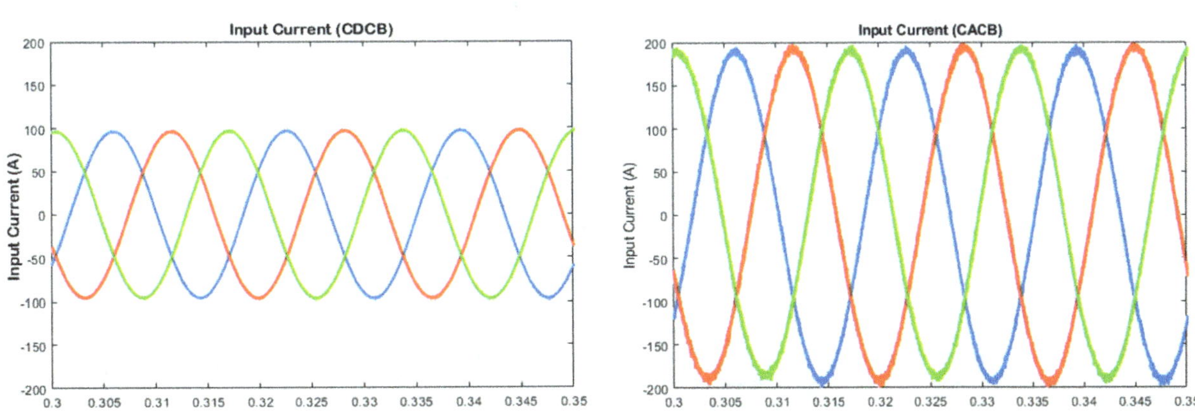

(a) (b)

In figure 4a comparison of SOC of battery during charging in both the architectures is given. Battery SOC is at 30% initially and charges up to 31.6% in 20 seconds. It means battery charges about 1.6% in 20 seconds. If this constant rate is maintained, the battery will get fully charged approximately in 21 minutes from 0% to 100%. In FCS, this time can be improved more by increasing the charging current. In most practical cases, battery charging is recommended when it is at around 30% and charged up to 90%. Then, it takes approximately 13 minutes to charge the battery by using the CDCB Architecture. Similarly, SOC of EV battery during charging in CACB architecture is shown in same Figure 4a and zoomed version is in Figure 4b. Battery SOC reaches only to approx. 30.007% from 30% in 20 seconds i.e. it would take approximately 33 hours to charge the battery from 30% to 90% in CACB architecture. Hence, a higher grid rating is required to increase the charging rate in the case of CACB architecture which is a serious concern.

Fig. 5a reveals the Input voltage (Phase-to-Phase) of the CDCB Charging station and Fig 5b shows the Input voltage (Phase-to-Phase) of the CACB Charging station. It is observed from the comparative analysis of both of these waveforms that voltage distortions are 9 to 10% higher (also depends upon the transformer configuration) in case of CACB charging station as compared to the CDCB charging station. Both of the waveforms are distorted, and the magnitude of distortions is seen through FFT analysis in the further part of the study.

Figs. 6a and 6b show the comparative analysis of Input current at PCC for both the architectures. It is observed that in CACB architecture, approximately 40-50% more current is drawn by 10 EVs. The current waveform of CACB is distorted as compared to CDCB, and the magnitude of distortion is observed in the next section with FFT analysis.

8.2 Case II (Comparative Power Quality Analysis)

In this section, comparative power quality analysis (harmonic-analysis) of both the architectures with different DT configurations is done. Fast Fourier Transform (FFT) analysis tool of MATLAB/Simulink® is used for this comparison. Different DT configurations are used to offer the best suitable configuration with the lesser harmonic component for both the current and voltage profiles.

Figure 7. (a) Grid Voltage Harmonics CDCB (THD_V = 4.59%) and (b) CACB (THD_V = 20.74%)

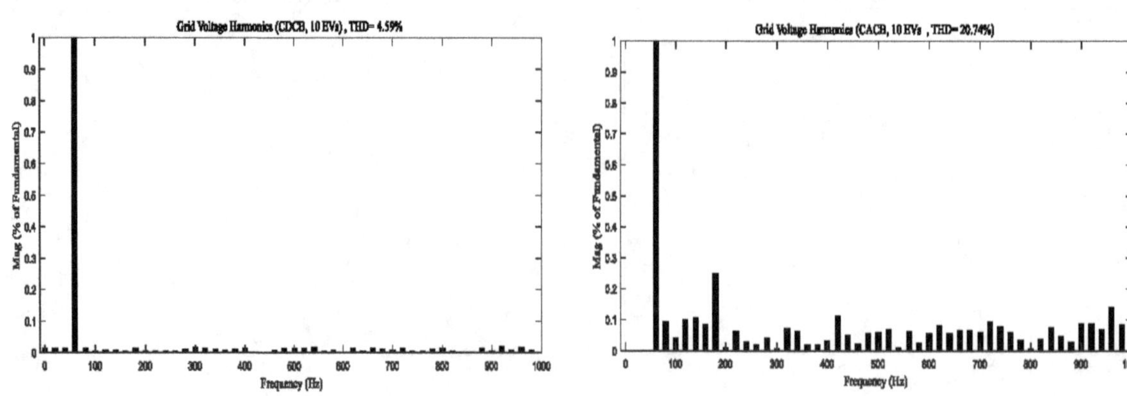

(a) (b)

Figure 8. (a) Grid Current Harmonics CDCB (THD_I = 1.12%) and (b) CACB (THD_I = 3.81%)

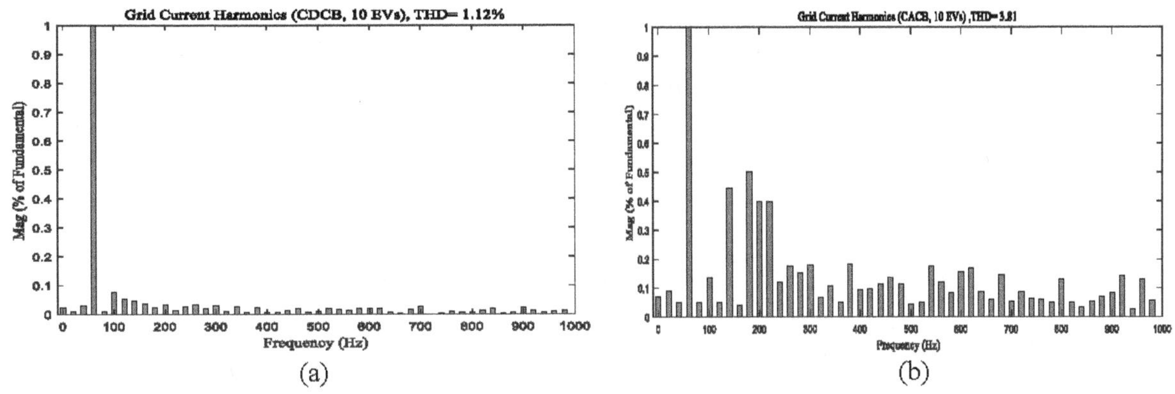

Figure 7 and figure 8 shows the comparison of FFT analysis of the DC and AC architectures, respectively, with DT having star-delta configuration.

From the FFT analysis, it is seen that the THD_V is higher than the recommended values in IEEE 519 standard in common AC bus configuration. In Table #.4, readings of THD_V, THD_I, and DC offset component is compiled for both fully and lightly loaded buses with different transformer configurations.

Table 4. Comparison table for different Transformer Configurations (Fully Loaded Buses and lightly loaded buses)

Harmonics during Battery Charging (G2V)						
			Fully Loaded Bus (10 EVs)		**Lightly Loaded Bus (5 EVs)**	
Type	**Transformer Connection Configuration**	**Harmonics Type**	**Percentage Harmonics (THD)**	**Off-set component**	**Percentage Harmonics (THD)**	**Off-set component**
Unipolar Common DC Bus Charging	*Star-Delta*	Voltage	4.74	0.005	2.50	0.010
		Current	1.18	0.032	0.83	0.030
	Delta-Delta	Voltage	4.89	0.008	2.49	0.008
		Current	1.13	0.039	0.83	0.034
	Star-Star	Voltage	4.93	0.008	2.48	0.007
		Current	1.13	0.039	0.83	0.034
	Delta-Star	Voltage	4.88	0.007	2.49	0.006
		Current	1.12	0.030	0.82	0.055
Unipolar Common AC Bus Charging	*Star-Delta*	Voltage	20.74	0.006	11.66	0.063
		Current	3.81	0.626	4.66	0.696
	Delta-Delta	Voltage	20.64	0.127	11.62	0.032
		Current	3.87	0.259	4.81	0.575
	Star-Star	Voltage	20.64	0.127	11.62	0.032
		Current	3.87	0.259	4.81	0.575
	Delta-Star	Voltage	21.07	0.120	11.76	0.071
		Current	3.85	1.045	4.83	0.320

It is evident from Tables #.4 that the Unipolar DC bus gives better results in case of power quality. Star-delta configuration of DT is with fewer harmonics and minimal value of DC offset for both the common AC and DC bus configurations. As per FFT analysis, the current harmonic percentage is very high in CACB charging infrastructure as compared to CDCB architecture. In terms of load, there is a considerable impact in the case of CACB charging. Substantial increase in voltage harmonics in case of a CACB structure has been observed, and it is unacceptable as per IEEE 519 standards. Large and expensive filters could be used as a possible solution to mitigate these surges.

8.3 Case III (Analysis During a Fault on the Bus (Dynamic State))

The dynamic state of both the architectures (Fully loaded state) is studied and compared in subsequent figures. The 3-phase fault state is formed for 3 cycles duration for both the architectures, as can be seen in Figs. 9 and 10 Power quality analysis is done to compare both the architectures during the dynamic state.

Figure 9. (a) Input Voltage (CDCB) during Fault; (b) Input Voltage (CACB) during Fault

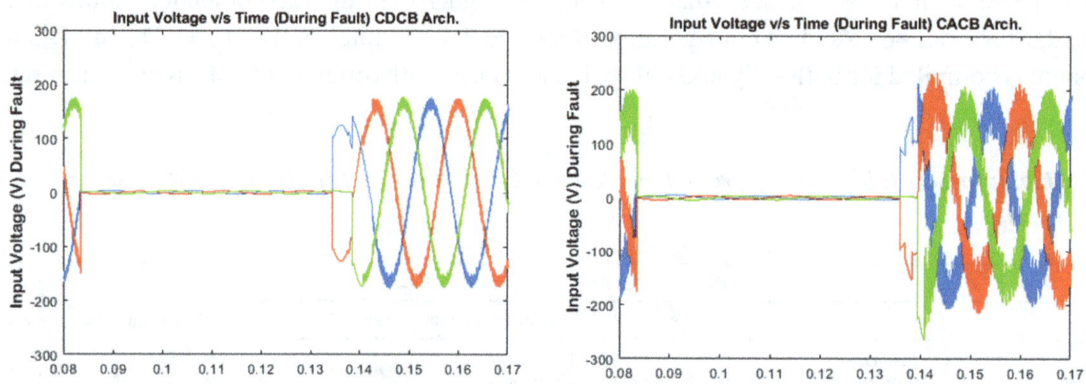

(a) (b)

Figure 10. (a) Input Current (CDCB) during Fault; (b) Input Current (CACB) during Fault

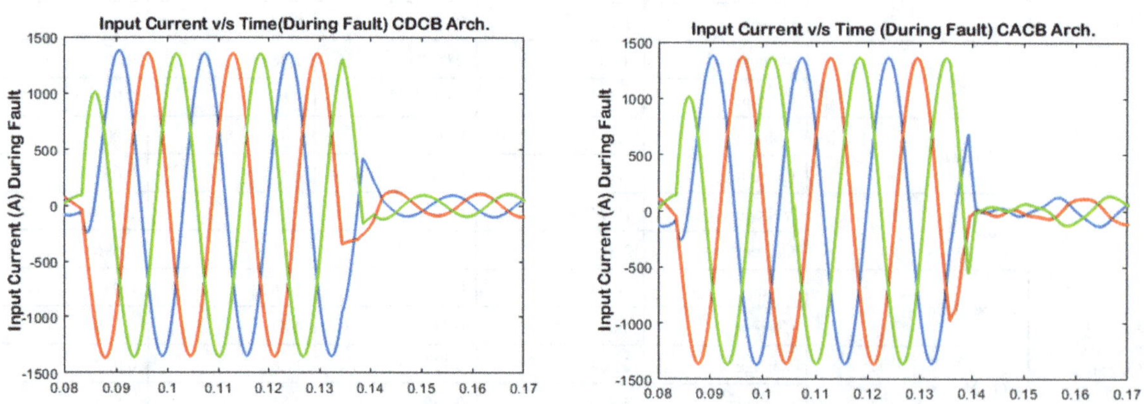

(a) (b)

In the above-shown figures, waveforms of input voltage at PCC and input currents at PCC during fault condition (3-Cycles) are shown for CDCB and CACB Architectures (10 EVs). It is clear from the comparative analysis of these figures that the CDCB charging station regains its normal state faster than the CACB station in the dynamic state. Harmonic spectra for both the architectures in a dynamic state is shown below in Fig. 11 and 12.

Figure 11. Impact of three phase fault fully loaded (for very small duration of 3-Cycles) on the Voltage waveforms (Star-Delta Configuration), #.11 (a) CDCB fully loaded (THD_V = 80.77%, DC Component = 4.576) and, #.11(b) CACB fully loaded THD_V = 110.54%, DC Component = 12.52)

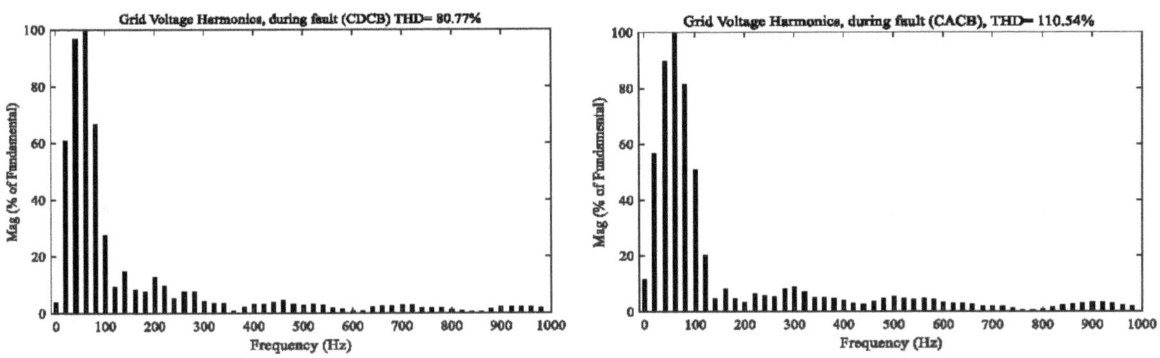

Figure 12. Impact of three phase fault fully loaded (for very small duration of 3-Cycles) on the Current waveforms (Star-Delta Configuration), #.12 (a) CDCB fully loaded (THD_I = 36.64%, DC Component = 4.116) and, #.12 (b) CACB fully loaded THD_I = 42.79%, DC Component = 1.819)

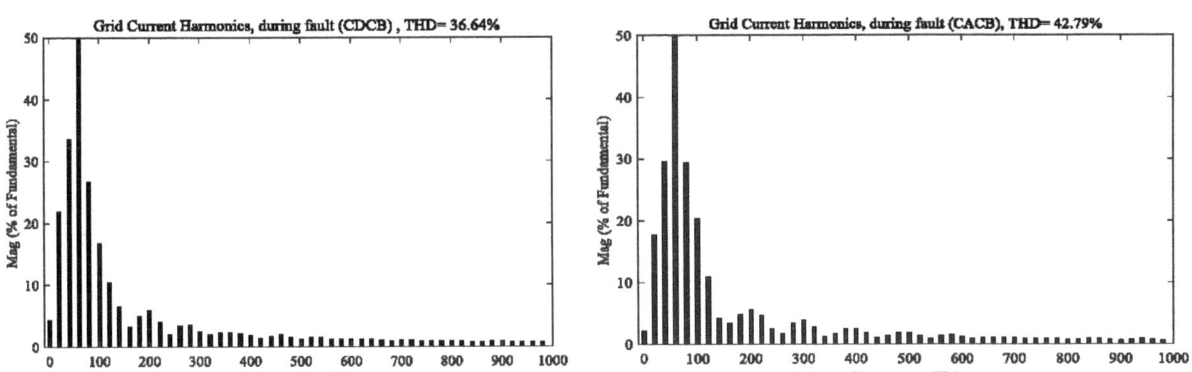

A significant increase in the THD_V and THD_I was observed during the dynamic state, which is damaging for connected equipment if fault condition persists for a longer duration. Proper protective devices are needed to safeguard the expensive infrastructure. Circuit breakers should be employed at key locations to protect the costly structure associated with the FCS.

9 CONCLUSION

A Unipolar CDCB and CACB architectures for EV FCS is proposed and simulated in this chapter. Both of these architectures are realized by using grid-connected charger/s through common DC/AC bus. These architectures are compared by using different transformer connection configurations to study the impacts on power quality.

It influences the life expectancy of DT and other associated equipment. THD_V and THD_I of both the architectures are compared, and conclusions are made as follows:

- Star-Delta (Y- Δ) configuration of DT is a suitable method of charging mode (Table #.4) as it gives minimal THD_V and THD_I in both the architectures. The reasons for lesser harmonics and better performance of Star-Delta (Y- Δ) configuration are as follows:
    - The Y- Δ connected DT has fewer difficulties with the third harmonic components due to the circulating currents in the delta windings. Delta connected secondary is also extra firm to unbalanced loads as it rearranges any imbalance in secondary.
    - In the Y- Δ transformer, neutral is available on the star side, which can be earthed to avoid most of the distortions on the primary side. It also reduces insulation costs as $V_L = \sqrt{3}\ V_{ph,}$ so phase voltage is lesser than the line voltage.
    - These reasons made Y- Δ configuration more stable in case of unbalanced loads and having lesser THD_V and THD_I.
- CDCB architecture is better than the CACB Architecture because of
    - Fewer transformation stages as there is only one main AC to DC converter as compared to the common AC bus.
    - Lesser THD_V and THD_I, and as a result, better power quality can be obtained.
    - The system efficiency of the common DC bus become more due to better power quality and improved power factor as compared to the CACB.
    - CDCB architecture is more stable in a dynamic state as compared to the CACB.
    - It can be implemented in coordination with distributed generation and grid coordination. EVs can be used as storage devices during idle time.
    - Lesser circulating currents in between converters which results in low DC component.
- There is an impact of the number of bays on power quality. THD rises with the growth in the number of EVs on the same rated system.

REFERENCES

Bangarraju, J., Rajagopal, V., & Jayalaxmi, A. (2014). Unit template synchronous reference frame theory based Control algorithm for DSTATCOM. *J.Inst. Eng. India Ser. B*, *95*(2), 135–141. doi:10.100740031-014-0087-y

Ben Jar & Watson. (2016). *Rapid EV chargers: Implementation of a charger*. Electric Power Engineering Centre (EPECentre), University of Canterbury.

BNEF. (n.d.). https://about.bnef.com/electric-vehicle-outlook/

Duffey, C. K., & Stratford, R. D. (1989). Update of Harmonic Standard IEEE-519: IEEE Recommended Practices and Requirements for Harmonic Control in Electric Power Systems. *IEEE Transactions on Industry Applications*, *25*(6), 1025–1034. doi:10.1109/28.44238

El Kharbachi, A., Zavorotynska, O., Latroche, M., Cuevas, F., Yartys, V., & Fichtner, M. (2020). Exploits, Advances and Challenges Benefiting Beyond Li-Ion Battery Technologies. *Journal of Alloys and Compounds*, *817*, 153261. doi:10.1016/j.jallcom.2019.153261

Erb, D. C., Onar, O. C., & Khaligh, A. (2010). Bi-directional charging topologies for plug-in hybrid electric vehicles. *Proc. 25th Annu. IEEE Appl. Power Electron. Conf.*, 2066–2072. 10.1109/APEC.2010.5433520

Expansion, challenges, and opportunities in the EV market and EVSE industry blog on e-mobility. (n.d.). https://circontrol.com/expansion-challenges-ev-market-evse-industry/

Francis, S. (n.d.). *Top 20 Electric Vehicle Charging Station Companies*. Robotics and automation news report. https://roboticsandautomationnews.com/

Fuchs, E., & Masoum, M. A. (2015). *Power Quality in Power Systems and Electrical Machines* (2nd ed.). Elsevier.

Globe Newswire. (n.d.). https://www.globenewswire.com/

Gray, M. K., & Morsi, W. G. (2015). Power quality assessment in distribution systems embedded with plug-in hybrid and battery electric vehicles. *IEEE Transactions on Power Systems*, *30*(2), 663–671. doi:10.1109/TPWRS.2014.2332058

IEEE Recommended Practice for Powering and Grounding Electronic Equipment-Red-line, in IEEE Std. 1100-2005 (Revision of IEEE Std. 1100-1999)- Redline ed., pp. 1-703, 2006.

Iyer, V. M., Gulur, S., Gohil, G., & Bhattacharya, S. (2018). Extreme fast charging station architecture for electric vehicles with partial power processing. *2018 IEEE Applied Power Electronics Conference and Exposition (APEC)*, 659-665. 10.1109/APEC.2018.8341082

Khan, W., Ahmad, F., & Alam, M. S. (2019, February). Fast EV charging station integration with grid Ensuring optimal and quality power exchange. *Elsevier International Journal of Engineering Science and Technology*, *22*(1), 143–152. doi:10.1016/j.jestch.2018.08.005

Miao, Y., Hynan, P., von Jouanne, A., & Yokochi, A. (2019). Current Li-Ion Battery Technologies in Electric Vehicles and Opportunities for Advancements. *Energies*, *12*(6), 1074. doi:10.3390/en12061074

Mojdehi & Ghosh. (2016). An On-Demand Compensation Function for an EV as a Reactive Power Service Provider. *IEEE Transactions on Vehicular Technology, 65*(6), 4572-4583.

Monteiro, V., Pinto, J. G., & Afonso, J. L. (2016, March). Operation modes for the electric vehicle in smart grids and smart homes: Present and proposed modes. *IEEE Transactions on Vehicular Technology*, *65*(3), 1007–1020. doi:10.1109/TVT.2015.2481005

Moradewicz, A. J. (2019). On / Off – board chargers for electric vehicles. *Przegląd Elektrotechniczny, 95*(2).

Naveen, G., Yip, I. H.-I., & Xie, Y. (2014). Modeling and protection of electric vehicle charging station. *Proc. of 2014 6th IEEE Power India International Conference (PIICON)*, 1-6.

Rajashekara, K. (2013, March). Present status and future trends in electric vehicle propulsion technologies. *IEEE Journal of Emerging and Selected Topics in Power Electronics, 1*(1), 3–10. doi:10.1109/JESTPE.2013.2259614

Sharma, Sood, Alam, & Shariff. (2020). Comparison of Common DC and AC Bus Architectures for EV Fast Charging Stations and Impact on Power Quality. *Elsevier eTransportation, 5*.

Shaukat, N. (2018, January). A survey on electric vehicle transportation within smart grid system. *Renew. Sustain. Energy Rev., 81*, 1329–1349. Advance online publication. doi:10.1016/j.rser.2017

Tan, L., Wu, B., Yaramasu, V., Rivera, S., & Guo, X. (2016). Effective Voltage Balance Control for Bipolar-DC-Bus-Fed EV Charging Station with Three-Level DC–DC Fast Charger. *IEEE Transactions on Industrial Electronics, 63*(7), 4031–4041. doi:10.1109/TIE.2016.2539248

Tran, V. T., Sutanto, D., & Muttaqi, K. M. (2017). The state of the art of battery charging infrastructure for electrical vehicles: Topologies, power control strategies, and future trend. *Proc. of AUPEC*, 1-6. 10.1109/AUPEC.2017.8282421

Van den Bossche, Turcksin, Omar, & Van Mierlo. (2016). Developments and Challenges for EV Charging Infrastructure Standardization. *World Electric Vehicle Journal, 8*, 1–7.

Watson & Watson. (2017). Impact of Electric Vehicle chargers on harmonic levels in New Zealand. *Proc. of 2017 7th IEEE Innovative Smart Grid Technologies (ISGT Asia 2017) annual premier smart grid technology meeting*.

Chapter 7
Optimal Charging Management of Microgrid-Integrated Electric Vehicles

Giulio Ferro
https://orcid.org/0000-0002-0641-6846
University of Genova, Italy

Riccardo Minciardi
University of Genova, Italy

Luca Parodi
University of Genova, Italy

Michela Robba
University of Genova, Italy

ABSTRACT

The relevance of electric vehicles (EVs) is increasing along with the relative issues. The definition of smart policies for scheduling the EVs charging process represents one of the most important problems. A discrete-event approach is proposed for the optimal scheduling of EVs in microgrids. This choice is due to the necessity of limiting the number of the decision variables, which rapidly grows when a small-time discretization step is chosen. The considered optimization problem regards the charging of a series of vehicles in a microgrid characterized by renewable energy source, a storage element, the connection to the main grid, and a charging station. The objective function to be minimized results from the weighted sum of the cost for purchasing energy from the external grid, the weighted tardiness of the services provided, and a cost related to the occupancy of the socket. The approach is tested on a real case study.

DOI: 10.4018/978-1-7998-6858-3.ch007

INTRODUCTION

The constant growth of Renewable Energy Systems (RESs) as a sustainable replacement for carbon-intensive fossil fuels for energy production is a key priority at the international level. Typically, RESs are intermittent and geographically distributed energy sources. Moreover, RESs enable the transition of the traditional electricity consumers to become prosumers actively engaged in energy markets through renewable production, storage and demand-side management. Additionally, new actors such as aggregators are present and electric vehicles (EVs) may represent an additional, intermittent and distributed load. In this framework, smart grid management and optimization technology solutions must become reality in order to guarantee the operation of the grid and the satisfaction of all connected users. To this end, Energy Management Systems (EMSs) based on optimization models are currently considered essential tools for the optimal management of an electrical grid characterized by different actors, distributed production, flexible loads, storage systems and EVs. Microgrids can be seen as a distribution grid at a small scale, which requires EMSs for the optimal management of production plants, storage systems, flexible loads, renewables, etc. When also EVs are present, the complexity of the optimization problem arises, and it is necessary to find new mathematical models and solution approaches. In fact, EVs' integration is particularly important to reduce greenhouse gas emissions in cities. Emissions can be reduced by using EVs when renewable resources are used to produce feeding the vehicles. As confirmed by the authors in (Ferrero et al., 2016; Shareef et al., 2016), mass distribution of EVs may be a good solution, but, unfortunately, their wide usage may have important drawbacks and cause technical problems. As an example, the power grid can be harmfully affected by uncontrolled charging, long charging times and interruptions and can become unstable for distributed and intermittent loads (Qian et al., 2015; Sbordone et al., 2015). Thus, it is necessary to define planning and management strategies that are able to ensure the operation of the grid and to satisfy EVs demand. Moreover, it is necessary to consider new technologies, such as Vehicle-to-Grid (V2G), Smart Charging (SC) and Vehicle-to-Building (V2B), allow the injection of power by EVs into the electrical grid and/or the modulation of the power during the charging process.

Finally, it should be considered that EVs could be charged in a much longer time because the charging station is not able to manage multiple vehicles through optimized power management strategies, thus creating long queues and users' dissatisfaction.

In this chapter, attention is focused on the optimal management of microgrids that include EVs and charging stations with multiple sockets.

In the literature, discrete-time decision models are used by several approaches that aim at the optimal scheduling of EVs and result in problems difficult to be solved, especially due to a large number of variables. Then, also heuristics and metaheuristics have been applied as well as decision architectures based on decentralized optimization. In the following, a discrete-event approach is used for the definition of an optimization model for the integration of EVs in the electrical grid. The approach is particularly advantageous because it allows reducing the number of variables and computational time. The considered case study includes a microgrid with a multiple sockets charging service station, photovoltaics, storage system, and a connection to the external grid. The resulting optimization model is (nonlinear) Mixed Integer Problem (MIP). This serious complication, with respect to the case considering the charge of one single EV, can still be managed when problem instances have moderate size, like those usually solved in real cases (perhaps in real-time). Anyway, the discrete-event approach counterbalances this complication, by considering a much lower number of time intervals, and consequently a significant reduction of the optimization problem size.

In summary, the contribution of the chapter are:

- A description of the role of optimization for EVs' scheduling in microgrids;
- A discrete event optimization model for multiple sockets charging stations;
- The application to a case study and discussion of results.

This chapter is organized as follows:

- Section II: discussion about the mathematical background on optimization and discrete event systems;
- Section III: state of art regarding EVs in the smart grid's context;
- Section IV: formalization of the optimization model;
- Section V: application of the model to a case study;
- Section VI: comments and remarks.

MATHEMATICAL BACKGROUND ON OPTIMIZATION

An optimization problem is composed of three main elements:

1. An objective function, i.e. a function which needs to be minimized or maximized usually related to some cost;
2. A set of decision variables whose values represent the decisions that should be applied on the basis of performance indicators (formalized in an objective function);
3. A set of constraints that are restrictions on the values that the variables can take, typically in the form of equality and inequality constraints.

Depending on the nature of these elements (linear, quadratic, non-linear, integer, mixed-integer, etc.) it is possible to differentiate the optimization problems.

An optimization problem has the general form:

$$\min_{x} = f(x) \tag{I}$$

s.t.

$$h(x) = 0 \tag{II}$$

$$g(x) \leq 0 \tag{III}$$

where $x \in R^n$ is the decision variable vector, $f(x):\mathbb{R}^n \to \mathbb{R}$ is the objective function to be minimized, $h(x):\mathbb{R}^n \to \mathbb{R}^{n_E}$ are functions for the equality constraints and $g(x):\mathbb{R}^n \to \mathbb{R}^{n_I}$ for the inequality constraints. The following observations are valid in this context:

- Any $x \in \mathbb{R}^n$ is called a solution, while a solution meeting (II) and (III) is called a feasible solution. A feasible solution minimizing (I) is called an optimal solution or a minimizer. The objective function value at the minimizer is the objective function optimal value;
- Minimizing $f(x)$ is equivalent to maximize $-f(x)$;
- The set of solutions meeting both $h(x)=0$ and $g(x)\leq 0$ determines the feasible region of the optimization problem.

Optimization-based control problems can be divided into different classes on the basis of the time discretization, i.e. continuous, discrete and discrete-event problems. In the first class, all the quantities are considered as continuous. In the second class, time is discretized in intervals with constant length. The last one considers the time divided into intervals which length is a variable itself.

Example

In the following, an example of a linear programming problem is provided

$$\min_{x,y} = (x + y - 30) \tag{IV}$$

$$5x + 10y \geq 250 \tag{V}$$

$$10x + 4y \leq 300 \tag{VI}$$

$$x \geq 20 \tag{VII}$$

$$y \geq 12 \tag{VIII}$$

$$x, y \in \mathbb{R} \tag{IX}$$

The optimal solution (obtained by means of a mathematical programming solver) is given by $x=20$, $y=15$ and the value of the objective function is 5.

In the case of smart grids, examples of decision variables are: size, number, and kind of available technologies to be used; the schedule of production plants and storage systems; in the case of interconnected microgrids, the power exchange between them.

Instead, possible objective functions could be the minimization of: operational cost, power losses in the distribution network, and CO_2 emissions.

Finally, constraints could be the maximum power production of the plants or maximum and minimum power exchanged with the external grid and with the storage system.

The main decision problems regarding microgrids are:

- Planning problems: the decision-maker (i.e., a Public Authority, project manager, etc.) must decide which technologies and the related size that have to be installed in a specific area to satisfy a known demand. These kinds of problems need to consider a yearly demand for energy and a time horizon equal to the lifetime of the technologies. The criteria for the choice are in general economic costs and environmental impacts such as greenhouse gas emissions.
- Operational management problems: when all plants and storage systems are already installed, the decision problem investigates how to schedule production and storage of power inside the microgrid during the day. Typically, the day before the schedule should be considered (i.e., 24 hours in advance). In this case, the microgrid's EMS receives data on the forecasted demands and power from non-controllable plants (i.e., production plants from renewable resources like wind turbines and PV plants) for the next day. According to such data, the optimal schedule is found, using an optimization model.
- Real-time control problems: during the day, the forecasted demands and renewables determined the day before could differ from the expected ones (due to the uncertainty of predictions). Thus, it is necessary to manage in real-time such discrepancies using specific algorithms (that, by the way, can be based on a simplified version of the models used for operational management

STATE OF THE ART AND CHALLENGES

The literature on smart grids is wide and presents very different cases and applications. One of the most interesting subjects directly connected to smart grids is represented by EMSs. An efficient EMS must be able to manage all the new technologies present in the controlled system. Among the last challenges, Demand Response (DR, i.e., the possibility for a traditional load to decrease the active power absorption for a certain period) can be considered as a key issue for cost reduction, for more effective integration of RESs and for reliable grid operations (Fontenot & Dong, 2019; Li et al., 2018; Srinivasan et al., 2016). In (Aliasghari et al., 2018b), the authors highlight how the implementation of EVs can be an advantageous option where the DR program is considered in microgrids or, as in (S. Wang et al., 2019), in power systems with multiple microgrids. Authors in (Amini et al., 2017), state that EVs and renewable resources should be coupled and managed together to have a reduction of emissions. Another example is given by (Asensio et al., 2018a, 2018b), where the authors jointly study energy storage systems (ESSs) and DR analyzing their impact and possible use in smart grids.

Among the survey papers, (Tan et al., 2016; Yang et al., 2015; Zheng et al., 2019) are focused on smart grids integrating EVs in their operations. In (Yang et al., 2015), scheduling methods to integrate plug-in electric vehicles are reviewed, examined and categorized based on different possible approaches (i.e., conventional optimization methods (e.g. linear, non-linear mixed-integer programming and dynamic programming), analytical scheduling, game theory, and meta-heuristic algorithms). Multi-objective and high-dimension problems represent a major problem resulting from the scheduling problems formalization for EVs which integrate intermittent renewable energy generation and stochastic charging behaviors. In (Tan et al., 2016) and (Zheng et al., 2019) the authors pay attention to the V2G technology. In particular, (Tan et al., 2016) examines it integrated into a smart grid, focusing on optimization-based approaches.

Another interesting paper is (Rahman et al., 2016), which reviews the use of optimization methods for the charging infrastructure related to plug-in hybrid electric vehicles (PHEVs) and EVs. It is assessed that optimization models specifically applied in solar-based charging stations are needed to obtain an optimal integration of EVs in the charging infrastructure while guaranteeing a reduction of emissions.

Some issues related to EVs are represented by the power peak associated with their charge. The power needed by the Charging Stations (CSs) is a very critical point while considering the planning of new plants due to their huge impact on the infrastructure. Studies like (Sachan & Adnan, 2018), state that considering the technical constraints imposed by the electrical grid is necessary to reduce load peaks and voltage violations. Then, it is decisive to consider the optimal charging process of those EVs. The authors in (Sachan & Adnan, 2018) propose a coordinated scheme for the evaluation of the optimal number of EVs' charges in a distribution network without recurring to the grid reinforcement or the infrastructure change. Other important aspects are considered in (Aujla et al., 2019): energy trading and pricing for electrical vehicles in smart cities. It proposes an energy-trading scheme which allows EVs to obtain energy from any available CS using a multi-leader multi-follower Stackelberg game and a multi-parameter pricing scheme.

Different methodological approaches have been considered in (Yang et al., 2015) formalizing different mathematical models. The most common approach, used for example in some recent papers as (Aliasghari et al., 2018a; Ferro et al., 2018; Ito et al., 2018; Sarikprueck et al., 2018), have formalized discrete-time setting, solved by means of mathematical programming techniques, characterized by a high number of variables with a consequent high computational time required. In (Atallah et al., 2018), the authors consider multiple charging outlets in supercharge facilities and compare different methods (centralized and decentralized) for the optimal scheduling of EVs. The centralized linear integer optimization model resulted in the best solution in terms of waiting time but characterized by low scalability. Instead, the (game-theoretic) distributed approach overcomes the scalability issue obtaining promising results. Another distributed game-theoretic approach is adopted by (Latifi et al., 2019) where the authors want to maximize grid efficiency, minimize customers' costs, and maximize the capacity for ancillary services. In (R. Wang et al., 2017) employing an architecture which is based on both centralized and decentralized optimization, the customers' charging preferences are considered through a model predictive control-based adaptive scheduling strategy. In (Liu et al., 2018) the authors propose day-ahead scheduling of EVs' charging based on an aggregative game model. The charging costs are minimized taking into account constraints representing the users' requirements. In (Zhang et al., 2018), the authors consider the optimal charging of a platoon of EVs by using a contract theoretic approach that maximizes the utilities of the CSs under constraints relative to the quality of service. It is an iterative algorithm whose convergence is demonstrated. In (Liu et al., 2019) an aggregator of EVs is considered and both the day-ahead electricity procurement and the real-time EVs charging management are managed through a predictive transactive control which minimizes the total operating cost, according to a quantification of the charging flexibility.

A hierarchical event-driven multi-agent system approach is proposed by authors in (Azar & Jacobsen, 2016) for coordinated scheduling of the charging process of EVs. However, as noted in (Ferro et al., 2019a), an event-driven approach present in the literature commonly consists of on-line scheduling algorithms that are able to perform a "correction" on a pre-existing schedule whenever a new vehicle requiring charging service arrives at the CS.

As showed in (Ferro et al., 2019b), compared to the discrete-time approach, discrete-event approaches for the management of EVs, and charging stations present a reduced number of variables and are considered promising. The method presented in this chapter consists of developing a predictive discrete-event optimization model for the optimal management of a microgrid where some EVs' charges must be scheduled. This is basically the same approach followed in (Ferro et al., 2019a), but it is now extended to the case in which several EVs may be charged simultaneously on different sockets.

FORMALIZATION OF THE OPTIMIZATION MODEL

A discrete-event approach is adopted in the formalization of the optimization problem and a series of events corresponding to vehicles' charging process, considering C_i the completion time of the *i-th* vehicle charging process (see Figure 1).

Figure 1. Time axis discretization

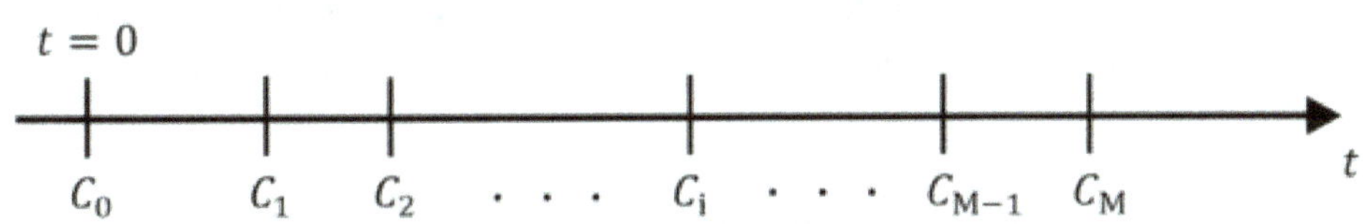

The order of the completion time instants is assumed to be given, according to an a priori defined criterion, as, for example, the arrival order of the customers.

Figure 2 presents the system architecture. We consider a microgrid (light-blue ellipse) where a charging station (with its charging units) is connected to a storage element and a renewable energy source. Then, the microgrid is connected to the main grid. The charging unit can provide simultaneous charging services to various vehicles. N is the maximum number of EVs that can be connected to the charging unit (the charging station has N sockets).

Figure 2. Schematic representation of the microgrid

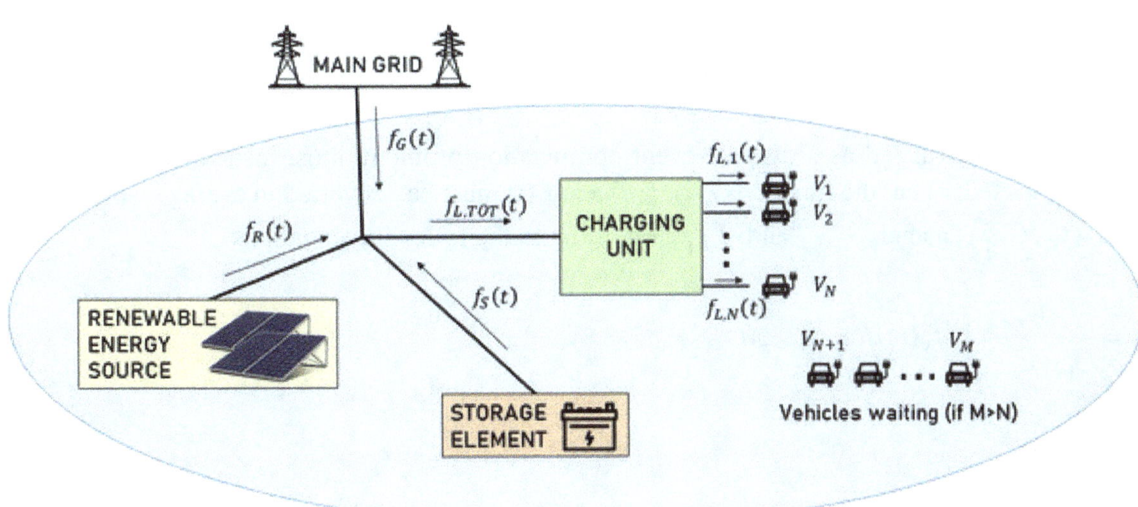

Decision Variables

As the model considers a discrete-event approach, the time intervals (C_i, C_{i-1}), i=1,…,M (let C_0=0 by definition), are decision variables. Regarding those time intervals, define:

- $f_{L,i,j}$= the constant (monodirectional) power flow delivered for charging of vehicle V_j, $j=i$, i+1,…, M (obviously $f_{L,j}(t)$=0 if V_j is not connected at time instant t);
- $f_{G,i}$= the constant power flow drawn from/injected into the grid. This flow is bi-directional and is considered as positive when power is drawn from the main grid;
- $f_{L,TOT,i} = \sum_{j=i}^{M} f_{L,i,j}$ = the total power flow delivered to the charging station;
- $f_S(t)$= power flow drawn (delivered) from (to) the storage element. This is a bi-directional flow, and, it is assumed to be positive when the power is drawn from the storage;
- $f_R(t)$= the (monodirectional) power flow coming from the renewable energy source;
- $y_{i,j}$= binary variable to indicate whether in a given time interval a certain vehicle is under charging or not (see Eq. (1));

$$y_{i,j} = \begin{cases} 1 & \text{if vehicle } V_j \text{ is under charging during time interval } (C_{i-1}, C_i) \\ 0 & otherwise \end{cases} \quad (1)$$

for i=1,…,M, j=1,…,M

- $tard_i$= the tardiness of the charging service provided to vehicle V_i and its definition is provided by (2).

$$tard_i = \max\{C_i - dd_i, 0\} \quad i=1,\ldots,M \quad (2)$$

where dd_i is the due date of the i-th vehicle whose meaning will be presented in the section regarding the parameters.

Evaluation of Average Productions

The problem is formalized as a discrete-event optimization problem, thus, instantaneous constraints cannot be imposed. Then, the functions $f_S(t)$, $f_G(t)$ and $f_R(t)$ must be converted in average values on time interval (C_i, C_{i-1}), namely, $\overline{f}_{S,i}$ and $\overline{f}_{R,i}$, according to the following equations

$$\overline{f}_{S,i} = \frac{1}{C_i - C_{i-1}} \int_{C_{i-1}}^{C_i} f_S(t)\,dt \quad i=1,\ldots,M \quad (3)$$

$$\overline{f}_{R,i} = \frac{1}{C_i - C_{i-1}} \int_{C_{i-1}}^{C_i} f_R(t)\,dt \quad i=1,\ldots,M \tag{4}$$

It is apparent that $\overline{f}_{R,i}$ is a function of C_{i-1}, C_i. Note that (3) is a definition, whereas (4) is a constraint to be inserted within the statement of the optimization problem. Like in (3), similar considerations can be done for $f_G(t)$.

In the solution of the optimization problem that we are going to formalize, only the average values $\overline{f}_{S,i}$, $i=1,\ldots,M$ will be determined (and not, of course, the function $f_S(t)$). These average values must be constrained to avoid physically infeasible power flows from/to the storage. Such a constraint should be imposed on the values of $f_S(t)$, for any value of t. But this is not possible, due to the discrete-event formalization of the model, and thus the bounds over $\overline{f}_{S,i}$ are to be established in order to allow the fulfillment of the physical constraint on $f_S(t)$ for any time instant.

In addition, note that we have imposed that functions $f_G(t)$, $f_{L,j}(t)$, as well as $f_{L,TOT}(t)$, are, in the solution of the optimization problem, piece-wise constant, and are represented by $f_{G,i}(t)$, $f_{L,j,i}(t)$ and $f_{L,TOT,i}(t)$. Obviously, this is a slightly restrictive condition, but it is necessary if we want to deal with parametric (and not functional) optimization problems.

Parameters

Among the parameters of this problem, BP(t) and SP(t) are the time-varying buying and selling prices from and to the main grid, respectively. These functions BP(t) and SP(t) are assumed as given.

Each vehicle V_i requiring charging service is characterized by the following parameters:

- Due date (dd_i): time instant at which the service should be completed;
- Deadline (dl_i): the time instant at which the service must be completed;
- Energy request (ER_i): amount of energy required for the charging service;
- Penalty tardiness coefficient (α_i): cost paid for a unit delay (with respect to the due date dd_i) regarding the completion of the service, for a unit of energy requested; α_i is expressed in $\left[\frac{€}{kWh \cdot h}\right]$;
- Occupancy cost (β): (fixed) cost of a socket per unit time, and is considered as independent from the value of the power flow for the specific considered time interval;
- η: minimum difference between two subsequent events.

The required energy must, in any case, be satisfied and no service can be denied.

All vehicles are assumed to be immediately available for service, thus there is no necessity of specifying any release time.

The considered optimization problem that refers to M vehicles that are already present within the system. In this chapter, the formalization of the problem will be provided in a form that includes the case $M \geq N$ as well as the case $M \geq N$.

Objective Function

The objective function that characterizes this optimal scheduling problem is presented in the following

$$\min \sum_{i=1}^{M} \left\{ FGCOST_i + \alpha_i \cdot tard_i \cdot ER_i + \sum_{j=1}^{M} \beta \cdot y_{i,j} \cdot (C_i - C_{i-1}) \right\} \tag{5}$$

where

$$FGCOST_i = \left\{ \left[\int_{C_{i-1}}^{C_i} BP(t)dt \right] f_{G,i}^{+} - \left[\int_{C_{i-1}}^{C_i} SP(t)dt \right] f_{G,i}^{-} \right\} i=1,\ldots,M \tag{6}$$

where:

- $f_{G,i}^{+} = \max\{f_{G,i}, 0\}$;
- $f_{G,i}^{-} = \max\{-f_{G,i}, 0\}$; $FGCOST_i$ takes into account both economic costs/revenues due to buying/selling power from/to the grid;
- $\alpha_i tard_i ER_i$ is the overall tardiness cost (where tardiness is "weighted" by the amount of the recharging request);
- $\beta \cdot y_{i,j} \cdot (C_i - C_{i-1})$ is a penalizing the occupation of a socket;
- BP(t) and SP(t) are perfectly known for all the time horizon of interest;
- $FGCOST_i$ results to be a function of decision variable C_i and C_{i-1}.

Constraints

The first class of constraints is related to the power balance in the microgrid. The corresponding instantaneous constraint is

$$f_{L,TOT}(t) = f_S(t) + f_G(t) + f_R(t)$$

where $f_{L,TOT}(t)$ is the function representing the total power flow delivered to vehicles at time t.

However, in the formalization as a discrete event optimization problem, instantaneous constraints cannot be imposed. To transform the latter constraint into one referring to time intervals, we must write,

$$f_{L,TOT,i} = \overline{f}_{S,i} + f_{G,i} + \overline{f}_{R,i} \; i=1,\ldots,M \tag{7}$$

Moreover, the following constraints must be introduced

$$y_{i,j} \bullet K - f_{L,i,j} \geq 0;\; i=1,\ldots,M,\; j=1,\ldots,M \tag{8}$$

to impose that when the power flow $f_{L,i,j}$ is greater than zero, the binary variable $y_{i,j}$ takes value 1. Note that K is the so-called "big M" constant whose value is arbitrarily chosen provided that it is considerably higher with respect to the possible values of the variables and parameters appearing in the statement of the problem.

The requirement that the entire energy request ER_i is satisfied for any vehicle V_j requires the introduction of the constraints

$$\sum_{i=1}^{j} f_{L,i,j}(C_i - C_{i-1}) = ER_i \quad j=1,\ldots,M \tag{9}$$

The dynamics of the storage element is represented by the discrete-event equation

$$x(C_i) = \mathrm{x}(C_{i-1}) - \overline{f}_{S,i}(C_i - C_{i-1}) \quad j=1,\ldots,M$$

that can be simply rewritten as

$$x_i = x_{i-1} - \overline{f}_{S,i}(C_i - C_{i-1}) \quad j=1,\ldots,M \tag{10}$$

where the initial state $x(C_0)=x(0)=x_0$ is known.

In the following all the remaining constraints are listed

$$f_{G,i} = f_{G,i}^{+} - f_{G,i}^{-} \quad i=1,\ldots,M \tag{11}$$

$$-F_{G,MAX} \le f_{G,i} \le F_{G,MAX}; \quad i=1,\ldots,M \tag{12}$$

$$0 \le f_{L,i,j} \le F_{L,j,MAX}; \quad i=1,\ldots,M,\ j=i,\ i+1,\ldots,M \tag{13}$$

$$0 \le f_{L,TOT,i} \le F_{L,TOT,MAX}; \quad i=1,\ldots,M \tag{14}$$

$$-F_{S,MAX} \le \overline{f}_{S,i} \le F_{S,MAX} \quad i=1,\ldots,M \tag{15}$$

$$x_{MIN} \le x_i \le x_{MAX}; \quad i=1,\ldots,M \tag{16}$$

$$C_i \ge C_{i-1} + \eta \quad i=1,\ldots,M \tag{17}$$

$$\sum_{j=1}^{M} y_{i,j} \le N \quad i=1,\ldots,M \tag{18}$$

$$tard_i \ge C_i - dd_i; \quad i=1,\ldots,M \tag{19}$$

$C_i \leq dl_i; \quad i=1,\ldots,M$ (20)

where all the variables, apart from $f_{G,i}$, $\overline{f}_{S,i}$, $FGCOST_i$, $i=1,\ldots,M$, are constrained to be nonnegative.

Equations (12)-(16) are introduced to consider physical constraints on the power flows. Constraints (17) are introduced to prevent the presence of zero-length time intervals. Constraints (18) are introduced to impose that the number of vehicles simultaneously under charge does not exceed the value N that is characteristic of the charging station. Constraints (19) are necessary to give significance to the term $tard_i$ appearing in (5). Indeed, constraints (19) are equivalent to this definition, since $tard_i$ in constrained to be nonnegative, and the structure of the cost to be minimized in (5) prevents that constraints (19) are satisfied by a strong inequality.

It is worth observing that in the formulation of the cost function to be minimized, no benefit is considered to form the sale of energy to customers (vehicles). This can be justified within two possible modeling frameworks. In the first one, it is assumed that charging service cannot be refused by the provider and that the selling prices (to customers) are fixed. In the second one, the vehicles are considered as a property of the company providing the charging service.

APPLICATION OF THE PROPOSED MODEL TO A CASE STUDY

In this section, a case study is used to apply the proposed approach. The number of EVs considered is 8 ($M=8$) and they must be charged in a grid-connected microgrid characterized by renewable power production plants, electrical storage, and an EV charging station (composed by 3 sockets, i.e. $N=3$).

By solving though the software Lingo the mixed-integer nonlinear optimization problem (MINLP), the optimal schedule of storage systems and EV charging is obtained. In Table 1, the values of the parameters of the elements of the microgrid for this case study are provided.

Table 1. System Parameters

Parameter	Value
x_{MAX}	125 [kWh]
x_{Min}	40 [kWh]
F_g^{MAX}	100 [kW]
$F_{S,MAX}$	36 [kW]
$f_{L,MAX}$	30 [kW]
$f_{L,TOT,MAX}$	75 [kW]
β	75 [kWh]

The renewable power production forecast $f_R(t)$ is available with a time discretization step equal to 15 minutes, over a whole day. To express $\overline{f}_{R,i}$ as a function of C_{i-1} and C_i, an interpolation of the available forecasts has been determined through aseventh order polynomial function, i.e.

$$f_R(x) = a_7x^7 + a_6x^6 + a_5x^5 + a_4x^4 + a_3x^3 + a_2x^2 + a_1x + a_0$$

The following values for the parameters of the above function are found (see Table 2) through parameters identification.

Table 2. Polynomial Approximation Parameters of Power Production from RES

Parameter	Value
a_7	2.978e-05
a_6	-1.572e-03
a_5	3.071e-02
a_4	-2.606e-01
a_3	7.742e-01
a_2	1.862e-01
a_1	1.358e+00

Table 3 reports the data for each EV for the basic Scenario (see below). The values represent time instants expressed in hours, starting from an initial time instant put equal to 0 (this means that they are absolute values since the optimization interval is from 9 AM to 12 PM).

Table 3. Polynomial Approximation Parameters of Power Production from RES

Vehicle	1	2	3	4	5	6	7	8
dd_i	1	1.5	1.8	2	2.4	2.8	3.3	3.5
dl_i	2	2.3	3	3.8	4.3	4.7	5.2	5.8
$ER_i[kWh]$	26	40	30	15	38	40	35	30
$\alpha_i[\epsilon / kWh \cdot h]$	0.1	0.1	0.1	0.1	0.1	0.1	0.1	0.1

Regarding buying prices, their variability is considered through a function resulting from a fourth-order polynomial interpolation. Then, the first term in equation (6) can be computed as functions of C_{i-1}, C_i by determining the definite integrals of the interpolating function given by the following fourth-order polynomial

$$f_P(x) = b_4x^4 + b_3x^3 + b_2x^2 + b_1x + b_0$$

where the corresponding coefficients are reported in Table 4.

Table 4. Polynomial approximation parameters of the time-varying energy buying price

Parameter	Value
b_4	4.47e-06
b_3	-3.91e-05
b_2	-7.99e-04
b_1	-6.94e-06
b_0	2.40e-01

To test the proposed method, three different scenarios have been considered:

- *Scenario 1*: constant energy prices and equal tardiness penalty coefficients for all vehicles;
- *Scenario 2*: constant energy prices and variable tardiness penalty coefficients;
- *Scenario 3*: time-varying energy prices and variable tardiness penalty coefficients.

Scenario 1

In the first scenario, the eight EVs are charged considering all of them equal in terms of tardiness penalty coefficient and constant energy buying prices. The result of the simulation is presented in Table 5, Figure 3, Figure 4, and Figure 5. Table 5 shows the time intervals and the tardiness results, while the Figures respectively present the charging schedule, the power coming from the main grid and the battery state of charge.

Table 5. Time interval and tardiness results

Vehicle	1	2	3	4	5	6	7	8
$C_i[h]$	0.87	1.42	1.43	2.00	2.40	2.90	3.37	4.37
$tard_i[h]$	0	0	0	0	0	0.10	0.07	0.87

Figure 3. Scenario 1: Gantt Diagram of the charging schedule

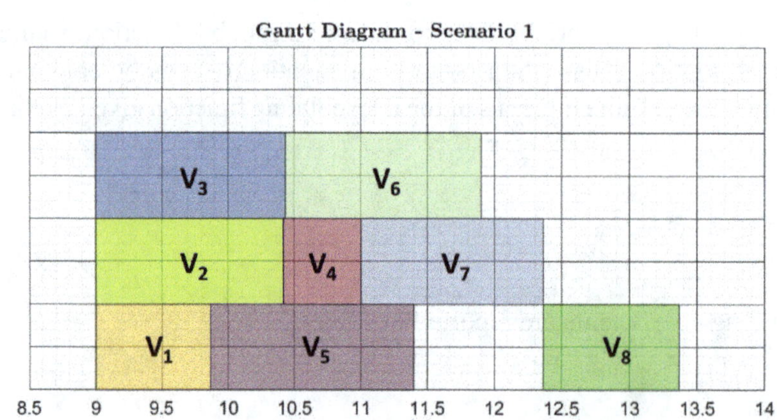

Figure 4. Scenario 1: Power coming from the main grid

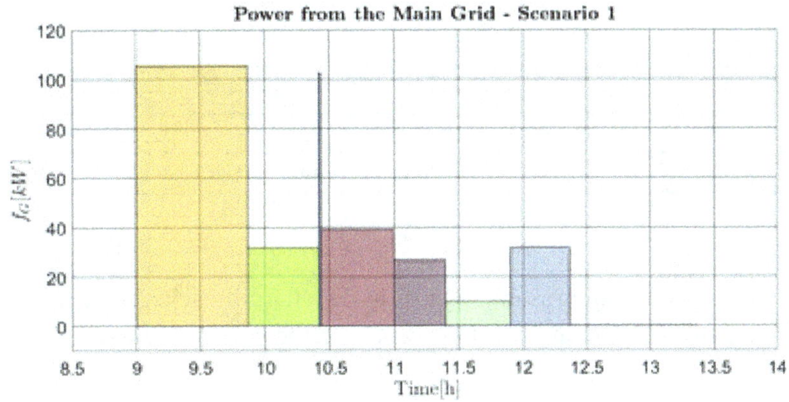

Figure 5. Scenario 1: State of Charge (SOC) of the battery

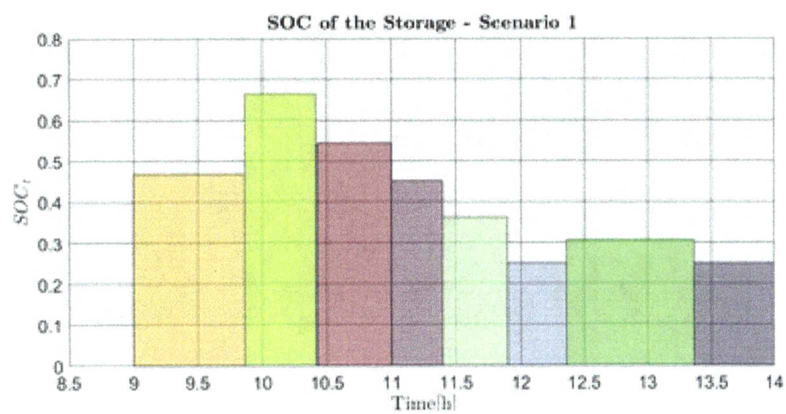

The solution regarding the first Scenario shows that the system allows a delay in the charge of the last EV because the power coming from the renewables is enough to compensate the tardiness cost because, by using that power, the contributions coming from the grid, is zero.

Scenario 2

In the second scenario, the eight EVs are charged considering different values of the tardiness penalty coefficient and constant energy prices. The values of αi adopted in this calculation are reported in Table 6.

Table 6. Scenario 2: Tardiness penalty coefficients

Vehicle	1	2	3	4	5	6	7	8
$\alpha_i[\epsilon / kWh \cdot h]$	0.1	0.2	0.2	0.1	0.1	0.2	0.1	0.2

The results regarding the time intervals and the tardiness are presented in Table 7.

Table 7. Scenario 2: Time interval and tardiness results

Vehicle	1	2	3	4	5	6	7	8
$C_i[h]$	0.87	1.38	1.39	2.20	2.40	2.80	3.47	3.60

The following Figures present the results regarding the charging schedule (Figure 6) and the corresponding power coming from the main grid (Figure 7) and the battery state of charge (Figure 8).

Figure 6. Scenario 2: Gantt Diagram of the charging schedule

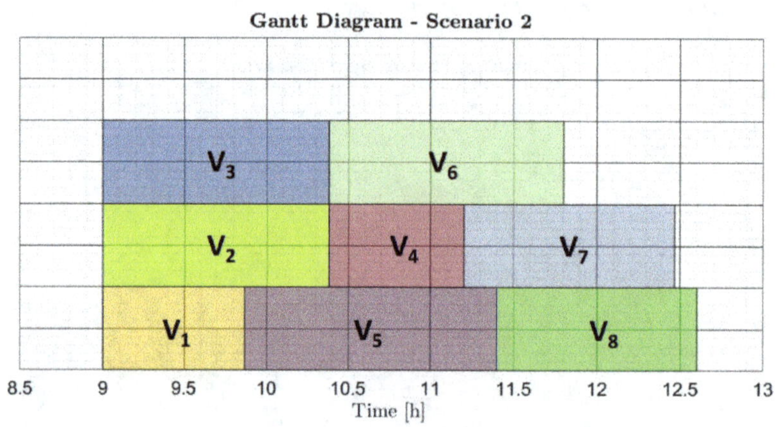

Figure 7. Scenario 2: power coming from the main grid

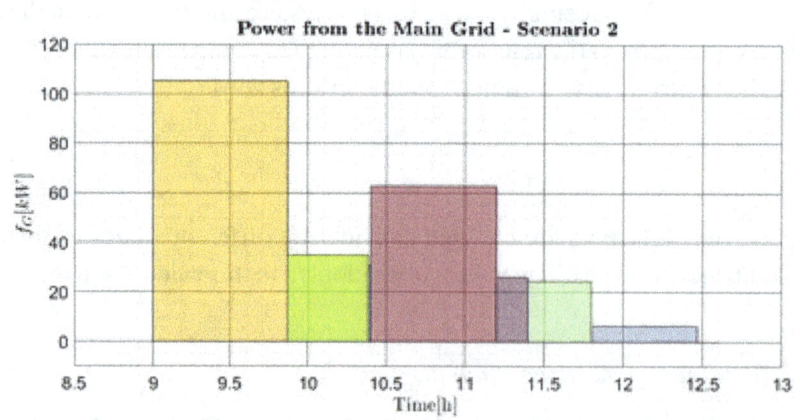

Figure 8. Scenario 2: State of Charge (SOC) of the battery

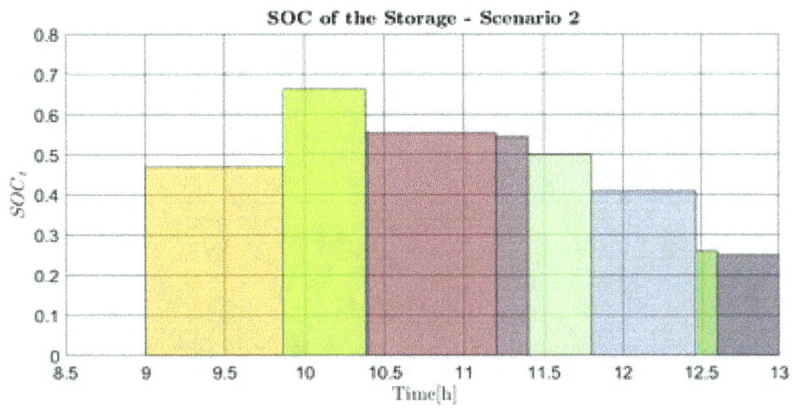

The results highlight how the tardiness penalty coefficient reduces the charging time and makes the charge scheduling solution more compact. The storage is charged in the first interval because of the purchase of a large amount of power from the main grid.

Scenario 3

In the third scenario, the eight EVs are charged considering different values of the tardiness penalty coefficient (the same of Scenario 2), and time-varying buying prices. The results are presented in Table 8.

Table 8. Scenario 3: time interval and tardiness results

Vehicle	1	2	3	4	5	6	7	8
$C_i[h]$	0.97	1.37	1.38	2.20	2.40	2.80	3.47	3.60
$tard_i[h]$	0	0	0	0.20	0	0	0.17	0.10

As in the previous cases, the three figures representing the charging schedule (Figure 9), the power exchange with the main grid (Figure 10) and the SOC of the battery (Figure 11), are presented.

From this Scenario, it is possible to see how the variability of the buying price affects the optimization result. Since the price is high in the first hours and then it decreases, in the first time interval the storage is discharged and the power bought from the main grid is minimum. Then, the battery SOC remains at its minimum value and the power exchange with the grid assumes a value more or less constant while the three sockets are busy and decreased when one of the sockets is emptied. Even if the buying price is decreasing, the tardiness of the last EV is constrained by its penalty coefficient which is higher than the previous interval.

Figure 9. Scenario 3: Gantt Diagram of the charging schedule

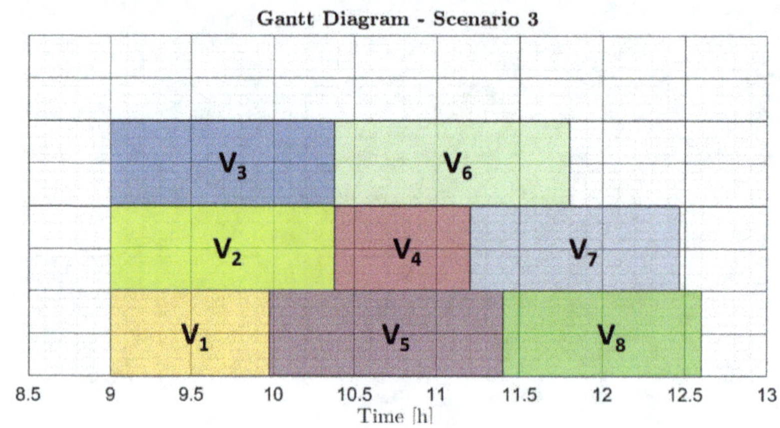

Figure 10. Scenario 3: power coming from the main grid

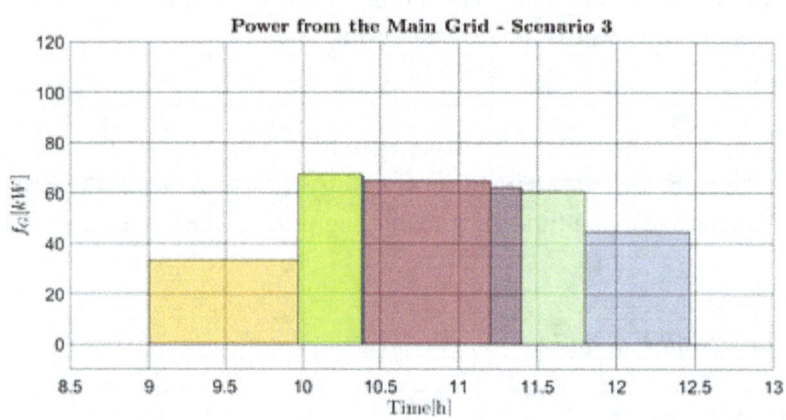

Figure 11. Scenario 3: State of Charge (SOC) of the battery

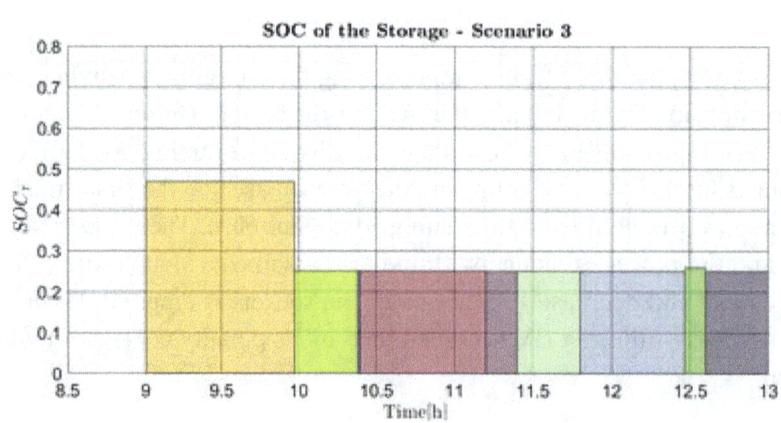

The three different scenarios highlight the main differences introduced when the variability of the energy buying prices and/or the urgency of an EV charge is considered. This leads to a different scheduling since the main effect given by the price variability is that, as expected, the power drawn from the grid diminishes. The main variation brought by the charge urgency is that some vehicles must be completed as fast as possible. This, with the assumptions regarding the sequencing of the charging services, leads to a more compact charge scheduling.

FUTURE RESEARCH DIRECTIONS

As a future development of this work it will be possible to model the non-linearities of the batteries for both vehicles and storage, and to investigate new approaches combining both the discrete time and discrete event models in order to reach more precise values of the power flows during the time while maintaining a restricted number of variables.

CONCLUSION

In this chapter, an optimization model based on a discrete event approach has been presented for the optimal EVs charging in microgrids. The proposed approach allows limiting the number of variables to be determined avoiding a time discretization of the optimization horizon. Moreover, it is possible to consider charging stations with multiple sockets and thus simultaneous charging of different EVs.

Future developments regard detailed modeling of the different phases in the charging station, a more detailed model for storage systems, and the inclusion of such a model in a polygeneration microgrid. Finally, a link with the transportation network could be established by jointly deciding on EVs scheduling, charging and path planning.

REFERENCES

Aliasghari, P., Mohammadi-Ivatloo, B., Alipour, M., Abapour, M., & Zare, K. (2018a). Optimal scheduling of plug-in electric vehicles and renewable micro-grid in energy and reserve markets considering demand response program. *Journal of Cleaner Production*, *186*, 293–303. doi:10.1016/j.jclepro.2018.03.058

Aliasghari, P., Mohammadi-Ivatloo, B., Alipour, M., Abapour, M., & Zare, K. (2018b). Optimal scheduling of plug-in electric vehicles and renewable micro-grid in energy and reserve markets considering demand response program. *Journal of Cleaner Production*, *186*, 293–303. doi:10.1016/j.jclepro.2018.03.058

Amini, M. H., Moghaddam, M. P., & Karabasoglu, O. (2017). Simultaneous allocation of electric vehicles' parking lots and distributed renewable resources in smart power distribution networks. *Sustainable Cities and Society*, *28*, 332–342. Advance online publication. doi:10.1016/j.scs.2016.10.006

Asensio, M., De Quevedo, P. M., Muñoz-Delgado, G., & Contreras, J. (2018a). Joint distribution network and renewable energy expansion planning considering demand response and energy storage-part I: Stochastic programming model. *IEEE Transactions on Smart Grid*, *9*(2), 655–666. doi:10.1109/TSG.2016.2560339

Asensio, M., De Quevedo, P. M., Muñoz-Delgado, G., & Contreras, J. (2018b). Joint distribution network and renewable energy expansion planning considering demand response and energy storage-part II: Numerical results. *IEEE Transactions on Smart Grid*, *9*(2), 667–675. doi:10.1109/TSG.2016.2560341

Atallah, R. F., Assi, C. M., Fawaz, W., Tushar, M. H. K., & Khabbaz, M. J. (2018). Optimal Supercharge Scheduling of Electric Vehicles: Centralized Versus Decentralized Methods. *IEEE Transactions on Vehicular Technology*, *67*(9), 7896–7909. doi:10.1109/TVT.2018.2842128

Aujla, G. S., Kumar, N., Singh, M., & Zomaya, A. Y. (2019). Energy trading with dynamic pricing for electric vehicles in a smart city environment. *Journal of Parallel and Distributed Computing*, *127*, 169–183. doi:10.1016/j.jpdc.2018.06.010

Azar, A. G., & Jacobsen, R. H. (2016). Agent-based charging scheduling of electric vehicles. *2016 IEEE Online Conference on Green Communications (OnlineGreenComm)*, 64–69. 10.1109/OnlineGreenCom.2016.7805408

Ferrero, E., Alessandrini, S., & Balanzino, A. (2016). Impact of the electric vehicles on the air pollution from a highway. *Applied Energy*, *169*(x), 450–459. doi:10.1016/j.apenergy.2016.01.098

Ferro, G., Laureri, F., Minciardi, R., & Robba, M. (2018). An optimization model for electrical vehicles scheduling in a smart grid. *Sustainable Energy. Grids and Networks*, *14*, 62–70. doi:10.1016/j.segan.2018.04.002

Ferro, G., Laureri, F., Minciardi, R., & Robba, M. (2019a). A predictive discrete event approach for the optimal charging of electric vehicles in microgrids. *Control Engineering Practice, 86*(September), 11–23. doi:10.1016/j.conengprac.2019.02.004

Ferro, G., Laureri, F., Minciardi, R., & Robba, M. (2019b). Optimal charging of electric vehicles in microgrids through discrete event optimization. *2019 18th European Control Conference (ECC)*, 916–921. 10.23919/ECC.2019.8795999

Fontenot, H., & Dong, B. (2019). Modeling and control of building-integrated microgrids for optimal energy management – A review. *Applied Energy*, *254*(July), 113689. doi:10.1016/j.apenergy.2019.113689

Ito, A., Kawashima, A., Suzuki, T., Inagaki, S., Yamaguchi, T., & Zhou, Z. (2018). Model Predictive Charging Control of In-Vehicle Batteries for Home Energy Management Based on Vehicle State Prediction. *IEEE Transactions on Control Systems Technology*, *26*(1), 51–64. doi:10.1109/TCST.2017.2664727

Latifi, M., Rastegarnia, A., Khalili, A., & Sanei, S. (2019). Agent-Based Decentralized Optimal Charging Strategy for Plug-in Electric Vehicles. *IEEE Transactions on Industrial Electronics*, *66*(5), 3668–3680. doi:10.1109/TIE.2018.2853609

Li, D., Chiu, W.-Y. Y., Sun, H., & Poor, H. V. (2018). Multiobjective optimization for demand side management program in smart grid. *IEEE Transactions on Industrial Informatics, 14*(4), 1482–1490. doi:10.1109/TII.2017.2776104

Liu, Z., Wu, Q., Huang, S., Wang, L., Shahidehpour, M., & Xue, Y. (2018). Optimal Day-Ahead Charging Scheduling of Electric Vehicles Through an Aggregative Game Model. *IEEE Transactions on Smart Grid, 9*(5), 5173–5184. doi:10.1109/TSG.2017.2682340

Liu, Z., Wu, Q., Ma, K., Shahidehpour, M., Xue, Y., & Huang, S. (2019). Two-Stage Optimal Scheduling of Electric Vehicle Charging Based on Transactive Control. *IEEE Transactions on Smart Grid, 10*(3), 2948–2958. doi:10.1109/TSG.2018.2815593

Qian, K., Zhou, C., & Yuan, Y. (2015). Impacts of high penetration level of fully electric vehicles charging loads on the thermal ageing of power transformers. *International Journal of Electrical Power & Energy Systems, 65*, 102–112. doi:10.1016/j.ijepes.2014.09.040

Rahman, I., Vasant, P. M., Singh, B. S. M., Abdullah-Al-Wadud, M., & Adnan, N. (2016). Review of recent trends in optimization techniques for plug-in hybrid, and electric vehicle charging infrastructures. *Renewable & Sustainable Energy Reviews, 58*, 1039–1047. doi:10.1016/j.rser.2015.12.353

Sachan, S., & Adnan, N. (2018). Stochastic charging of electric vehicles in smart power distribution grids. *Sustainable Cities and Society, 40*, 91–100. Advance online publication. doi:10.1016/j.scs.2018.03.031

Sarikprueck, P., Lee, W.-J., Kulvanitchaiyanunt, A., Chen, V. C. P., & Rosenberger, J. M. (2018). Bounds for Optimal Control of a Regional Plug-in Electric Vehicle Charging Station System. *IEEE Transactions on Industry Applications, 54*(2), 977–986. doi:10.1109/TIA.2017.2766230

Sbordone, D., Bertini, I., Di Pietra, B., Falvo, M. C., Genovese, A., & Martirano, L. (2015). EV fast charging stations and energy storage technologies: A real implementation in the smart micro grid paradigm. *Electric Power Systems Research, 120*, 96–108. doi:10.1016/j.epsr.2014.07.033

Shareef, H., Islam, M. M., & Mohamed, A. (2016). A review of the stage-of-the-art charging technologies, placement methodologies, and impacts of electric vehicles. *Renewable & Sustainable Energy Reviews, 64*, 403–420. doi:10.1016/j.rser.2016.06.033

Srinivasan, S., Palmie, G., Srinivasan, S., Palmieri, G., Del Vecchio, C., & Glielmo, L. (2016). Demand Side Management for heating controls in Microgrids. *IFAC-PapersOnLine, 49*(1), 611–616. doi:10.1016/j.ifacol.2016.03.123

Tan, K. M., Ramachandaramurthy, V. K., & Yong, J. Y. (2016). Integration of electric vehicles in smart grid: A review on vehicle to grid technologies and optimization techniques. Renewable and Sustainable Energy Reviews. doi:10.1016/j.rser.2015.09.012

Wang, R., Xiao, G., & Wang, P. (2017). Hybrid Centralized-Decentralized (HCD) Charging Control of Electric Vehicles. *IEEE Transactions on Vehicular Technology, 66*(8), 6728–6741. doi:10.1109/TVT.2017.2668443

Wang, S., Gangammanavar, H., Eksioglu, S. D., & Mason, S. J. (2019). Stochastic Optimization for Energy Management in Power Systems with Multiple Microgrids. *IEEE Transactions on Smart Grid, 10*(1), 1068–1079. doi:10.1109/TSG.2017.2759159

Yang, Z., Li, K., & Foley, A. (2015). Computational scheduling methods for integrating plug-in electric vehicles with power systems: A review. *Renewable & Sustainable Energy Reviews, 51*, 396–416. doi:10.1016/j.rser.2015.06.007

Zhang, K., Mao, Y., Leng, S., He, Y., Maharjan, S., Gjessing, S., Zhang, Y., & Tsang, D. H. K. (2018). Optimal Charging Schemes for Electric Vehicles in Smart Grid: A Contract Theoretic Approach. *IEEE Transactions on Intelligent Transportation Systems, 19*(9), 3046–3058. doi:10.1109/TITS.2018.2841965

Zheng, Y., Niu, S., Shang, Y., Shao, Z., & Jian, L. (2019). Integrating plug-in electric vehicles into power grids: A comprehensive review on power interaction mode, scheduling methodology and mathematical foundation. *Renewable & Sustainable Energy Reviews, 112*, 424–439. doi:10.1016/j.rser.2019.05.059

ADDITIONAL READING

Cao, Y., Huang, L., Li, Y., Jermsittiparsert, K., Ahmadi-Nezamabad, H., & Nojavan, S. (2020). Optimal scheduling of electric vehicles aggregator under market price uncertainty using robust optimization technique. *International Journal of Electrical Power and Energy Systems, 117*(September), 105628. doi:10.1016/j.ijepes.2019.105628

Cassandras, C. G., & Lafortune, S. (2008). Introduction to discrete event systems. In Introduction to Discrete Event Systems. doi:10.1007/978-0-387-68612-7

Ebrahimi, J., Abedini, M., & Rezaei, M. M. (2020). Optimal scheduling of distributed generations in microgrids for reducing system peak load based on load shifting. *Sustainable Energy. Grids and Networks, 23*, 100368. doi:10.1016/j.segan.2020.100368

Li, Y., Yang, Z., Li, G., Mu, Y., Zhao, D., Chen, C., & Shen, B. (2018). Optimal scheduling of isolated microgrid with an electric vehicle battery swapping station in multi-stakeholder scenarios: A bi-level programming approach via real-time pricing. *Applied Energy, 232*(October), 54–68. doi:10.1016/j.apenergy.2018.09.211

Pinedo, M. L. (2008). Scheduling: Theory, algorithms, and systems. In Scheduling: Theory, Algorithms, and Systems. doi:10.1007/978-0-387-78935-4

Smith, D. K., & Bertsekas, D. P. (1996). Dynamic Programming and Optimal Control. Volume 1. *The Journal of the Operational Research Society, 47*(6), 833. Advance online publication. doi:10.2307/3010291

Wang, Y., Huang, Y., Xu, J., & Barclay, N. (2017). Optimal recharging scheduling for urban electric buses: A case study in Davis. *Transportation Research Part E, Logistics and Transportation Review, 100*, 115–132. doi:10.1016/j.tre.2017.01.001

KEY TERMS AND DEFINITIONS

Charging: Process in which energy is injected into the battery of the electric vehicles.

Discrete Event Model: Model characterized by a discrete sequence of events which determine a variation in the state of the system.

Electric Vehicles: Vehicles characterized by an electric motor and a battery to store the energy.

Microgrid: A small electrical network usually attached to a centralized national grid but able to function independently.

Optimization: Selection of a best element, with regard to some criterion, from some set of available alternatives.

Scheduling: Process of controlling, arranging, and optimizing work and workloads in process.

Tardiness: Measure of a delay in executing certain operation.

Chapter 8
Environmental Impact of EV Batteries and Their Recycling

Zeeshan Ahmad Arfeen
The Islamia University, Bahawalpur, Pakistan

Rabia Hassan
Institute of Business Management, Karachi, Pakistan

Mehreen Kausar Azam
https://orcid.org/0000-0001-9676-5312
Institute of Business Management, Karachi, Pakistan

Mohammad Pauzi Abdullah
Universiti Technologi Malaysia, Malaysia

ABSTRACT

Electrifying transportation is one of the biggest keys to solving the looming climate crisis. The demand for electric vehicles (EV) is booming in the last five years and will increase in the coming years. In this modern age, where EV is the finest means of transportation due to null exhaust gases, there is a dire need to think about ways of recycling and reusing those batteries associated with EVs. In this context, it is estimated that post-vehicle battery packs application will be crossed from 1.4 million to 6.8 million by the year 2035. Numerous researches have been done on the re-purposing and safe disposal of EV batteries. However, presently, Lithium-ion batteries (LiBs) are the optimal choice for electric transportation due to greater energy density, compact size, and extended life cycles. Nonetheless, the trade-off between re-purposing and disposal of LiBs is substantial for the protection of the environment and human health. Regrettably, Lithium-ion battery recycling percentage is only 3% currently whereas its revival is negligible.

DOI: 10.4018/978-1-7998-6858-3.ch008

INTRODUCTION

Electric vehicles have substantial advantages contrasted with internal combustion-based engine vehicles due to their eco-friendly nature. EVs are preferred as they are more economical to maintain and run as well as they add more value for life because of their negligible air contamination, less reliance on oil as well as no poisonous substances emerge which can destroy the ozone layer (Arfeen et al., n.d.). Furthermore, electric engines and batteries add to the vitality of electric-vehicle. Many studies observed that the materials utilized in batteries are perilous for the earth. The mining and preparing of items such as lithium, copper, and nickel require much exuberance and this can lead to the emission of harmful substances. In the chapter, it comprehensively discussed how the second-life batteries, which already completed their useful life could be reconditioned. Furthermore, if they need to be disposed of what are the best ways of doing so to opt.

Battery Re-Purposing and Clearance for Sustainable Society

Living in a period of sustainability, which underpins the crucial endeavouring to improve the natural wellbeing and personal satisfaction for the network. As reusing the paper, cardboard boxes, and utilize containers of spaghetti sauce, however, what is demanded by electric vehicle battery after it has lost its dynamism? So analysts of the world are dealing with the territories where they can create ways on which these EV batteries which are disposed of can be reused. EVs sales are skyrocketing. As given in **Figure 1** underneath, the best five EV automotive organizations for the year 2019.

Figure 1. Sale of top 5 EV manufacturing companies in 2019. (Irle, 2019)
Short ALT Text for a photo, p. 2 ():
The top 5 manufacturers of Electric Vehicles (EVs) sales data for the year 2019 depicts a sharp increase in the sales of EVs

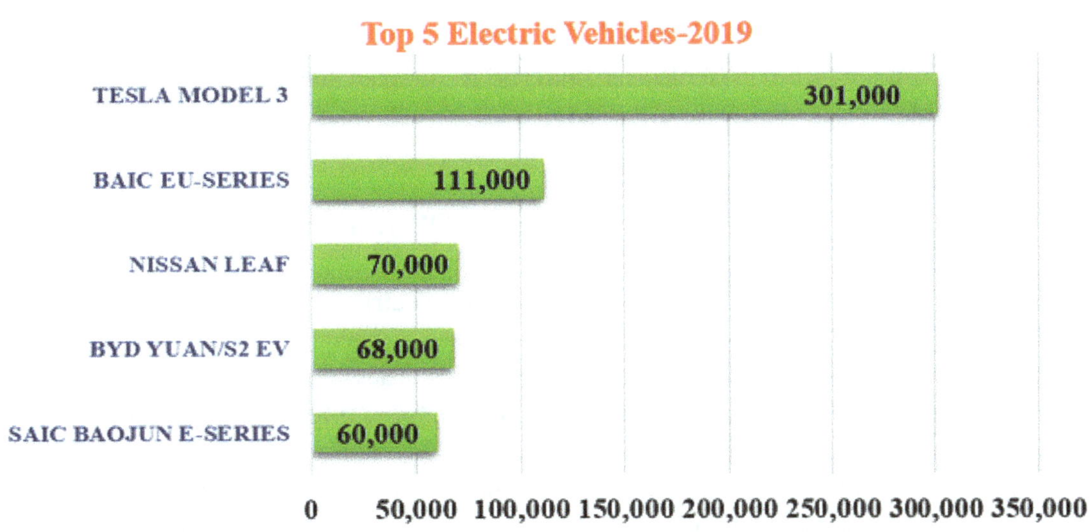

As an electric vehicle sales increase, the number of batteries that will get access for a second life outside of the vehicle increases. EV batteries are a gauge to surpass what might be compared to about 3.4 million packs by 2025, contrasted and around 55,000 of every 2018 as expressed in Bloomberg Businessweek. Does the inquiry rise that where will these batteries land? Various roads have been inquired about for the repurposing of the EV battery and a portion of the major are examined in detail in this chapter. **Figure 2** signified the battery's production volume increases with a fall of net cost yearly.

Figure 2. Sales of EV increases with falling battery cost. (Slowik et al., 2016)
Short ALT Text for a photo, p. 2 [28 words]:
The influx in the purchase of Electric Vehicles is impacting the cost of Li-ion batteries as a higher volume of EVs production results in the lower production cost of batteries

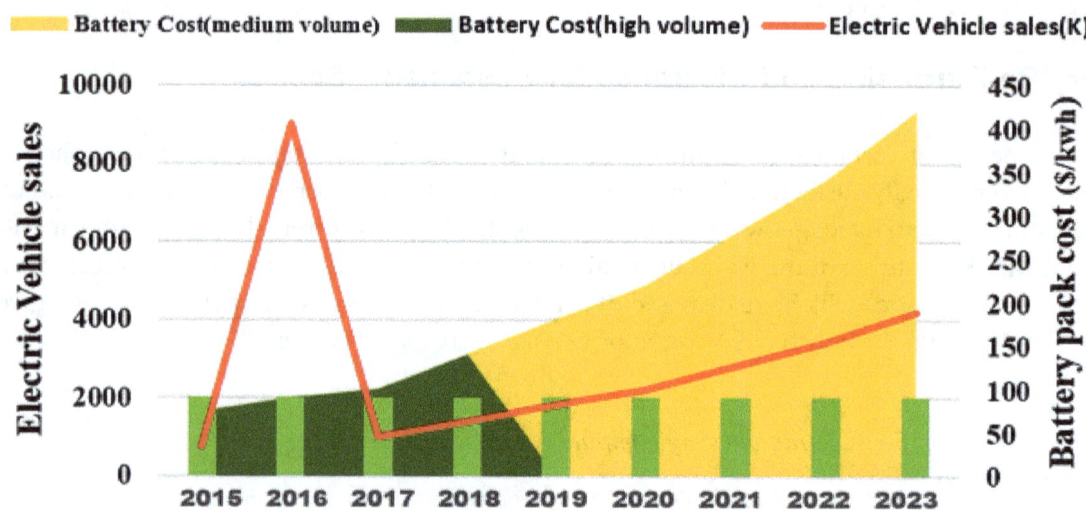

There are several manufacturers of electric vehicles. A compact summary of the types of plug-in EVs is given in **Table 1**.

1 DELAYING RECYCLING THROUGH REPURPOSING

Three kinds of procedures are right now accessible for reusing electric vehicle batteries. In the first place, the batteries can experience a refining procedure, accessible on an enormous scale for different kinds of batteries, including lithium-particle and nickel-metal batteries. These batteries are encouraged into the smelter to recuperate significant metals. Remaining materials, for example, lithium is lost to slag which can be used for the manufacturing, not accessible for a wide range of batteries, including isolating parts through different physical and synthetic procedures, including hydrometallurgical advances, and afterward recouping any battery-grade materials directly. Finally, transitional procedures, which include extricating exclusively perilous battery segments toward the end of the battery life, which is useful in limiting the measure of dangerous substances that advance into the earth. **Table 2** gives a very precise view of the merits and demerits which can be attained by resuing the EV batteries (Better World Solutions, 2019).

Table 1. Types of plug-in Electric Vehicles. (Bhatti & Salam, 2018; Lü et al., 2020; Slowik et al., 2016)

Types of EVs	EV Models	Prominent Features
Plugin EV (Gasoline + battery)	Toyota Prius, Ford Fusion, Audi A3 E-Tron; BMW 530Le	It is powered by dual-fuel similar to Hybrid, EV with another option of plug-in charging with any electrical outlet. The energy stored in battery packs supplied power to an electric motor or by an internal combustion engine that runs on conventional fuel or regenerative braking. It employs both in the charge sustaining mode as well as a charge depleting mode of operation.
Hybrid, EV (Gasoline + battery)	Honda Civic Hybrid, Ford Fusion Hybrid, Toyota Prius Hybrid, LEXUX NX	HEVs cannot be charged from outside outlets like the power grid. The battery bank charged by IC engine default built-in and by regenerative braking, i.e. by transforming its kinematic energy into chemical energy that deposited in the banks of the battery.
Extended Range EV/PEV	Chevrolet, Cadillac	This is an example of a merger between PHEV and BEV, which has a more efficient system for fuel and fewer emissions of toxic nature
Battery, EV (Battery)	Nissan Leaf, BMW i-3, Mitsubishi I MiEV, Tesla Roadster Model S, Ford Focus Electric, Benz EQC	Zero-emission is achieved in this type by the merger of batteries which can be recharged by an electrical outlet. It works in the charge depleting states of operation require enough power battery packs and high power. As the fixed cost is high which does not break even quickly because of short mileage, the charging duration also long makes this option less feasible.
Fuel Cell EV, (FC +SC/battery)	Nissan Motor	It accumulates hydrogen as a fuel that is produced from natural gas. The energy in this case hydrogen, provisions of a fuel cell within combustions with the oxygen from around to produce a flow of electrons which is due to electrolysis in the reverse direction, thus resulting in heat and water as byproducts of this chemical reaction

Short ALT Text for the table, p. 3 [12 words]:
Three major types of EVs include plugin EV, Hybrid EV and extended EVs

Table 2. Pros and Cons of reconditioned EV batteries.

Sr.No	Pros	Cons
1	Making their repurposing for a second and even third life much progressively significant to expand their monetary and ecological incentive before reusing.	The reusing lithium-particle batteries involve expense and potential waste.
2	The batteries can be used for stationary storage, which is monetary favorable and thus saving on the initial amount of the product. This battery property is another factor that will be making EVs popular.	New batteries are used for stationary storage at the moment which has a high density and has a bigger life than the batteries which have served their first life. At the moment companies are not issuing warranty to batteries that are used.
3	The stability of the grid and the ability of the grid to integrate renewable energy is aided by the secondary storage.	The size and shape, as well as the performance of the EV batteries, differ. The batteries perform under different climatic conditions and stressors and for different car models.
4	As per the costing, this is the most economical system where the battery packs are of the uniform constitution.	When the battery packs are of the non-uniform constitution the output has to be regulated by using software that is non-economical. Research is being conducted to attain performance by incorporating old batteries with new ones, but it will take time and resources to reach the desired results.

Short ALT Text for the table, p. 4 [14 words]:
The repurposing of Electric Vehicles (EVs) comes with several advantages and disadvantages

Figure 3 Sheds light on the EV batteries are currently used and how this can be further utilized in second life.

Figure 3. Possible vehicle battery life. (Catton et al., 2019)
Short ALT Text for a photo, p. 5 ():
Lithium-ion batteries primary applications include usage in Electric Vehicles whereas they can be repurposed in several ways

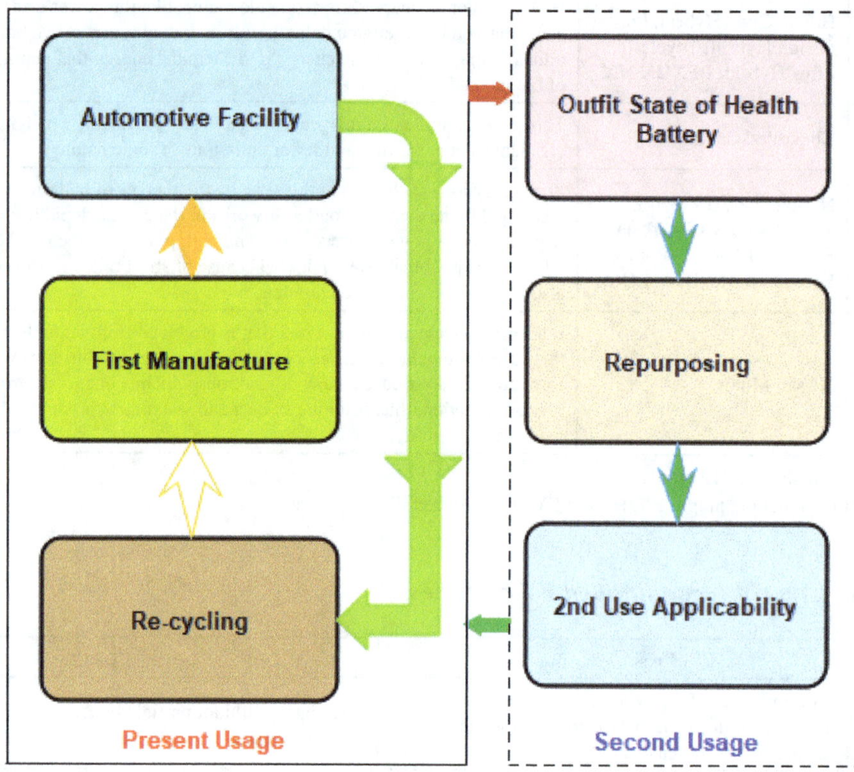

1.1 Re-Purposing

Synthetic batteries in electric vehicles give a quick and enormous force supply. Most flow module electric vehicles – characterized as vehicles that attach into the lattice for a few or the entirety of their capacity – use lithium-particle batteries [7]. These batteries, though in various configurations, normally utilized inconvenient customer hardware, for example, phones and PCs. In contrast their higher vitality in per unit mass comparative with other electrical vitality stockpiling frameworks. Nevertheless, the specific science of vehicle batteries regularly contrasts from buyer hardware, batteries, just as from one another relying upon the automaker. In general, batteries in vehicles have a higher all-out force limit and size.

Automakers lean toward lithium-particle batteries since they convey predominant execution in both force and vitality thickness, permitting them to accomplish a high weight-to-execution proportion. Moreover, most segments of lithium-particle batteries can be recycled. Most of the industries are using LiBs because of their economic and ecological perspectives, re-using the LiB batteries which are no longer to be used in their prime application field is preferred (Falk et al., n.d.). Participants at the assembling noticed that expanded assembling and economies of scale have discounted costs and broadened their helpful life cycle, with an anticipated value diminishes of around seven-to-eight percent for each year. **Table 3** depicts the mechanical design of the battery for different brands of EVs.

Table 3. Common zero-emission vehicle battery packs. (Canals Casals & Amante García, 2016)

Battery Types	Cell Design	Company	Cells/Modules	Energy in kWh
Tesla Model S	Cylindrical	Panasonic	7,104 /16	85-large pack
Nissan Leaf	Pouch	NEC	192/48	24
Chevy Volt	Prismatic	LG Chem	288/9	16.5

Short ALT Text for the table, p. 6 ():
Design of cells used by the top three manufacturers of Electric Vehicles (EVs) along with the energy produced by them

Rather than reusing them promptly, the huge number of batteries that will be leaving electric vehicles in the coming years could be repurposed, prompting a surge of reasonable batteries that can give vitality stockpiling administrations to clients, utilities, and matrix administrators. Since second-life batteries will hold critical limits, they might be appropriate for different clients and lattice applications, especially whenever amassed for mass vitality stockpiling (K Environmental, 2019; Pagliaro & Meneguzzo, 2019).

2 ECONOMIC ASPECTS

Electricity is usually generated by nuclear or hydroelectric means in developing countries whereas developing and underdeveloped countries are still stuck with hydroelectric resources. A country's demand for electricity is never-ending, increases every day as population and industry are on the rise, and most of the appliances use electrical energy to operate. As we enter the digital age the electricity demand has increased many folds. A nuclear supply can only be operated on a level load, as it is not feasible to turn down or increase the level easily. To cater to these demands, non-renewable energy sources need to be used. During the peak energy situation, the produced energy could be stored somewhere else. Thus, at this point, reusing the EV batteries is not a choice but a necessity.

Reused batteries for EVs, HEVs, and PHEVs present a superb, financially perceptive alternative for vitality stockpiling applications. The organizations involved reusing lithium-particle batteries to form a circular economy are currently working with battery creators to embrace effectively disassembled item structures, and are preparing them to take-up the new green science forms for the green recuperation of all esteemed battery segments.

From a business stance, the risk is marginalized for investments by putting the resources into an enormous number of little applications than few huge applications following the age-old saying of "Don't put all your eggs in one basket". Smaller applications can incorporate various private customers and a wide assortment of business clients, including telecommunications organizations requiring 5 to 10 EV battery packs each, light business structures requiring 10 to 15 packs each and nourishment dissemination focuses requiring 30 to 40 packs each. The batteries can be used in several applications. A few of them are discussed below.

2.1 Identifying Domestic Demand

Energy storage systems (ESS) are capturing the market by storm. It works by reusing the stationary Li-Ion Electric vehicle battery as energy storage devices. It helps to counter the effects of supply and demand gaps. It can be used to power households, industries by improving the economy and decreases greenhouse gas emissions.

Now ESS is an established system that is controlled centrally. The demand side used it for deployment at the micro-grid side. The system is currently catering to the efficient usage of renewable energy resources, other resources such as power demand shortfalls and standby operator as the weather conditions change, the usage of peak demand rate factors and climate changes. However, the efficiencies are not up to the standard and the high cost of installations makes the system lagging and gives room for further research.

2.2 Identifying Industrial demand

The impact of solutions provided by ESS in the areas of the smart grid is quite visible. The utility grid problems can be stabilized by using ESSs. The ESS optimizes the quality of the grid as it maintains the power as constant. Renewable Energy Sources are used to balance the load curve and the frequency is controlled. The capability of the transmitting lines can be enhanced as well as the voltage fluctuations are mitigated by using them (Vazquez, Lukic, Galvan, Franquelo, & Carrasco, 2010).

The industry needs systems that provide capacity with a high rating and the electric wires that are used should be of high grade as well to handle the rise in voltage and current. This is provided by centralizing power generation systems.

3 STANDARDS FOR REUSING EVS BATTERIES

The organization which is currently working for the EV is known as Electric Vehicles and the Environment Informal Working Group (EVE IWG) which was formed in March 2012. This organization is currently working on environmental issues related to EVs (Electric Vehicle Regulatory Reference Guide, 2019).

New guideline in China presently considers EV creators answerable for the recuperation of batteries, expecting them to set up reusing channels and administration outlets where used batteries can be gathered, put away and moved to reuse organizations. Before the finish of Feb 2019, 393 carmakers, 44 rejected vehicle destroying ventures, 37-course use endeavours, and 42 reusing undertakings had just joined the new discernibility stage to follow starting point and proprietors of disposed of batteries (Battery Recycling as a Business, 2019).

Worldwide battery reusing prerequisites are either missing totally or where they exist; they vary a lot, as well as the profundity of inclusion. This presents another interesting area of research on a worldwide premise because of the unpredictable idea of practices, and mentalities towards reusing EV batteries around the world. What is prescribed is that assets be assigned to assess the benefit of creating producing for-recyclability necessities (Electric Vehicle Regulatory Reference Guide, 2019).

4 ENVIRONMENTAL IMPACTS OF ELECTRIC VEHICLE BATTERIES | EVBS

The design of LiBs depends on the selection of active elements to be used for cathode whereas these materials having a substantial impact on the health of homo-sapiens the surrounding environment. These impacts are due to the design procedure of materials, the use of energy, and the way they are extracted from the surface of the earth. The elements which are needed for the manufacturing of LiBs contribute more towards metal degradation and greenhouse gasses (GHGs). An Environmental Protection Agency (EPA) study found that the batteries which used Aluminum have higher rates of contributing to ozone depletion (Amarakoon et al., 2013). Environmental risks arising from lithium-ion batteries must be addressed while execution of recycling and waste strategies. Lithium-ion batteries contain a high proportion of hazardous heavy metals which is harmful to the environment as well as for the human body (Tchounwou et al., 2012). In 2005, the used LiBs were gathered and the result showed that from 4000 tons of LiBs, 1100 tons of heavy metals and around and above 200 tons of toxic materials were produced, toxic materials in lithium-ion batteries include nickel, copper, organic and lead chemicals (Ordoñez et al., 2016). LiBs categorized as a class nine dangerous materials. As LiBs have the hazardous, especially the possibility of rapid discharge, in this case, the batteries generate avalanche currents, which cause overheating, and risk of explosion. Therefore, lithium-ion batteries might be very hazardous when transported, particularly by air flights (Hazard Class 9 - Hazard Class Labels Explained, 2019).

The energy demand for the manufacturing of LiBs has estimated at approximately 90 Mega-Joule/kg and 12.5 kg of carbon dioxide per kg. The main contribution towards the health and environmental effects start with the development of raw materials accompanied by battery manufacturing, supply, uses, services, recycling, and waste management (McManus, 2012).

4.1 Raw Material Manufacturing Effects

Extraction, casting procedure, smelting, leaching, filtering are the main steps used to convert the metal and its elements in the particular material used for the batteries. In the production of LiBs for electric vehicles, the aluminum alone produces 2-3 kg of CO2 equivalent/kg of battery. The manufacturing of electrode like Lithium Manganese Oxide ($LiMn_2O_4$) generates about 800-1000 kg of carbon dioxide equivalent/kg of battery. However, the manufacturing of electrolytes such as dimethyl carbonate and Lithium hexa-fluoro phosphate ($LiPF_6$) generates 100-500 kg of CO_2 equivalent/kg of battery. The manufacturing of such raw materials has high threats to the environment (Meshram et al., 2019).

4.2 Battery Manufacturing Effects

Hazardous air emissions, solid waste, and water pollution can generate damage to health and the environment during the manufacturing of LiBs. The automotive industry has surged the demand for LiB, with the dire negligence of greenhouse gas emissions (GHS). It is reported that three categories of cathode materials are mostly used in the manufacturing of LiBs that are mainly responsible for GHS emission. It is reported in (Hao et al., 2017), the manufacturing of 28 kWh battery, the cathode materials of cobalt oxide, lithium nickel manganese, lithium-ion manganese oxide, and lithium iron phosphate produced GHG emissions approximately 2912 $Kg.CO_2.eq$, 2705 $Kg.CO_2.eq$ and 3061 $Kg.CO_2.eq$ respectively.

4.3 Thermal Gas Emission

The major thermal run-away effects are gas and heat emissions, which are highly flammable. The structure of the batteries and cells typically involve protections to release the gas without generating a risk of cells or batteries exploding. Likewise, non-flammable material like plastics is also avoided to prevent the additional participation of plastic combustion to the generated heat. **Table 4** showed the emission of gases throughout the lithium-ion battery thermal lifecycle.

Table 4. Lithium-ion battery thermal run-away emission of gases. (K Environmental, 2019)

Compound	Concentration (percentage)
Carbon monoxide	40
Hydrogen	30
Carbon dioxide	20
Hydrocarbons	7
Hydrogen Fluoride HF	less than 3

Short ALT Text for the table, p. 9 [14 words]:
The concentration of gases that are emitted during the thermal lifecycle of the li-ion battery.

4.4 Chemical Hazards

Lithium-ion battery leaking is equal to hydrofluoric acid production. Since the electrolyte is fluid, therefore there is a possibility that this fluid may be a leak inside the battery and make contact with water and air. Two chemical reactions may cause the hydrofluoric acid production, the electrolyte ions of PF_6^- in water presence and that PF_6^- ions combustion. PF_6^- ions hydrolysis take place in the existence of water in basic and acidic (PH level between 1- 12) medium whereas hydrolysis production is not favorable. Through this reaction hydrofluoric acid produced. The hydrofluoric acid may cause serious chemical accidents when it interacts with the eyes or skin.

5 BATTERY DISMANTLING AND HANDLING HEALTH HAZARDS

The batteries of electric vehicles available in a wide variety in terms of chemistry and forms. To recycle batteries of an electric vehicle, the first important thing is sorting and classifying chemistry of the battery through inspection after that forward the batteries for dismantling safely towards workbenches. Trained technicians dismantle the packs of batteries to the module or cell, circuitry, wiring, and pieces of assembly segregated from individual cells. Disassembling through the manual process is performed by manual and power equipment as cells and packs of the battery are constructed with complexity due to that it arduous to devise cost-effective End-of-Life (EOL) process plan for electric vehicle batteries. Safety training is vital for the personnel who are responsible for the dismantling of these items and the instructions for dismantling should be given by the manufacturers.

Lithium-ion battery disassembly is a complicated, costly, and lengthy process because of the extensive range of battery composition, components, and potential hazards due to high voltage and chemical products in cells of the battery. All of these elements must be taken into consideration while planning the dismantling process and appropriate workstations. A lithium-ion battery includes toxic chemicals and by-products which are either destructive or by the reaction with other materials they produced toxic, heat-generating or flammable chemicals. The detail of the main components, material incompatibility, and their health effects are summarized in **Table 5** below.

Table 5. Components detail of lithium-ion battery. [21]

Components	Examples	Effects on Health	Incompatibility	Non- inflammable/ Flammable
Cathode (Negative)	Lithium Cobalt ($LiCoO_2$); Lithium Manganate ($LiMn_2O_4$); Lithium Phosphate ($LiFePO_4$)	Respiratory, skin, Eye and gastrointestinal irritant; possible carcinogens	Nil, but elude of extreme temperature and fire	Non- inflammable
Anode (Positive)	Graphite (C)	In solid form no effect (evade dust)	Nil, but elude of extreme temperature and fire	Non- inflammable
Salt Electrolyte	Lithium Hexa-fluoro phosphate ($LiPF_6$)	Causes respiratory, eye, skin and gastrointestinal	Agents of oxidizing; Water; strong acids	Non- inflammable
Solvent Electrolyte	EC; DMC; PC; DEC	Respiratory, skin, Eye and gastrointestinal irritant;	Agents of oxidizing; alkalis and acids;	Flammable
Separator	Polyethylene; Polypropylene	Nil	Nil, but elude of extreme temperature and fire	Non- combustible

Short ALT Text for the table, p. 11 [18 words]:
Health effects of the materials used in the manufacturing of Li-ion batteries along with their examples are discussed.

The component of the cathode has a significant risk of toxicity. Therefore, the probability of health effects is termed as low as the elements are solids and a high temperature is used to melt them. If these elements are disclosed they will involve the breaking of a battery with the chemical component separation or vaporization. In a battery of lithium-ion, the cathode and anode are covered by toxic substances such as the most hazardous chemicals are solid electrolyte interface layer, which is coated on anode and cathode, electrolyte salt, carbon anode, and electrolyte solvent.

5.1 Lithium-ion Battery Landfill

The reuse of LiBs is done in fewer countries, while in other countries LiBs were landfill or thrown away. Discarded batteries have eventually reached landfills and saturated, this may also harm the environment [22]. Greenhouse gas emissions from organic components like electrolyte solvent, plastic, and paper are produced during landfill processing. Besides, heavy metals can be leached to the atmosphere and can enter water or soil based on the landfill state. Landfilling possesses more danger to the environment in comparison to the recycling of the batteries.

5.2 Impact of Recycling on the Environment

Growing waste required more land space for disposal, as these batteries contain harmful chemicals so ultimately these chemicals return to the environment. However, with the comparison to other batteries, the LiB contains small toxic materials, so recycling of batteries may have a less impact on the environment. Particularly by recycling lithium-ion batteries, the considerable reduction in energy consumption, GHS emission, and the use of natural resources are found in comparison to the manufacturing of new LiBs.

Additionally, by recycling, we can retrieve lithium and other materials. In particular, the use of lithium is on the rise due to the increasing need for LiBs which are used in electric vehicles. Around 25% of the total amount of Lithium is used by industries battery manufacturing industries for EVs and this number will surge exponentially shortly [23].

5.3 Recycling of Electric Vehicles (EV) Battery

The demand for Electric Vehicles (EV) is rising and will be increased in the coming years. It is estimated that post-vehicle battery packs application will be increased from 1.4 to 6.8 million by 2035 [24]. LiBs is a suitable choice for electric vehicles; secondly, the disposal of LIBs is vital to conserve the environment and health of homo-sapiens. Regrettably, the LiBs recycling percentage is only 3%, whereas its revival is negligible. It is estimated that if the rate of recycling will not increase, the requirement of lithium transcends by 2023 [25]. **Figure 4** summarizes that after the manufacturing of lithium-ion battery its life is approximately 8-10 years. When the battery completes its life, it should be sent towards service and retail centers where these batteries can be repaired. If the product has the end of life, then it should be sent towards the recycling process from where raw materials can be obtained to manufacturing new ones [26].

6 ENVIRONMENTAL ASPECTS OF REUSE

Assuming that the second life of batteries of EV may reduce the strain of producing a new battery, or eventually reduce the usage of resources. To adequately evaluate the environmental aspects of using secondary batteries, the environmental impacts of EV batteries should be empirically measured over the entire life cycle. Contemplating the reuse of nearing retirement of electric vehicle batteries, the complete environmental effect of the production, primary reuse, and recycling of electric vehicle batteries should be dealt with consideration in the analysis of the life of the batteries. Reusing an EV retired lithium-ion battery in a stationary system might reduce Global Warming Potential (GWP) by 15% when compared to non-stationary use. It has been reported that LiB's refurbishment may reduce GWP effects by 12 to 46% compared with the similar battery functionality of Pb-acid [27]. The production phase of the battery pack produced the majority of Greenhouse Gas emissions by approximately 40% when compared with both phases of usage and re-usage (31% and 26%) thus contributing to Greenhouse Gas emissions significantly [28].

Figure 4. The lifecycle of a Lithium-ion battery.
Short ALT Text for a photo, p. 12 [12 words]:
Electric vehicle Lithium-ion battery segregation of reuse and recycling, after completion of its lifetime, extract its raw material and manufacture new battery

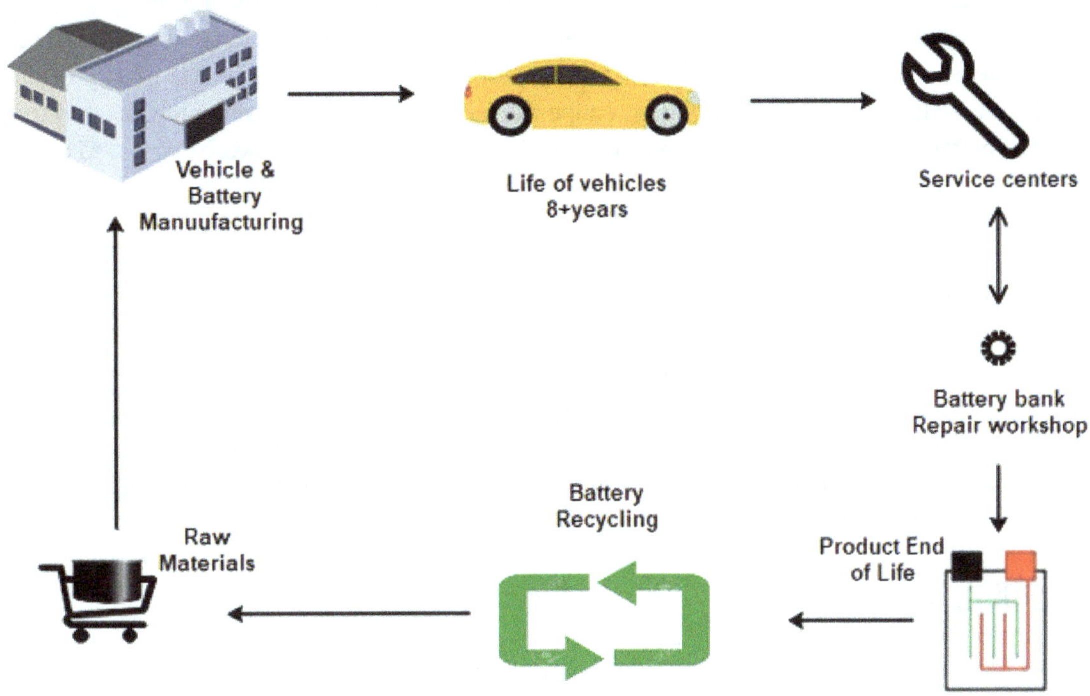

7 ENVIRONMENTAL ASPECTS OF RECYCLING

When electric vehicle batteries are decommissioned, the overall impact on the environment must be evaluated while considering the processes of battery recycling, which include collection, transport, dismantling, and recycling. Transporting lithium-ion battery to recyclers have a slight contribution to the environment as compared to manufacturing and usage phases. However, there is a need for adequate transport to minimize or exclude the environmental effects of the process. Research has shown that truck and rail transport is capable of reducing Greenhouse Gas emissions related to transportation by about 23% to 45%. Besides, they combined geospatial models and life cycle analysis to recognize the anticipated infrastructure needs, categorize emissions and material flows and evaluate human health risk from the lithium-ion battery recycling system in California [29]. These findings indicate that developing of facilities more than two for dismantling would probably reduce investment returns. Optimum recycling technologies and multimodal supply chain collection and transport will help in dealing with the economic and environmental challenges of LiBs production.

Pyrometallurgical methods of recycling can minimize Greenhouse Gas emissions and primary energy utilization by 23% and 6% to 56%, respectively, compared to virgin production [29]. Hydrometallurgy recycling provides enormous electrical energy savings and slight air pollution; furthermore, pyrometallurgical recycling has significant benefits concerning water utilization. The significant environmental challenges regarding the processes of recycling are waste material landfills, plastics incineration, and electricity consumption, specifically in the procedures of smelting which releases a lot of energy. The hydrometallurgical processes' major impact on climate change is landfill, while plastics incineration is the major impact of pyrometallurgical processes. Mechanical and hydrometallurgical techniques may retrieve more materials consuming lesser energy than the pyrometallurgical technique.

The resources required for battery recycling is just 387.4 MJ for the manufacturing of 1kg of cathode material, whereas, the manufacturing of new battery production required 795.4 MJ of 1kg of cathode material. The recycling of batteries may also recuperate 51% of the valuable resources used during the production of new batteries while nuclear energy consumption and fossil fuel were decreased by 57.25% and 45.3% respectively [30].

8 RECYCLING

Recycling of the electric vehicle batteries will be the major concern in upcoming decades, whereas in the manufacturing of EV batteries different chemical materials are used which makes the recycling process more complex. Lithium-ion battery packs have a complex cell, module, and structure due to cylindrical or prismatic cells associated with a series-parallel configuration such as wire bonding, welding, and mechanical appending [31].

8.1 Recycling Methods

In each cell of lithium-ion, batteries require a broad spectrum of materials which results with a more complex recycling process, including

- Lithium, copper, manganese, aluminum, zinc, cobalt, steel
- Components of plastic: Polyethylene terephthalate, polypropylene
- Graphitic carbon
- Sulphuric acid electrolyte, solvent
- Fiberglass
- A management system of battery/coolant

The flow process of lithium-ion battery recycling as depicted in **Figure 5.**

During recycling, all of the above materials must be separated from one another. The following are the various recycling methods/processes currently used for Lithium-ion batteries.

8.2 Mechanical Procedure | MP

A mechanical method is used to split components of electric vehicle batteries where electrolyte securely separates to cut down the metals so that the process becomes easier. Batteries can be fed into equipment through a conveyer and shredded components of the battery into small pieces then moved towards recycling.

Figure 5. Lithium-ion battery-recycling procedures.
Short ALT Text for a photo, p. 14 [14 words]:
Dismantle and disassemble batteries can go through any of these three processes where materials like alloy, Li, Ni found for battery manufacturing.

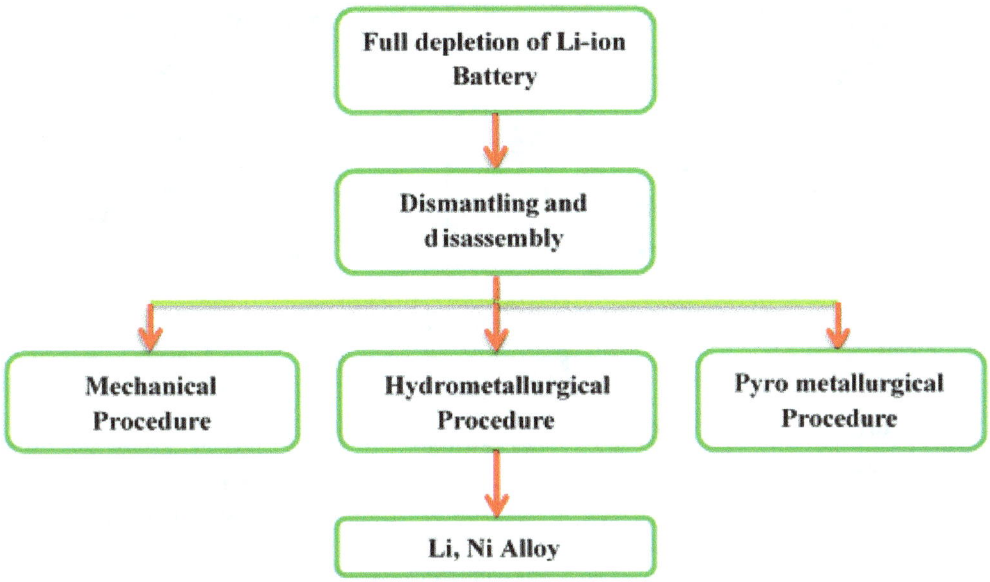

8.3 Pyro Metallurgical Procedure | PM

This is the process of extreme temperature, which includes two operations. Lithium-ion batteries firstly kindle in the smelter at which substances are smashed down and other materials like separators and plastic burned away. Then by carbon curtailment new alloys are produced, metal alloys further extracted to retrieve materials [32]. During this process, costly metals like copper (Cu), nickel (Ni), and cobalt (Co) are retrieved with maximum competence. The electrolyte, plastic, and anode are oxidized and provide the energy for the procedure. The major advantages of this process are (i) a mature and simple process (ii) reduction in size and sorting out is not mandatory- a composition of LiB and Nickel-metal hydride batteries which can be recovered (iii) Production is comprised of basic building blocks which might be used to synthesize materials of the cathode by various chemistries. On the other hand, the main drawbacks are i) carbon dioxide formation and higher power consumption during the procedure of smelting ii) alloy needs more refining due to which the cost of recycling will increase iii) many products in Lithium-ion batteries like graphite, plastic and Aluminium were not retrieved, the procedure recuperates Cobalt and Nickel from the material of cathode and Copper from anode current collector, which accounts for only 30% w.r.t. Lithium-ion batteries for electronics (iv) the business strategy might not work effectively for electric vehicle batteries, because of the lower concentration of cobalt. Furthermore, the aim of the industries is cobalt reduction or eventually to produce cobalt-free cathode materials. Recently, the Pyro-metallurgical process exists extensively in the industry due to its naivety and higher efficiency. There is no standardized procedure for pyro-metallurgical recycling of the batteries and every current method is peculiar to the concerned company. Given **Figure 6** portrayed the whole recycling processes.

Figure 6. Recycling methods.
Short ALT Text for a photo, p. 16 [16 words]:
The three recycling methods which include pyrometallurgy, Hydrometallurgical and direct recycling, through these methods, manufacturing companies can retrieve different components and materials

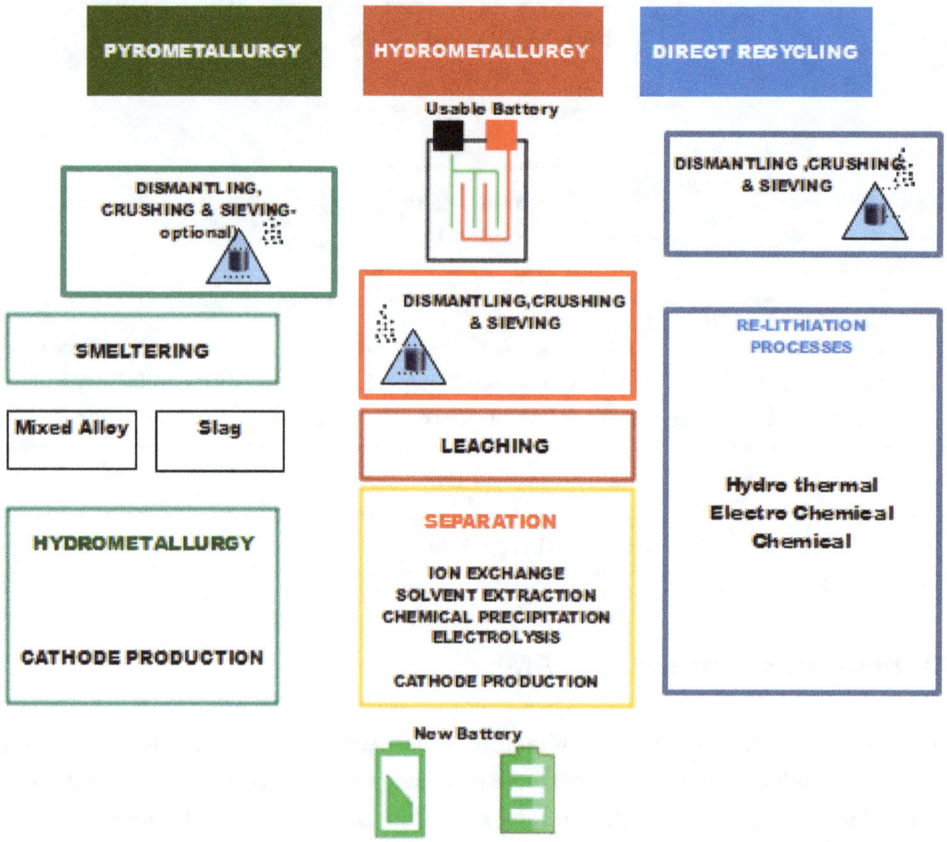

8.4 Hydrometallurgical Procedure | HP

In the hydrometallurgical procedure, material retrieval is obtained through aqueous chemistry, by leaching method and later concentration and filtration are done. Lithium-ion batteries, solution ions are segregated using different technologies like solvent extraction, ion exchange, electrolysis, and chemical precipitation [33]. In this process, various steps involved in recycling. The following steps are for dismantling, crushing and sieving after that the leaching process takes place where metals in current EOL of lithium-ion batteries are dissolved by leaching process and the latter leachate go through by further process to isolate the metal ions. The last step is the separation of different solvent and production of the cathode from where one can develop a new battery. The major hydrometallurgical process benefits are (i) generation of high purity materials (ii) most constituents of a LiB can be retrieved (iii) the operation of the hydrometallurgical process is at a lower temperature (iv) the emission of CO_2 is lesser in the comparison of a pyro metallurgical process [34]. The major disadvantages of this process are (i) sorting requires which need more storage space due to which cost increases and process complexity arise (2) difficulty in the segregation of elements like nickel, iron, cobalt, aluminum, and copper from the solution as they have similar in properties, this again results in higher costs (3) water waste treatment and associated costs which associated with them.

8.5 Direct Recycling Procedure | DRP

DRP method of recovery which suggested for the direct processing and recovery of active lithium-ion battery material while maintaining the structure of its origin [35]. In this procedure, battery components are separated with the use of methods such as magnetic separation, physical separation, and thermal process to evade active materials chemical breakdown. Active materials are then refined and defects are retrieved by the hydrothermal or re-lithiation process. Cathodes might be a combination of two or more active materials, the segregation of these materials is difficult. The major advantages of this process involve: (i) relatively easy procedure (ii) after regeneration of active materials they can be used directly (iii) in the comparability of hydrometallurgy and pyro metallurgy process, the recycling process generates fewer emissions and secondary pollutants [20]. The major disadvantages of this process are (i) involve pre-processing/comprehensive sorting for active chemical substances (ii) untested technology, which to date only exists at the laboratory scale (iii) significant susceptibility to changes in the data stream (iv) this process is not flexible because the material is not purified, and therefore cycle might not be acceptable to the cathode chemistry.

9 LITHIUM-ION BATTERIES RECYCLING BEST PRACTICES

The best practices for recycling LIBs in the industries are discussed below

9.1 Umicore Company

Umicore Company exploits the pyro metallurgical procedure for the recycling of LIBs by using the method of ultra-high temperature. This company able to handle 7000 metric tons/year, the process of ultra-high temperature creates an alloy, Copper (Cu), Nickel (Ni), and Cobalt (Co) and generates slag for construction materials.

However, the further steps of segregation or purification involved the technique of hydrometallurgical, which includes leaching, extraction of solvents, and precipitation for the production of new cathode substances like nickel-manganese-cobalt (NMC) and Lithium cobalt oxide (LCO). The lithium had existed previously in the slag, which mainly used for building [36] but Umicore recently illustrates that by doing the further process, the Lithium-ion battery slag can be converted into Li-recovery flow sheets with external collaborators [37]. **Figure 7** showed the systematic process.

9.2 Retrieve Technologies

In North America, the company Retrieve Technologies utilizes a hydrometallurgical procedure to recycle lithium-ion batteries. They manually dismantled the lithium-ion battery pack to the cell level or module. After it fed into a lithium brine flooded hammer-mill for crushing to lower the battery reactivity, neutralization of electrolyte, and avoid the emissions of gas [39]. This company follows a three-stream process of materials which include plastic fluff, metal solids, and metal-enriched liquid. The metal-enriched liquid contains mainly lithium-ion, precipitated and purified and solid metals might contain copper (Cu), Aluminium (Al), and black mass. The solid metals are provided to downstream smelters who are interested in nickel (Ni) and Cobalt (Co) contents. On the other hand, plastic may be disposed of or recycled. However, lithium recovered as lithium-carbonate [40, 41].

Figure 7. Umicore Recycling Process.[38]
Short ALT Text for a photo, p. 18 [32 words]:
Recycling method of Umicore Company, end-of-life batteries go for the disassembling process after that smelter and granulation process took place, in granulation stage construction material, rare earth oxides and alloy were extracted

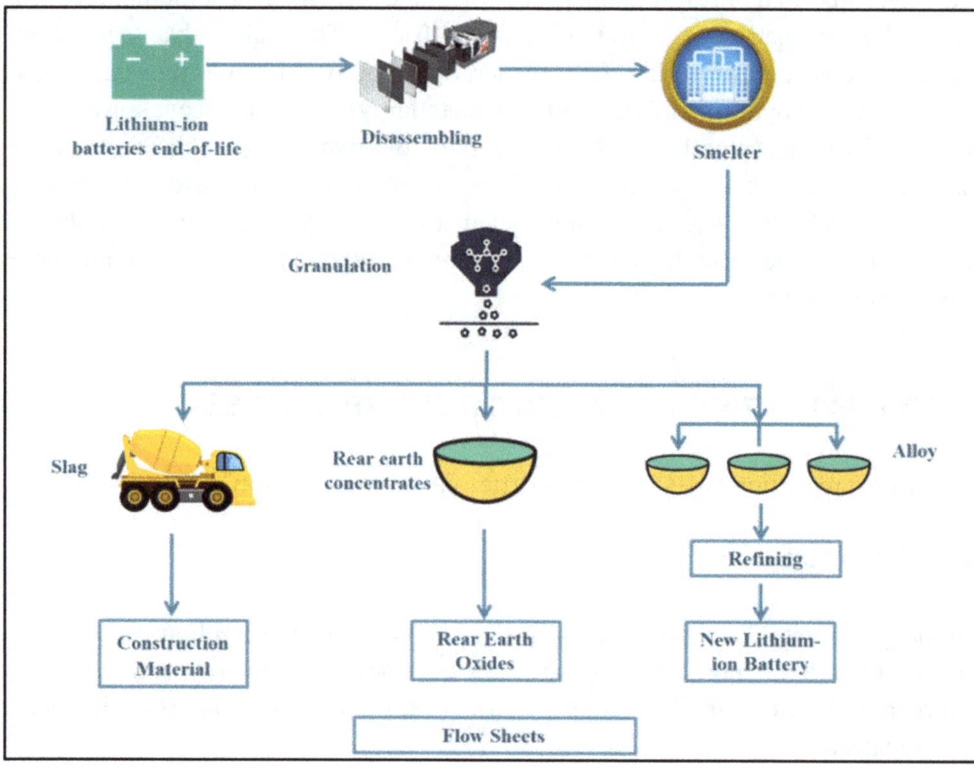

9.3 Onto Technology

Onto Technology recycled lithium-ion batteries by using the direct recycling method where they obtain the electrode from lithium-ion batteries. After that cathode, the material is obtained by combining and detaching the electrode from the current collector in an aqueous alkaline solution. The devalued cathode content might revive and possible to use in the development of cells by using hydrothermal and additional heat treatment.

10 SAFETY INDICATORS

Safety precautions are the most important aspect of lithium-ion batteries while dismantling, sorting, reorganizing and recycling the retired batteries and must require proper management to avoid risks.

11 DISMANTLING AND STORAGE

The lithium-ion batteries must be reassigned for reuse and after cessation, the module and battery pack must be dismantled with the elimination of the exterior circuit, removal battery scarp, and data sensors. The remaining ability, immensely high voltage, an exterior circuit, and reliability of connectivity need the dismantling of demolished batteries must be carried out in an open area with safety precautions. Disassembling processes might also prevent permanent impairment to the battery. To achieve non-destructive and secure dismantling, it is suggested to dismantle the module or pack level instead of cell level soft connection modules. Dismantling at the cell level is not viable both economically and technically. To minimize safety risks, transport and storage of demolished batteries should evade transportation of long-distance, long term storage, fire sources, rain exposure, and other collisions.

11.1 Reorganizing and Screening

Decommissioned batteries cause several risks like flatulence, maligned casing, leakage, insulation failure, and short circuit. Batteries screening out with long service life and good performance is a difficult task maintaining the protection of decommissioned batteries. Traditional screening techniques are based primarily on cells that are singled out, comprising assessment and testing for performance and appearance through nano-CT to evaluate the structure of three dimensional qualitatively; to create a model of electrochemical and to evaluate the electro-chemical external parameters performance. Some technologies have recently been developed, like contact-type ultrasonic, non-destructive screening technology that can quickly and correctly determine the cell's State of Health and State of Charge by correlating ultrasonic signals with a performance of electrochemical and non-destructive and non-contact screening methodology. Detection procedures are fairly successful for battery modules and might make visible Li-battery development sans the battery removal module, hence lowering the risk of short circuits and "dead Li". Furthermore, at this point research has not come forward with any technology or equipment that can detect and view battery modules.

The modules of batteries have to be rearranged after the screening, depending upon database compiling demand details for secondary usage, functional details, and test requirements comprising of the capacity, residual cycle life, internal resistance, and material system. The battery management system of decommissioned batteries seems more systematic when concerning about safety. Thus new packs of the battery must consider the following factors while reorganization to minimize the safety risk of separate integration of decommissioned batteries: Optimizing the module model connection to enhance flexibility; implementing intelligent technology of time-division; equipping an effective temperature monitoring system and current equalizers and the module protection through high-voltage and thermal monitoring.

12 TECHNOLOGICAL INITIATIVES

Research is simultaneously being done in three major avenues to curb the issue of used batteries in the current scenario.

Composition: Different compositions such as Lithium-sulphur batteries and lithium-air batteries are being tested for performance and cost in comparison to LiBs.

Available resources: The reserves of Lithium will become scarce as they are being consumed at a very high rate since its demand increases exponentially with the increase in EVs.Presently we are expecting a exceed in the lithium cost if the battery recycling is not made efficient and economical.

Recycling methods: It has been observed that the methods of hydrometallurgical and mechanical procedures are giving better results in the recollection of materials as well as it is low energy seeking procedures in comparison to pyro-metallurgical procedure. The Lithium-ion Battery Recycling Initiative (LiBRi) is teaming with different companies and are working on ways to recollect lithium from the slag content. Research is underway to recover the maximum amount of lithium during the dismantling of the batteries

13 CONCLUSION

Researchers are experimenting with different combinations of chemicals to replace lithium and cobalt. The alternatives can be the materials that are found in abundance and are less toxic. Nevertheless, if new batteries are not efficient or more costly than lithium, they could wind up negatively affecting the earth severely in general. If recycling of EV batteries is taken as a business practice; key areas to be considered can be a system to collect expired batteries, storing them, making logistical arrangements, etc.

14 RECOMMENDATIONS AND FUTURE DIRECTIONS

Further research can be done on developing a supply chain model of closed-loop for the batteries which can have properly marked and labelled pieces right at the manufacturing unit. This will help in the retrieval of these batteries after their end life. Also, sustainability can be applied to the whole manufacturing procedure for LiBs batteries which help in reusing the batteries to achieve zero carbon emissions.

REFERENCES

Ahmadi, L., Young, S. B., Fowler, M., Fraser, R. A., & Achachlouei, M. A. (2017). A cascaded life cycle: Reuse of electric vehicle lithium-ion battery packs in energy storage systems. *The International Journal of Life Cycle Assessment*, *22*(1), 111–124. doi:10.100711367-015-0959-7

Amarakoon, Smith, & Segal. (2013). *Application of life-cycle assessment to nanoscale technology: Lithium-ion batteries for electric vehicles*. Academic Press.

Arfeen, Khairuddin, Munir, Azam, Faisal, & Arif. (n.d.). En route of electric vehicles with the vehicle to grid technique in distribution networks–Status and technological review. *Energy Storage*.

Battery Recycling as a Business. (2019). Available: https://batteryuniversity.com

Better World Solutions. (2019). Available: https://www.betterworldsolutions.eu/repurpose-ev-batteries/

Bhatti, A. R., & Salam, Z. (2018). A rule-based energy management scheme for uninterrupted electric vehicles charging at constant price using photovoltaic-grid system. *Renewable Energy*, *125*, 384–400. doi:10.1016/j.renene.2018.02.126

Canals Casals, L., & Amante García, B. (2016). Assessing electric vehicles battery second life remanufacture and management. *Journal of Green Engineering*, *6*, 77–98. doi:10.13052/jge1904-4720.614

Ciez, R. E., & Whitacre, J. F. (2019). Examining different recycling processes for lithium-ion batteries. Nature Sustainability, 2, 148-156. doi:10.103841893-019-0222-5

Das, A., Li, D., Williams, D., & Greenwood, D. (2018). Joining technologies for automotive battery systems manufacturing. *World Electric Vehicle Journal*, *9*(2), 22. doi:10.3390/wevj9020022

Dewulf, J., Van der Vorst, G., Denturck, K., Van Langenhove, H., Ghyoot, W., Tytgat, J., & Vandeputte, K. (2010). Recycling rechargeable lithium ion batteries: Critical analysis of natural resource savings. *Resources, Conservation and Recycling*, *54*(4), 229–234. doi:10.1016/j.resconrec.2009.08.004

Electric Vehicle Regulatory Reference Guide. (2019). Available: https://www.unece.org/

Falk, Nedjalkov, Angelmahr, & Schade. (n.d.). *Applying Lithium-Ion Second Life Batteries for Off-Grid Solar Powered System—A Socio-Economic Case Study for Rural Development*. Academic Press.

Gaines, L. (2014). The future of automotive lithium-ion battery recycling: Charting a sustainable course. *Sustainable Materials and Technologies*, *1*, 2–7. doi:10.1016/j.susmat.2014.10.001

Gaines, L., & Nelson, P. (2009). Lithium-ion batteries: possible materials issues. *13th international battery materials recycling seminar and exhibit,* 16.

Gaines, L., Sullivan, J., Burnham, A., & Belharouak, I. (2011). Life-cycle analysis of production and recycling of lithium ion batteries. *Transportation Research Record: Journal of the Transportation Research Board*, *2252*(1), 57–65. doi:10.3141/2252-08

Hao, H., Mu, Z., Jiang, S., Liu, Z., & Zhao, F. (2017). GHG Emissions from the production of lithium-ion batteries for electric vehicles in China. *Sustainability*, *9*(4), 504. doi:10.3390u9040504

Hazard Class 9 - Hazard Class Labels Explained. (2019). Available: https://www.general-data.com/about/blog/hazard-class-9-hazard-class-labels-explained

Heelan, J., Gratz, E., Zheng, Z., Wang, Q., Chen, M., Apelian, D., & Wang, Y. (2016). Current and prospective Li-ion battery recycling and recovery processes. *JOM*, *68*(10), 2632–2638. doi:10.100711837-016-1994-y

Hendrickson, T. P., Kavvada, O., Shah, N., Sathre, R., & Scown, C. D. (2015). Life-cycle implications and supply chain logistics of electric vehicle battery recycling in California. *Environmental Research Letters*, *10*(1), 014011. doi:10.1088/1748-9326/10/1/014011

Irle, R. (2019). *Global BEV & PHEV Sales for 2019*. Available: www. EV-volumes.com

K Environmental. (2019). Research Study on Reuse and Recycling of Batteries Employed in Electric Vehicles. *Energy API.*

Li, X., Mo, Y., Qing, W., Shao, S., Tang, C. Y., & Li, J. (2019). Membrane-based technologies for lithium recovery from water lithium resources: A review. *Journal of Membrane Science*, *591*, 117317. doi:10.1016/j.memsci.2019.117317

Lithium-Ion. (2020). Available: https://www.retrievtech.com/lithiumion

Lü, X., Wu, Y., Lian, J., Zhang, Y., Chen, C., Wang, P., & Meng, L. (2020). Energy management of hybrid electric vehicles: A review of energy optimization of fuel cell hybrid power system based on genetic algorithm. *Energy Conversion and Management*, *205*, 112474. doi:10.1016/j.enconman.2020.112474

McManus, M. C. (2012). Environmental consequences of the use of batteries in low carbon systems: The impact of battery production. *Applied Energy*, *93*, 288–295. doi:10.1016/j.apenergy.2011.12.062

Meshram, P., Mishra, A., & Sahu, R. (2019). Environmental impact of spent lithium ION batteries and green recycling perspectives by organic acids–A review. *Chemosphere*, 125291. PMID:31896181

Ordoñez, J., Gago, E., & Girard, A. (2016). Processes and technologies for the recycling and recovery of spent lithium-ion batteries. *Renewable & Sustainable Energy Reviews*, *60*, 195–205. doi:10.1016/j.rser.2015.12.363

Pagliaro, M., & Meneguzzo, F. (2019). Lithium battery reusing and recycling: A circular economy insight. *Heliyon*, *5*(6), e01866–e01866. doi:10.1016/j.heliyon.2019.e01866 PMID:31245638

Popper, Lanfrankie, & Caro. (2019). *The (Potential) Value of Labeling in the Lithium Ion Battery Supply Chain*. Available: https://blogs.anderson.ucla.edu/

Recycling Process. (2020). Available: https://csm.umicore.com/en/recycling/battery-recycling/our-recycling-process

Richa, K., Babbitt, C. W., Nenadic, N. G., & Gaustad, G. (2017). Environmental trade-offs across cascading lithium-ion battery life cycles. *The International Journal of Life Cycle Assessment*, *22*(1), 66–81. doi:10.100711367-015-0942-3

Rohr, S., Müller, S., Baumann, M., Kerler, M., Ebert, F., Kaden, D., & Lienkamp, M. (2017). Quantifying uncertainties in reusing lithium-ion batteries from electric vehicles. *Procedia Manufacturing*, *8*, 603–610. doi:10.1016/j.promfg.2017.02.077

Shi, Y., Chen, G., Liu, F., Yue, X., & Chen, Z. (2018). Resolving the Compositional and Structural Defects of Degraded LiNi x Co y Mn z O2 Particles to Directly Regenerate High-Performance Lithium-Ion Battery Cathodes. *ACS Energy Letters*, *3*(7), 1683–1692. doi:10.1021/acsenergylett.8b00833

Slowik, P., Pavlenko, N., & Lutsey, N. (2016). *Assessment of next-generation electric vehicle technologies.* White paper, International Council on Clean Transportation.

Sonoc, A., Jeswiet, J., & Soo, V. K. (2015). Opportunities to improve recycling of automotive lithium ion batteries. *Procedia CIRP*, *29*, 752–757. doi:10.1016/j.procir.2015.02.039

Stephens, D., Shawcross, P., Stout, G., Sullivan, E., Saunders, J., & Risser, S. (2017). *Lithium-ion battery safety issues for electric and plug-in hybrid vehicles*. National Highway Traffic Safety Administration, Report No. *DOT HS*, *812*, 418.

Swain, B. (2017). Recovery and recycling of lithium: A review. *Separation and Purification Technology*, *172*, 388–403. doi:10.1016/j.seppur.2016.08.031

Tchounwou, P. B., Yedjou, C. G., Patlolla, A. K., & Sutton, D. J. (2012). Heavy metal toxicity and the environment. In Molecular, clinical and environmental toxicology. Springer. doi:10.1007/978-3-7643-8340-4_6

Weisman, J. (2019). *Tesla LIB and PV Recycling Programs*. Available: www2.calrecycle.ca.gov

Yang, J., Gu, F., & Guo, J. (2020). Environmental feasibility of secondary use of electric vehicle lithium-ion batteries in communication base stations. *Resources, Conservation and Recycling*, *156*, 104713. doi:10.1016/j.resconrec.2020.104713

Zeeshan Ahmad Arfeen, M. P. A. (2020). *Energy Storage Usages: Engineering Reactions, Economic-Technological Values for Electric Vehicles – A Technological Outlook*. International Transactions on Electrical Energy Systems.

Chapter 9
Multi-Level Inverters Interfacing Electric Vehicle Charging Stations With Microgrid for Vehicle-to-Grid (V2G) Applications

Mohd Rizwan Khalid
Aligarh Muslim University, India

Adil Sarwar
Aligarh Muslim University, India

Ibrahim Alsaidan
College of Engineering, Qassim University, Buraydah, Saudi Arabia

ABSTRACT

Multi-level inverters (MLI) are power electronic converters that convert DC power to AC power with high power-quality of output voltage waveforms. These MLI are the main streamline converters for integrating dc power of EV with microgrids. Thus, the recent interest of researchers is to investigate the MLI with a lower number of active and passive switch counts which could integrate the DC power of EV to the microgrid with the boosting ability. This chapter discusses various topologies of MLI for the integration of the DC power of EV to the grid for vehicle-to-grid (V2G) applications. MLI converts DC power to AC power with high quality of output voltage waveform. Thus, the recent interest of researchers is to investigate the MLI with a lower number of active and passive switch counts which could integrate the DC power of EV to the microgrid. Also, MLI must be capable of boosting the voltage level to meet the grid requirements. The aim of this chapter is to discuss the various topologies of MLI for the integration of DC power of EV to the grid for vehicle-to-grid (V2G) applications.

DOI: 10.4018/978-1-7998-6858-3.ch009

1 INTRODUCTION

The present environmental state of earth is degrading day-by-day due to emission of greenhouse gasses and particulate matters by the energy generation and transportation sector. These harmful emissions cause climate change and other health problems. To reduce these, the usage of renewable energy sources was introduced, like solar photovoltaic, wind, fuel cell, etc. Recently, integration of dc power of electric vehicles (EV) into microgrid is the emerging topic for the researchers. On integration dc power of EVs into microgrid, the reliability of grid is increased. The technology requires dc/ac converters which have the ability of converting dc power to ac power with high power quality, i.e. low contents of lower order harmonics. Figure 1 Block diagram of microgrid shows the block diagram for integration of renewable energy resources power along with EVs' dc power into grid.

The development of multi-level inverter (MLI) was first reported in 1981 (Nabae & Akagi, 1981). The main purpose of the MLI is to synthesize output voltage waveform like a staircase. This is

Figure 1. Block diagram of microgrid

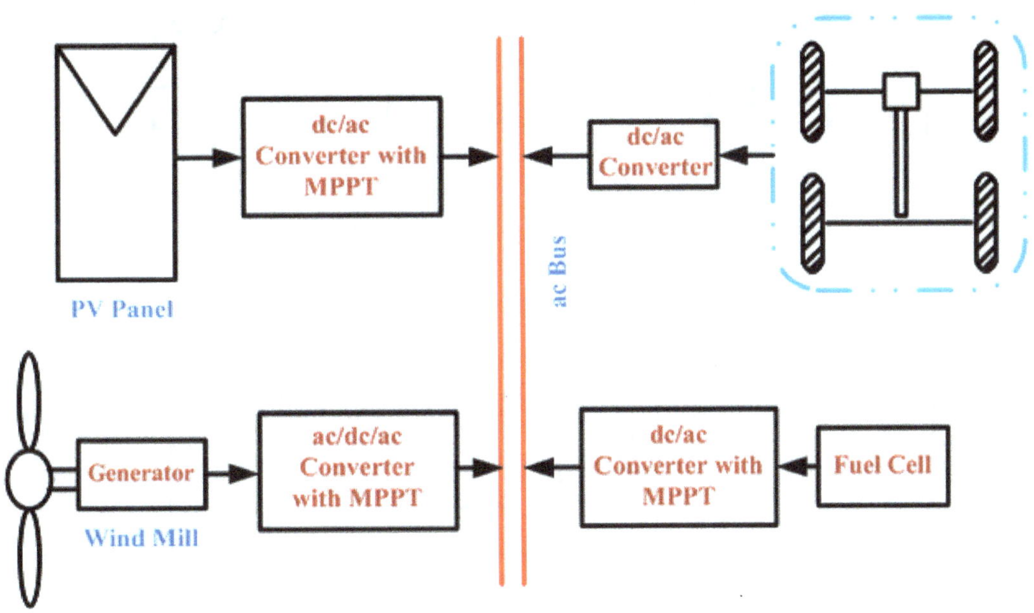

achieved by using several separate dc voltage sources at the input side. The low order harmonics in the staircase shaped voltage waveform are low; thus, the total harmonics distortion (THD) is minimum which further reduces electromagnetic interference (EMI) and the voltage stress on the switches (Shalchi Alishah, Sabahi, Nazarpour et al, 2014). In the literature there are three classical topologies namely (a) cascade H-bridge inverter (CHB), (b) flying capacitor clamped inverter (FC), and (c) neutral point clamped inverter (NPC) as shown in Figure 2. These topologies have simple control strategies and exclusive characteristics which makes their practical realization easy and simple. These MLI finds application in FACTS, induction heating, integrating renewable energy in microgrid, vehicle-to-grid (V2G) applications, UPS, variable frequency drives and small to large scale industries. To obtain large number of steps in the output voltage waveform, the required active switches increases this in turn makes the circuitry of MLI complex. At higher voltage levels, NPC and FC suffers the problem of capacitor voltage unbalancing. Cascaded H-bridge is most feasible and flexible among conventional topologies and utilizing a lesser number of components, however it requires large number of separate dc sources, which make MLI bulkier and costlier (Roy et al., 2019).

Figure 2. MLI (a) cascaded H-bridge (b) diode clamped and (c) flying capacitor

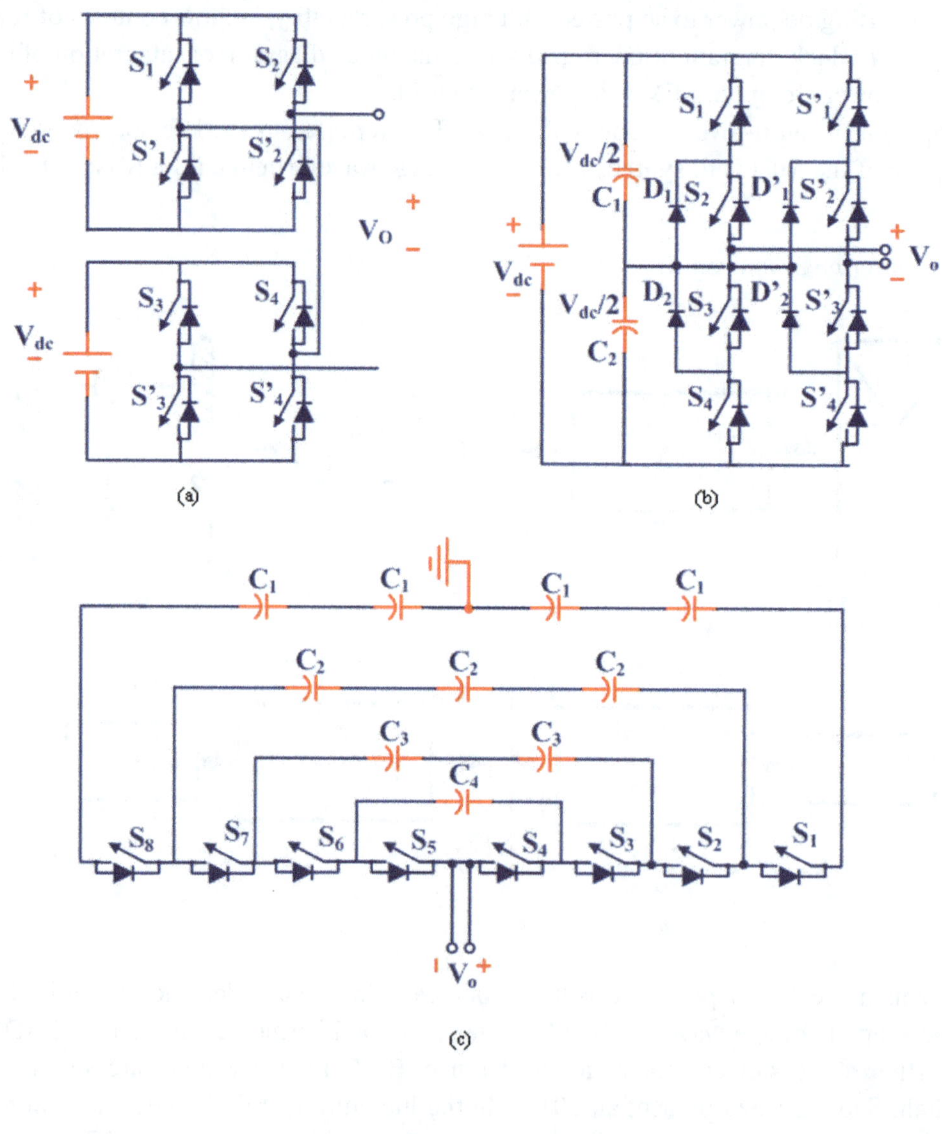

Initially, the requirement of large dc voltage sources was a big problem as this contributes to major portion of their size and overall cost. To overcome the issue the use of dc capacitor was introduced, which focused the research on balancing the capacitor voltage as the charging and discharging property of capacitor usually disturbs the voltage levels in the output voltage waveform. The requirement of more gate driver circuits and switches was also one major drawback, which further focused the research for the development of novel topologies with reduced active switches. The classical topologies and other topologies facing the problems discussed above were replaced by switched capacitor multilevel inverter (SCMLI) introduced by O.C. Mak and A. Ioinovici in 1998 (Mak & Ioinovici, 1998).

2 MULTI-LEVEL INVERTER (MLI) TOPOLOGIES

MLI topologies with less number of switches are proposed in the literature (Alishah et al., 2017; Babaei & Hosseini, 2009; Kangarlu et al., 2012; Shalchi Alishah, Nazarpour, Hosseini et al, 2014; Waltrich & Barbi, 2010). In (Babaei & Hosseini, 2009) a symmetrical converter topology having the same magnitude of dc voltage sources was presented that requires less active switches as compared to symmetric CHB topology. Moreover, to synthesize high voltage levels, it requires large active switches. The voltage stress on active switches is high at higher voltage levels. In (Kangarlu et al., 2012), (Shalchi Alishah, Nazarpour, Hosseini et al, 2014) symmetric topologies were presented requiring lesser switching devices as compared to topology presented in (Babaei & Hosseini, 2009) but they were not ideal for high voltage applications because of high blocking voltage across switches at higher levels. To overcome this problem a symmetric hybrid topology was proposed in (Shalchi Alishah, Sabahi, Nazarpour et al, 2014). In (Waltrich & Barbi, 2010), (Alishah et al., 2017) symmetric cascaded topologies were presented in which several basic units are cascaded to generate a high number of levels. At higher voltage levels, the symmetric topologies require large number of isolated dc sources, gate drivers, active/passive switches and capacitors.

To overcome these drawbacks, asymmetric topologies were proposed which utilizes dc voltage sources of different magnitude. In (Alishah et al., 2015)-(Wang et al., 2017), asymmetric MLI topologies were proposed. In (Alishah et al., 2015), a cascaded asymmetric MLI topology was proposed in which several basic units were connected in a cascaded manner for generating high voltage levels. The basic unit includes various dc voltage sources with bidirectional switches. This topology requires a high number of active switches and dc power supplies to generate higher levels which boost up the size and cost of MLI.

A cascaded asymmetric MLI topology was proposed by the author in (Boora & Kumar, 2017) based on the basic unit provided in (Babaei et al., 2014), to overcome the problems associated with (Alishah et al., 2015) i.e. to reduce the number of dc power supplies, active switches, voltage stress across switches and associated driver circuits. In (Banaei & Salary, 2011), a semi-cascaded MLI topologies utilized both in symmetric and asymmetric configuration were presented with lesser number of dc voltage sources, active switches, and associated driver circuits to generate higher voltage levels but they suffer from the high total standing voltage (TSV) and high conduction losses. The author in (Banaei et al., 2014), had proposed a cascaded MLI topology which is obtained by reconfiguration of semi-cascaded topologies proposed in (Boora & Kumar, 2017) and (Banaei & Salary, 2011) and it has low TSV and conduction losses as compared to these topologies for same numbers of active switches, dc voltage sources and output voltage levels.

In (Alishah & Hosseini, 2014) and (Alishah et al., 2013), cascaded switched-diode (CSD) based MLI topologies were utilized and an alteration technique for power generation from renewable energy sources were proposed, which produces output voltage waveform of high steps by utilizing fewer number of devices in contrast to the cascaded half-bridge topology. But these topologies do not have a path for reverse output load current due to which these topologies suffer from high voltage spikes in output voltage under R-L load which results in deterioration of power quality. A novel cascaded switched-diode

(CSD) topology was proposed by the author in (Wang et al., 2017), to overcome the problem of high voltage spikes in output voltage waveform as mentioned in (Alishah & Hosseini, 2014) and (Alishah et al., 2013), by providing a path for reverse output current and hence improved the power quality.

Topologies discussed above suffer from voltage boosting capability and unbalanced voltage problem in capacitors due to which these were not suitable for the applications that utilize low input dc voltage sources, such as renewable energy resources systems.

Some auxiliary circuits such as impedance network or forehead type boost converter were used along with MLI to achieve the voltage boosting capability (Abdullah et al., 2014). To overcome the problem of capacitor voltage unbalancing; a complex control algorithm had been discussed in (Khazraei et al., 2012), (Shukla et al., 2012). The cost, size and complexity of the control mechanism are enhanced due to the need of auxiliary circuits to compensate the voltage boosting incapability and unbalancing problems of capacitor voltage.

To overcome the voltage boosting inabilities and voltage unbalancing problems associated with conventional and above mentioned MLI topologies, switched-capacitor-multilevel-inverter (SCMLI) are proposed. Figure 3 shows different topologies for single-/three-phase SCMLI. Due to their unique characteristics these have emerged as a prominent solution to above discussed problems. In SCMIs, capacitors are utilized as substitute of dc voltage sources.

However, SCMLIs have following merits and demerits:

1. They require less number of switches and driver circuits.
2. They need a lesser number of dc power supplies because capacitors act as alternate dc sources.
3. They do not require any auxiliary circuits to overcome the problems of capacitors voltage unbalancing or self-voltage boosting inabilities and due to which the size and cost of SCMLIs are diminished as compared to conventional topologies.
4. Boosting the input voltage without using a bulky transformer and inductor.
5. In SCMLIs, the blocking voltage across switches becomes high at high boosting factor which in results the requirement of high rating switches leads to enhance the overall cost of the inverter.

Figure 3a. Topologies for SCMLI: single-phase

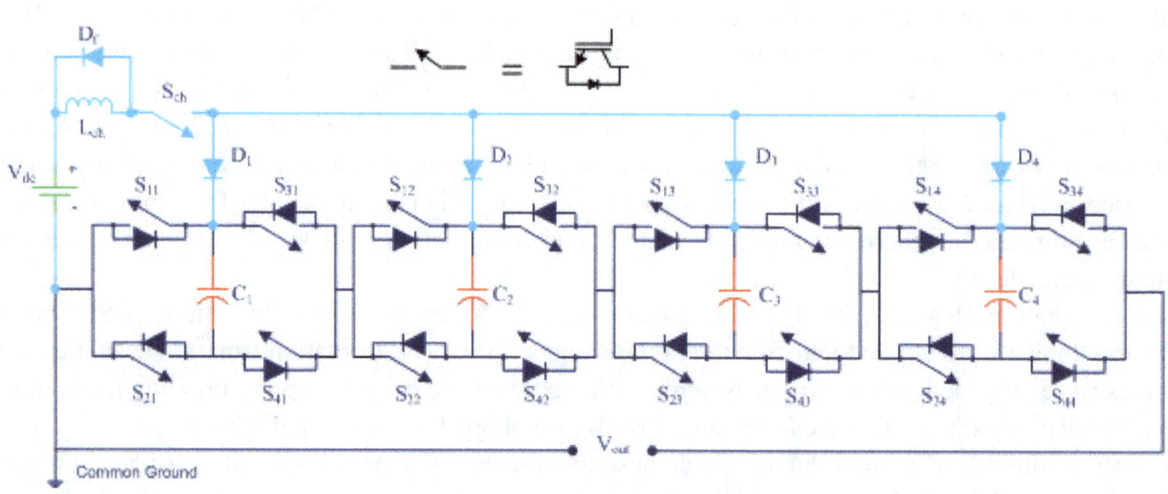

Figure 3b. Topologies for SCMLI: single-phase

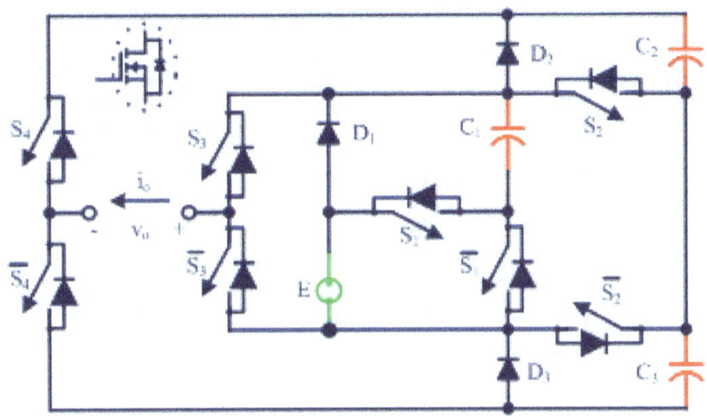

Figure 3c. Topologies for SCMLI: single-phase

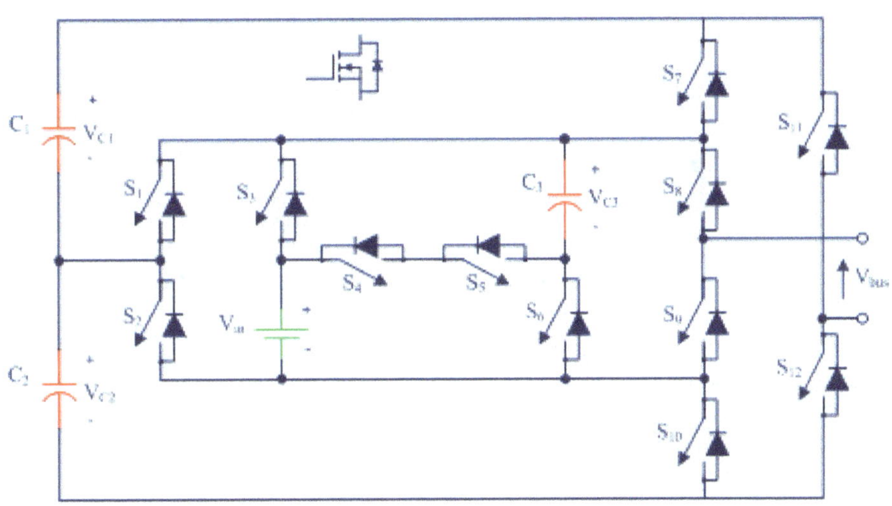

Figure 3d. Topologies for SCMLI: single-phase

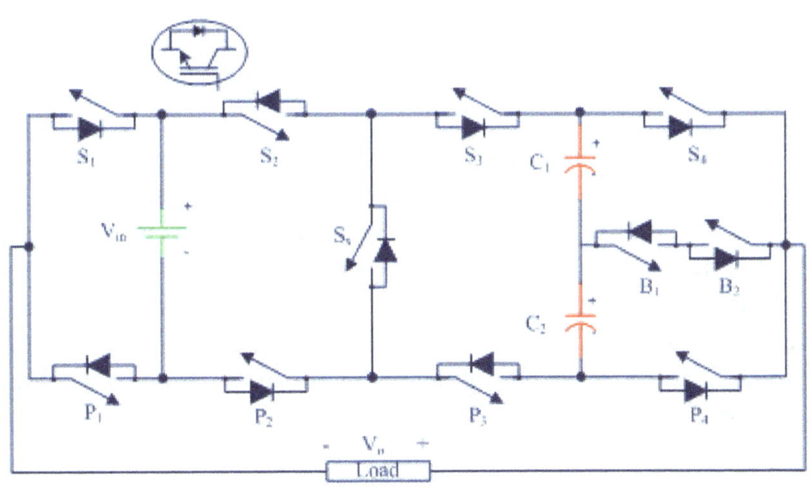

Figure 3e. Topologies for SCMLI: single-phase

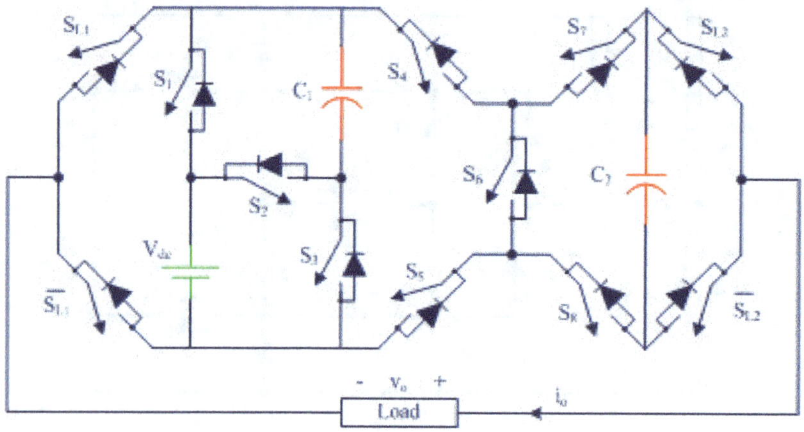

Figure 3f. Topologies for SCMLI: single-phase

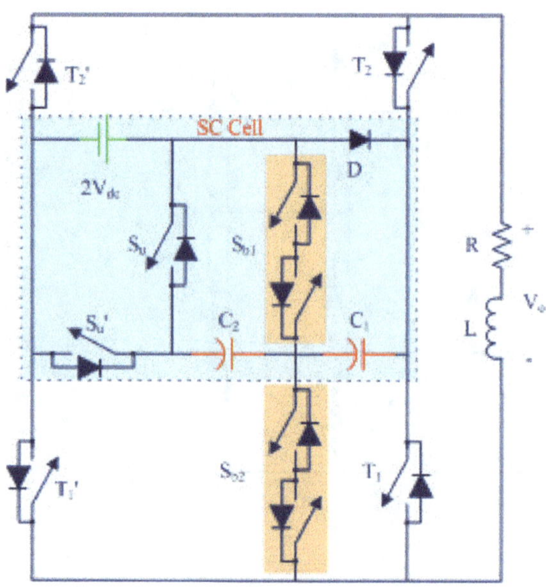

Figure 3g. Topologies for SCMLI: single-phase

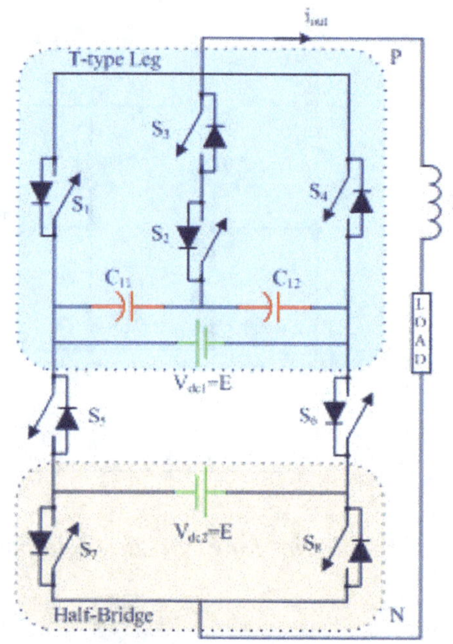

SCMLIs were first proposed by O.C. Mak and A. Ioinovici in 1998, which consists of a basic unit of switched capacitor (SC) that includes 2 active switches, 1 capacitor and 2 diodes (Mak & Ioinovici, 1998). Based on SC units, SCMLIs topologies had been proposed in (Marco & Chau Chanand, 2007) and (Axelrod et al., 2005), however, these topologies are not capable to generate the output voltage levels beyond 13 and 7 respectively. An SCMLI using a series-parallel conversion had been proposed in (Hinago & Koizumi, 2012), and it can be extended to generate the high number of voltage levels. But it requires a significant number of components at the high number of voltage levels which enhances the cost and size of the inverter.

Figure 3h. Topologies for SCMLI: three-phase

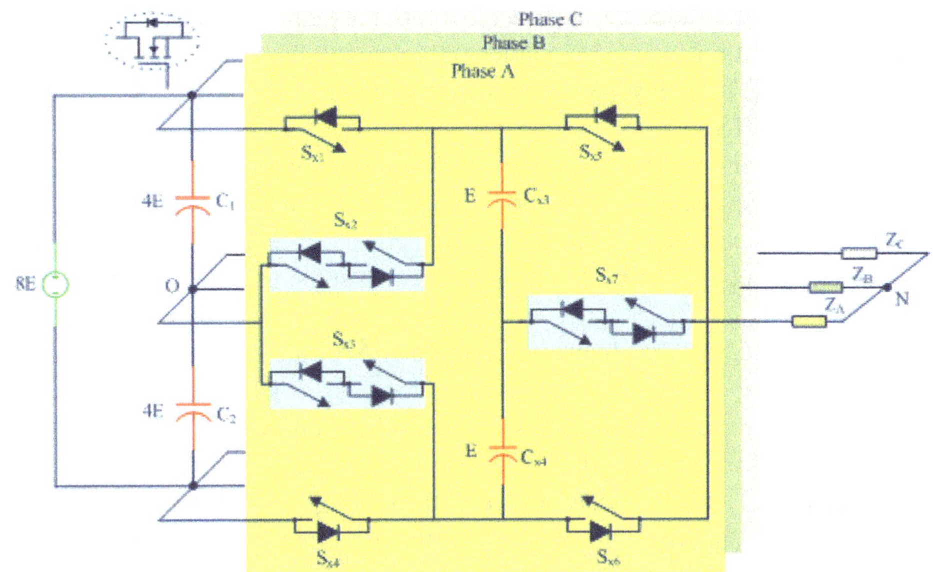

A novel step up SCMLI topology consisting of single dc voltage source has been presented in (Ye et al., 2014), to reduce the number of components required at a higher voltage level. It includes a SCC and a full-bridge converter. SCC and full-bridge converters are used as level generator and polarity generator, respectively. SCC consists of various units connecting in a cascaded manner and each unit consists of a switch, two diodes, and a single capacitor. In this topology, the output voltage level is a function of the number of units connected in SCC. By varying the number of units in SCC, voltage levels can be varied from low to high values. This topology generates higher output voltage levels from a single dc voltage source and also has voltage boosting ability along with capacitor voltage balancing property.

A cascaded SCMLI topology had been presented in (Hosseini et al., 2015), especially for maximum power point tracking (MPPT) in solar photovoltaic (PV) systems. This topology requires less active devices for generating higher output voltage levels and thus produces lower harmonic components.

To further increase the output voltage levels with a reduced number of devices the authors in (Ngo et al., 2018), proposed an SCMLI topology with a single dc voltage source which involves an H-bridge and an SCC. However, topologies presented in (Ye et al., 2014) and (Ngo et al., 2018) have high voltage stress across H-bridge active switches due to which it cannot be used in high voltage applications. To overcome these problems, a SCMLI having a single dc voltage source was proposed in (Taghvaie & Rezanejad, 2018). This topology has several SC units and each unit is a bipolar module consists of five switches, one capacitor. These units are connected in such a manner that the capacitors are charged in parallel and discharged in series. In this topology, H-bridge is not required both level generator and polarity generator. The voltage across each switch is equal to supply voltage which leads it to become a good option for high voltage applications, but to generate the higher levels it needs significant number of switches and capacitors which thus, enhance the size and cost of the overall converter. In (Barzegarkhoo et al., 2018), a novel single-phase SCMLI topology has been presented which generates nine levels of output voltage with reduced active switches and dc voltage sources. This topology includes a SC cell having 2 capacitors, 1 dc voltage source, 1 diode and 4 active switches. To generate a higher output voltage levels, extended structure was also presented in which several nine-level structures having individual SC cell in the packed form were added. This extended topology can be utilized in symmetrical as well as in asymmetrical configuration and the magnitude of isolated voltage source can be taken in binary or trinary design. However, this topology has the drawback of the requirement of high magnitude isolated dc voltage sources and high TSV.

In (Liu et al., n.d.), a seven-level SCMLI topology having low voltage stress has been proposed. To generate a high number of levels, a level extended structure was also presented which is obtained by connecting several seven-level topologies in a cascaded manner and each cell has its separate dc voltage source as in (Barzegarkhoo et al., 2018). This topology can also be used in the symmetrical as well as asymmetrical configuration.

In (Nakagawa & Koizumi, 2017), a hybrid nine-level SCMLI topology was proposed with a single dc voltage source and four capacitors. The design of this topology is just like a step-down operation due to which it is not able to generate higher output voltage levels. A boost type nine-level SCMLI having a single dc voltage source was proposed in (Nakagawa & Koizumi, 2019), and it is obtained by rearranging the location of capacitors and dc voltage source of the topology which was proposed in (Nakagawa & Koizumi, 2017), and replacing the third capacitor with a floating dc voltage source to make it fit for step-up operation. In (Nakagawa & Koizumi, 2019), the levels extended structure of quadruple boost type topology has been presented in which SC cells having four switches and capacitors are cascaded. This topology can generate higher levels with a single dc voltage source and requires a lower number of switching components. To further reduce the number of switching components with a single dc voltage source, the levels extended structure of quadruple boost type topology has been presented in (Nakagawa & Koizumi, 2019), in which pairs of SC cells were cascaded. Each pair of SC cells includes six switches and four capacitors.

3 MODULATION TECHNIQUES FOR MLI

In MLIs different modulation techniques are employed for controlling and regulating the output voltage. Switching losses and THD are the major parameters that are considered for the selection of modulation strategy for a particular topology of MLI. Figure 4 shows the classification of MLI modulation techniques. Modulation strategies which are widely used are discussed below.

3.1 Pulse Width Modulation

Pulse Width Modulation (PWM) is among the widely used modulation strategy. It is important to select a suitable modulation technique to optimize the output. In the initial stage of the development of the power electronic semi-controlled switches, the output voltage was controlled using firing angle delay called phase control. The phase control introduces distortion in the output current of the converter. Harmonic pollution of the system results in the faulty operation of system protection relays and EMI with communication systems (Trzynadlowski, 2016). Pulse width modulation scheme mitigated the above problems and as a consequence size of filters is reduced, and better spectral characteristics of the converter came into picture.

Figure 4. Different modulation techniques for MLIs

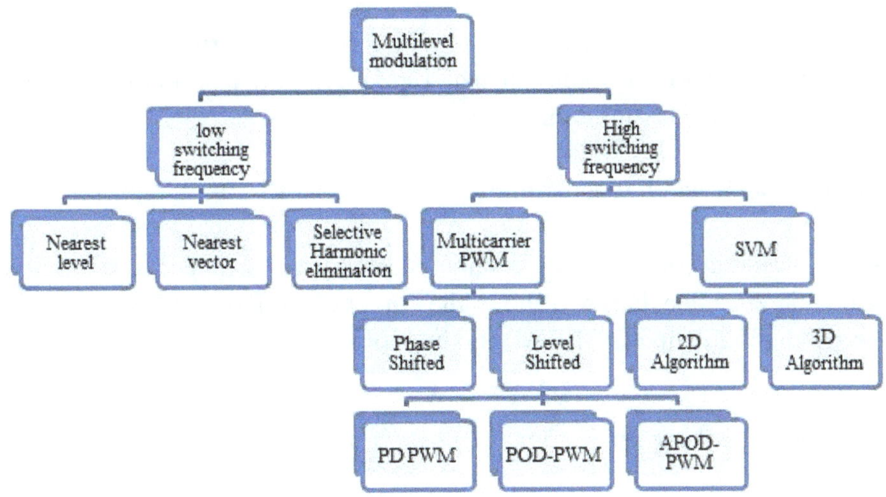

3.1.1 Phase shifted PWM

Phase-shifted PWM (PSPWM) is extension of conventional PWM techniques. In PSPWM, a reference signal (sinusoidal wave) is compared with multiple carrier waves (triangular signal) having the same maximum value with high frequency. But the adjacent carrier waves are shifted by some appropriate angle. The *m-1* carrier signals are compulsory for the generation of *m* voltage levels and these carrier waves are shifted by an angle of θ=(360/(m-1)). It is mainly used in FC and cascaded H bridge inverter. Figure **5** shows phase shift PWM techniques for generation of five-level output voltage.

Figure 5. Five-level output voltage using phase-shift PWM

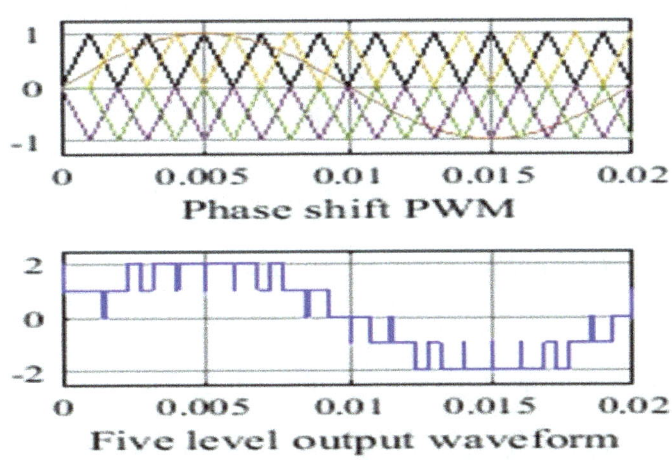

3.1.2 Level Shifted PWM

Level-shifted PWM (LSPWM) is the extension of bipolar PWM for multilevel inverters. It is based upon the amplitude shifts between carriers. In this each carrier is related to a particular voltage level (Angulo et al., 2007). These are further classified into three groups.

1. Phase disposition-pulse width modulation:

In this technique all the carriers are in the same phase but are arranged vertical above and below from the horizontal reference line. Figure 6 shows the output of the five-level with the help of the phase disposition-pulse width modulation (PDPWM) modulation scheme.

Figure 6. Five-level output voltage using phase disposition PWM

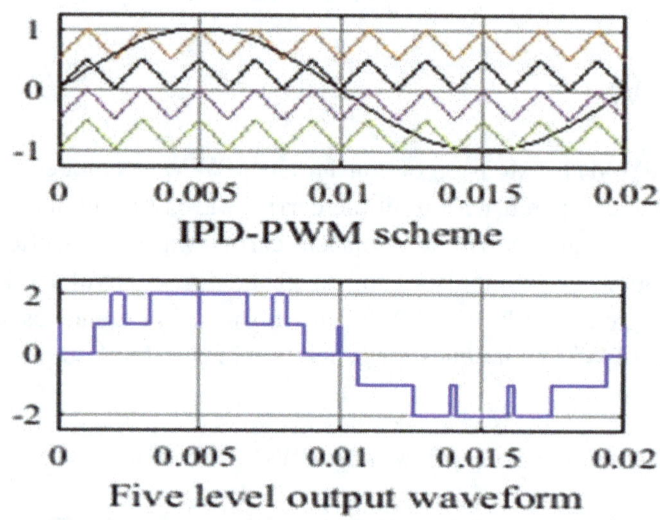

2. Phase opposite disposition-pulse width modulation:

When all positive carriers having the same frequency are in the same phase with vertical to each other and negative carriers (below zero reference line) are 180 degrees out of phase then this is called phase opposition disposition (PODPWM) technique. Figure 7 shows five levels of output voltage using this modulation scheme.

3. Alternate phase opposition disposition-pulse width modulation:

In this modulation technique, all the adjacent carriers are out of phase by 180°. Figure 8 shows five-levels of output voltage using this modulation scheme.

Figure 7. Five level output voltage using phase opposite disposition PWM

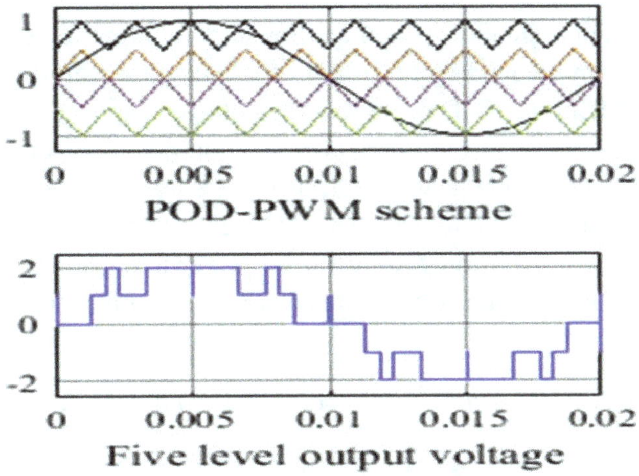

Figure 8. Five-level output voltage using phase opposite disposition PWM

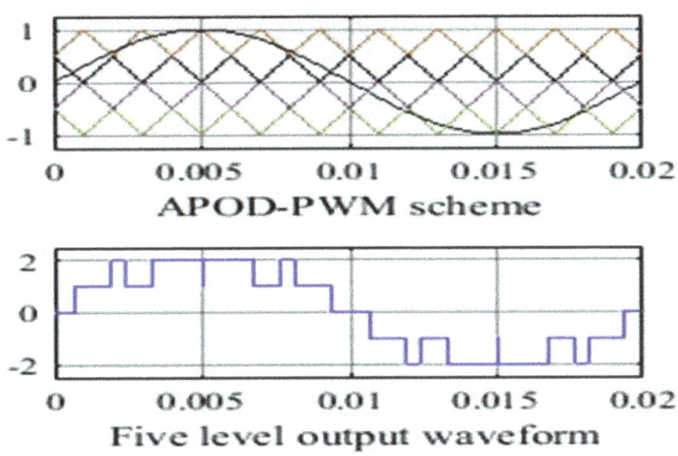

3.2 Selective Harmonic Elimination

Converters used in high power applications have significant switching losses. The traditional carrier PWM method is not suitable for them. They are operated at less than 1 kHz, but this introduces lower order harmonics in the output voltage. In such case selective harmonic elimination technique is used for the elimination of undesired lower-order harmonics (Karaca, 2015), (Sreedhar, n.d.).

4 OUTPUT VOLTAGE WAVEFORMS OF DIFFERENT MLI

Different MLI are compared in term of the power quality of their output waveforms and their simulation results for the same are shown below. Figure 9 shows the voltage waveform of line-to-line voltage of 5-level capacitor clamped MLI. The THD present in the voltage is approximately 41.95%.

Line-to-line voltage waveform of asymmetrical MLI is shown in Figure 10, which contains very low THD of 4.69%. This MLI have adequate performance in terms of power quality and is suitable and efficient for integrating EV dc power into grid. This MLI gives better performance in terms of power quality and can be used efficiently for integrating EV dc power into grid. THD present in 3-level MLI is usually high (67.69%), its line-to-line voltage waveform is shown in figure 11.

Figure 9. Line-to-line voltage waveform of 5-level capacitor clamped MLI

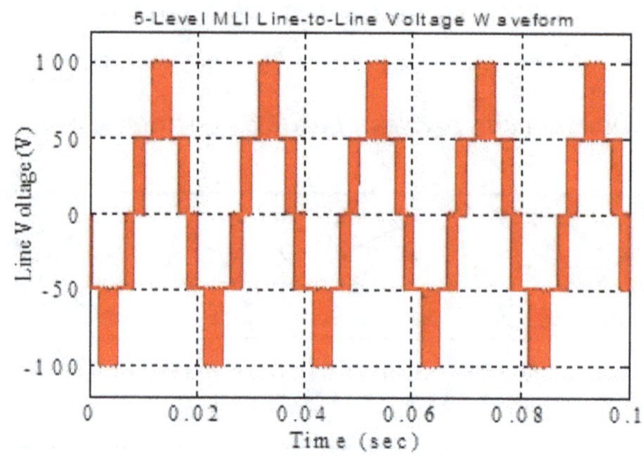

Figure 10. Line-to-line voltage waveform of asymmetrical MLI

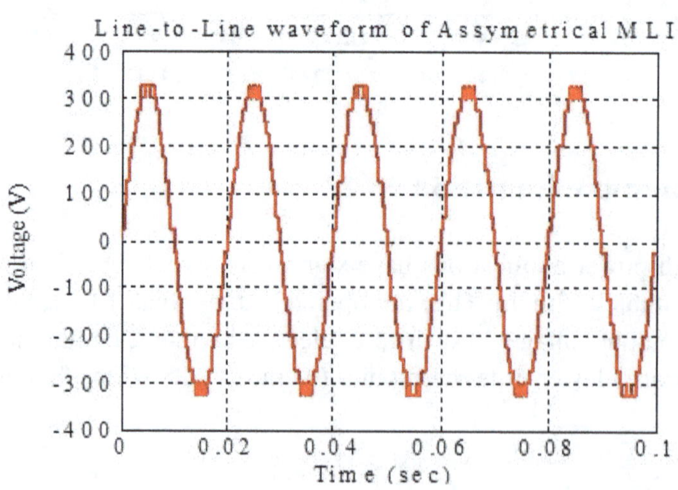

Figure 11. Line-to-line voltage waveform of asymmetrical MLI

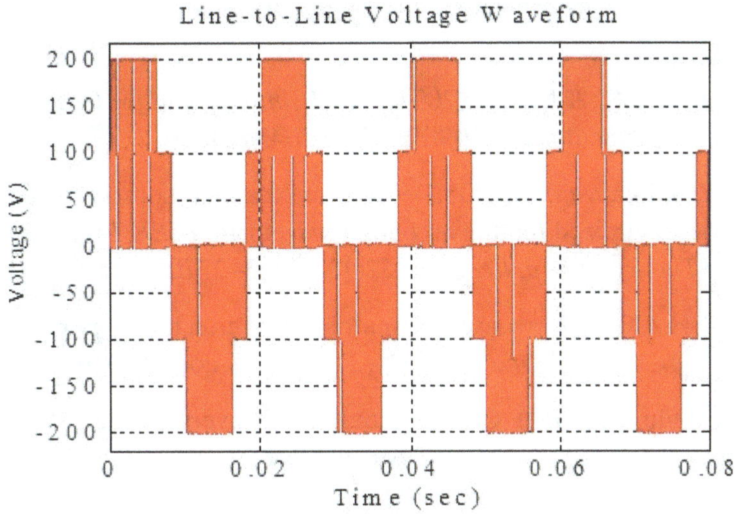

5 CONCLUSION

This chapter is focused on the need of multi-level inverters (MLI) for integration of dc power of EV into grid. The various topologies for the MLI are presented and are also discussed in detail, along with their key features. Various modulation techniques used for obtaining required power quality in the voltage waveform of MLI are also discussed. In the last part of the chapter the simulation results of the voltage waveforms are shown which are being obtained from MATLAB®/Simulink based MLI models.

REFERENCES

Abdullah, R., Rahim, N. A., Sheikh Raihan, S. R., & Ahmad, A. Z. (2014, October). Five-level diode-clamped inverter with three-level boost converter. *IEEE Transactions on Industrial Electronics*, *61*(10), 5155–5163. doi:10.1109/TIE.2013.2297315

Alishah, D. N. R. S., & Hosseini, S. H. (2014). Novel topologies for symmetric, asymmetric, and cascade switched-diode mutilevel converter with minimum number of power electronic components. *IEEE Transactions on Industrial Electronics*, *61*(10), 5300–5310. doi:10.1109/TIE.2013.2297300

Alishah, R. S., Hosseini, S. H., Babaei, E., & Sabahi, M. (2017). Optimal design of new cascaded switch-ladder multilevel inverter structure. *IEEE Transactions on Industrial Electronics*, *64*(3), 2072–2080. doi:10.1109/TIE.2016.2627019

Alishah, R. S., Nazarpour, D., & Hosseini, S. H. (2015). Reduction of power electronic elements in multilevel converters using a new cascade structure. *IEEE Transactions on Industrial Electronics*, *62*(1), 256–269. doi:10.1109/TIE.2014.2331012

Alishah, R. S., Nazarpour, D., Hosseini, S. H., & Sabahi, M. (2013). New hybrid structure for multilevel inverter with fewer number of components for high-voltage levels. *IET Power Electronics*, *7*(1), 96–104. doi:10.1049/iet-pel.2013.0156

Angulo, M., Lezana, P., Kouro, S., & Mar, S. (2007). *Level-shifted PWM for Cascaded Multilevel Inverters with Even Power Distribution*. doi:10.1109/PESC.2007.4342382

Axelrod, B., Berkovich, Y., & Ioinovici, A. (2005, December). A cascade boost switched-capacitor-converter two-level inverter with an optimized multilevel output waveform. *IEEE Transactions on Circuits and Systems. I, Fundamental Theory and Applications*, *52*(12), 2763–2770. doi:10.1109/TCSI.2005.852205

Babaei, E., & Hosseini, S. H. (2009). New cascaded multilevel inverter topology with minimum number of switches. *Elsevier J. Energy Convers. Manage*, *50*(11), 2761–2767. doi:10.1016/j.enconman.2009.06.032

Babaei, E., Laali, S., & Alilu, S. (2014). Cascaded multilevel inverter with series connection of novel H-bridge basic units. *IEEE Transactions on Industrial Electronics*, *61*(12), 6664–6671. doi:10.1109/TIE.2014.2316264

Banaei, M. R., Jannati Oskuee, M. R., & Khounjahan, H. (2014, May). Reconfiguration of semi-cascaded multilevel inverter to improve systems performance parameters. *IET Power Electronics*, *7*(5), 1106–1112. doi:10.1049/iet-pel.2013.0277

Banaei, M. R., & Salary, E. (2011). Verification of new family for cascade multilevel inverter switch reduction of components. *Journal of Electrical Engineering & Technology*, *6*(2), 245–254. doi:10.5370/JEET.2011.6.2.245

Barzegarkhoo, R., Moradzadeh, M., Zamiri, E., Madadi Kojabadi, H., & Blaabjerg, F. (2018). A New Boost Switched-Capacitor Multilevel Converter with Reduced Circuit Devices. *IEEE Transactions on Power Electronics*, *33*(8), 6738–6754. doi:10.1109/TPEL.2017.2751419

Boora, K., & Kumar, J. (2017). General topology for asymmetrical multilevel inverter with reduced number of switches. IET Power Electronics, 10(15), 2034-2041. doi:10.1049/iet-pel.2016.1011

Hinago, Y., & Koizumi, H. (2012, February). A Switched-Capacitor Inverter Using Series/Parallel Conversion With Inductive Load. *IEEE Transactions on Industrial Electronics*, *59*(2), 878–887. doi:10.1109/TIE.2011.2158768

Hosseini, S. H., Alishah, R. S., & Gharehkoushan, A. Z. (2015). Enhancement of extracted maximum power from partially shaded multi-string PV panels using a new cascaded high step-up DC-DC-AC converter. *2015 9th International Conference on Electrical and Electronics Engineering (ELECO)*, 644-648. 10.1109/ELECO.2015.7394619

Kangarlu, M. F., Babaei, E., & Laali, S. (2012). Symmetric multilevel inverter with reduced components based on non-insulated dc voltage sources. *IET Power Electronics*, *5*(5), 571–581. doi:10.1049/iet-pel.2011.0263

Karaca, H. (2015). GA Based Selective Harmonic Elimination for Multilevel Inverter with Reduced Number of Switches (Vol. I). Academic Press.

Khazraei, M., Sepahvand, H., Corzine, K. A., & Ferdowsi, M. (2012, February). Active capacitor voltage balancing in single-phase flying-capacitor multilevel power converters. *IEEE Transactions on Industrial Electronics*, *59*(2), 769–778. doi:10.1109/TIE.2011.2157290

Khoukha, C. H., & Madjid, B. E. (2007). Multilevel selective harmonic elimination PWM technique in the nine level voltage inverter. *2007 International Conference on Computer Engineering & Systems*, 387-392. 10.1109/ICCES.2007.4447075

Kolomyjski, W., Malinowski, M., & Kazmierkowski, M. P. (2006). *Adaptive space vector modulator for three-level inverter-fed induction motor. 9th IEEE Int. Work. Adv. Motion Control.*

Lai & Peng. (1996). Multilevel converters-a new breed of power converters. *IAS '95. Conf. Rec. 1995 IEEE Ind. Appl. Conf. Thirtieth IAS Annu. Meet.*, *3*(3), 2348–2356.

Liu, J., Zhu, X., & Zeng, J. (n.d.). A Seven-level Inverter with Self-balancing and Low Voltage Stress. IEEE Journal of Emerging and Selected Topics in Power Electronics. doi:10.1109/JESTPE.2018.2879890

Mak, O. C., & Ioinovici, A. (1998, April). Switched-capacitor inverter with high power density and enhanced regulation capability. *IEEE Transactions on Circuits and Systems. I, Fundamental Theory and Applications*, *45*(4), 336–347. doi:10.1109/81.669056

Marco, S. W., & Chau Chanand, K. T. (2007, December). A New Switched-Capacitor Boost-Multilevel Inverter Using Partial Charging. *IEEE Transactions on Circuits and Wystems. II, Express Briefs*, *54*(12), 1145–1149. doi:10.1109/TCSII.2007.905352

Meshram, P. M., & Borghate, V. B. (2015, January). A Simplified Nearest Level Control (NLC) Voltage Balancing Method for Modular Multilevel Converter (MMC). *IEEE Transactions on Power Electronics*, *30*(1), 450–462. doi:10.1109/TPEL.2014.2317705

Nabae, I. T., & Akagi, H. (1981). A New Neutral-Point-Clamped PWM Inverter. *IEEE Transactions on Industry Applications*, *IA-17*(5), 518–523. doi:10.1109/TIA.1981.4503992

Nakagawa, Y., & Koizumi, H. (2017). A hybrid nine-level inverter with series/parallel conversion. *Proc. IEEE Int. Symp. Circuits Syst.*, 1-4. 10.1109/ISCAS.2017.8050652

Nakagawa, Y., & Koizumi, H. (2019, July). A Boost-Type Nine-Level Switched Capacitor Inverter. *IEEE Transactions on Power Electronics*, *34*(7), 6522–6532. doi:10.1109/TPEL.2018.2876158

Ngo, B.-B., Nguyen, M.-K., Kim, J.-H., & Zare, F. (2018). Single-phase multilevel inverter based on switched-capacitor structure. *IET Power Electronics*, *11*(11), 1858–1865. doi:10.1049/iet-pel.2017.0857

Roy, T., Sadhu, P. K., & Dasgupta, A. (2019, November). Cross-Switched Multilevel Inverter Using Novel Switched Capacitor Converters. *IEEE Transactions on Industrial Electronics*, *66*(11), 8521–8532. doi:10.1109/TIE.2018.2889632

Shalchi Alishah, R., Nazarpour, D., Hosseini, S. H., & Sabahi, M. (2014). New hybrid structure for multilevel inverter with fewer number of components for high-voltage levels. *IET Power Electronics*, *1*(7), 96–104. doi:10.1049/iet-pel.2013.0156

Shalchi Alishah, R., Sabahi, M., Nazarpour, D., & Hosseini, S. H. (2014). Novel multilevel inverter topologies for medium and high-voltage applications with lower values of blocked voltage by switches. *IET Power Electronics*, *7*(12), 3062–3071. doi:10.1049/iet-pel.2013.0670

Shukla, A., Ghosh, A., & Joshi, A. (2012, November). Control of dc capacitor voltages in diode-clamped multilevel inverter using bidirectional buck-boost choppers. *IET Power Electronics*, *5*(9), 1723–1732. doi:10.1049/iet-pel.2012.0237

Sreedhar. (n.d.). *Genetic Algorithm Based Selective Harmonic Elimination Technique For Multilevel Inverters With Unequal Voltage Sources*. Academic Press.

Taghvaie, J. A., & Rezanejad, M. (2018). A Self-Balanced Step-Up Multilevel Inverter Based on Switched-Capacitor Structure. *IEEE Transactions on Power Electronics*, *33*(1), 199–209. doi:10.1109/TPEL.2017.2669377

Trzynadlowski, A. M. (2016). Introduction to Modern Power Electronics. John Wiley & Sons.

Waltrich, G., & Barbi, I. (2010). Three-phase cascaded multilevel inverter using power cells with two inverter legs in series. *IEEE Transactions on Industrial Electronics*, *57*(8), 2605–2612. doi:10.1109/TIE.2010.2043040

Wang, F. (2002). Sine-triangle versus space-vector modulation for three-level PWM voltage77 source inverters. *IEEE Transactions on Industry Applications*, *38*(2), 500–506. doi:10.1109/28.993172

Wang, L., Wu, Q. H., & Tang, W. (2017, December). Novel Cascaded Switched-Diode Multilevel Inverter for Renewable Energy Integration. *IEEE Transactions on Energy Conversion*, *32*(4), 1574–1582. doi:10.1109/TEC.2017.2710352

Ye, Y., Cheng, K. W. E., Liu, J., & Ding, K. (2014, December). A Step-Up Switched-Capacitor Multilevel Inverter With Self-Voltage Balancing. *IEEE Transactions on Industrial Electronics*, *61*(12), 6672–6680. doi:10.1109/TIE.2014.2314052

Chapter 10
Battery Swapping Station:
A Potential Solution to Address the Limitations of EV Charging Infrastructure

Furkan Ahmad
Center for Automotive Research and Tribology (CART), Indian Institute of Technology, Delhi, India

Essam A. Al-Ammar
King Saud University, Saudi Arabia

Ibrahim Alsaidan
College of Engineering, Qassim University, Saudi Arabia

ABSTRACT

State-of-the-art research to solve the grid congestion due to EVs is focused on smart charging and using (centralized, de-centralized, vehicle-to-grid) stationery energy storage as a buffer between times of peak and off-peak demand. On the other hand, the charging of EVs introduces new challenges and opportunities. This can prove to be beneficial for the EV aggregator as well as to consumers, regarding the economy. Also, EV as distributed storage makes the grid more steady, secure, and resilient by regulating frequency and the spinning reserve as backup power. However, the charging time and range anxiety lead to peak challenges for the use of EVs. In this chapter battery swapping station (BSS) as solution to the EV charging station is discussed.

1 INTRODUCTION

Issues like climate change, energy security, and declining fossil fuel reserves have caught the attention of the present world economies (Singh et al. 2020). The transportation sector is a major contributor to these issues. With the transportation sector being a cornerstone of the economic development of a country, efforts are being done to make it sustainable in the long run. A part of the solution is to replace the polluting and fossil fuel-based fleet with environmental friendly Electric Vehicles (EVs). Apart from reducing emissions, EVs offer reduced dependency on foreign economies for procurement of fossil

DOI: 10.4018/978-1-7998-6858-3.ch010

fuel (oil), thus increasing the energy security of a country. Apart from the economic and environmental benefits, EVs have numerous technical advantages also. One of the unique advantages is the technology of Vehicle-to-Grid (V2G), in which EVs act as dynamic Energy Storage System and assist the grid with services like peak load shaving, load profile regulation, active and reactive power support, spinning reserve, energy backup, frequency regulation etc. However, mass penetration of EVs also increase the load on the grid. The present practice of uncoordinated charging leads to overloading of the grid and increase in peak load as the charging period of EVs generally coincide with the evening peak period of a residential area. Implementation of smart EV charging strategies offer multiple advantages. Not only the battery life of EV is enhanced but helps in regulating voltage and frequency (Asaad et al. 2017b, a; Shariff et al. 2019). The smart charging strategies comprises of optimization of one or many of following objective functions viz.:

- Integration of renewable based generation
- Minimization of peak load, power and energy cost
- Aggregator profit maximization
- Load profile regulation and demand side management
- Operating cost reduction
- Increase in grid reliability

According to (Amjad et al. 2018), the following factors determines the design of smart EV charging strategy:

Duration of charging: The period and duration of charging (amount of energy consumption) is generally related to EV owner's choice and incentives for charging during non-peak load period.

Mobility of EV: The parking and moving of EV also decides the location, time and duration of charging.

Charging methods: For EV charging, the centralized or decentralized methods are commonly used. Based on power level, fast charging (higher power) and slow charging is employed.

EV to Grid connectivity: The unidirectional or bidirectional capability of V2G connection determines whether the EV will act only as a load or as both energy source and load.

EV Fleet Characteristics: Charging of a large EV fleet during peak load period drastically affect local grid voltage profile. Using demand side management to shift residential and commercial EV fleet charging to off-peak periods increases the dependability of grid.

EV Battery State of Charge: The state of charge (SOC) of EV batteries are estimated to properly schedule the charging of EVs. Information pertaining to batteries like remaining power, charging level (fast or slow), depth of discharge, operating temperature etc. are communicated to central entity for this purpose.

Power requirements of charging station: The fast (or commercial) charging station require comparatively higher power (higher current and voltage) compared to slow (or residential) charging station. Proper optimization of EV charging results in elimination of fluctuations in grid voltage and current.

2 PEAK LOAD SHAVING

The conventional way of dealing with peak load is capacity addition or use of small capacity backup sources like gas and diesel-based plants. However, these methods are uneconomical due to underutilization of total capacity as peak period occurrs for only a short duration compared to whole day. Other disadvantages are high operating cost, damage to equipments and high emissions (Ahmad and Alam 2017, 2018a, b, 2019; Amrr et al. 2018). As a result, the peak load shaving is preferred to deal with these disadvantages. The benefits of peak load shaving are:

Power Quality Enhancement: Imbalance of demand and supply results in voltage fluctuations, brownouts and blackouts (Grigsby 2017). Peak load shaving enhances the overall power quality of supply.

Efficient Energy Management: As load factor (measure of energy consumption variability) is inversely proportional to peak load power, so peak load shaving techniques result in high load factor which reduces the energy cost. Also, peak load shaving reduces high peak current flow thereby reducing high I2R losses of transmission line (Kalkhambkar et al. 2016).

Cost Reduction: Typically, power plants are designed to accommodate peak load. However, since peak load period only occurs for a brief duration, so power plants are under utilized for most of the time. Peak load shaving increases system efficiency and utilization, reduces on capital expenditure, fuel and maintenance cost. (Malinowski and Kaderly 2004) lists further economic benefits of peak load shaving.

Grid Reliability Enhancement: Using Energy Storage Systems (batteries and EVs) for mitigating the peak load period enhances the reliability of the grid.

Consumer Cost Reduction: To meet the peak demand, less efficient generators (gas or diesel based) are used which increase the cost of electricity generation at peak period. This cost is ultimately borne by consumers. By shifting usage of their non-essential loads to off-peak period, the consumers can save on their electricity bill and enjoy other non-economical benefits like uninterrupted supply and increased power quality.

Emission Reduction: Extra amount of fuel is consumed in frequent start-stop operation of peak load generators which produces more emission. Generally, peak load generators are of high polluting nature (gas or diesel based). Peak load shaving reduces variability in load and ensures reduced emission operation of power plants. For peak load shaving, three strategies are most common:

- Using Energy Storage System
- Using Electric Vehicles
- Demand Side Management

3 ESTIMATED COST OF CHARGING INFRASTRUCTURE

Total investment cost required for the establishment of charging infrastructure for EVs includes

- The cost of equipments to be used,
- Installation costs
- Operation and maintenance costs.

Further, the establishment cost incorporates the expense of common works, exchange costs in regards to dispersion framework administrator consent, and other related expenses relying upon factors like the prerequisite of another network association or up-degree of the current association. Notwithstanding, with an expansion in infiltration of EVs in the following barely any years, the quantity of EV chargers will increment, and henceforth the hardware cost is relied upon to diminish.

- For semi-private/semi-open spots where low or medium force level chargers are required, cost shifts between 500 € (Rs 36,431) and 1200 € (Rs 87,435).
- For open spots where high force level chargers are required, the establishment cost is generally higher extending between 2400 € (Rs 1, 74,871) to 3600 € (Rs 2, 62,306).
- For that activity and support cost is to be included, which might be taken to be 10% of the absolute establishment cost (counting gear cost)

4 CHALLENGES IN USING EV

The EV increases the difficulty of distribution network planning. New requirements as power request and the format of charging stations should be thought of, which add to the intricacy of system arranging. Improved quality supplies with high evaluations are required in the appropriation system to encourage the interconnection of the charging foundation. It requires a conveyance transformer with a bigger limit and appropriation line of the bigger cross area to stay away from issues like over-burdening, voltage deviation, and so on. It might prompt a diminishing in the economy of circulation framework activity. As the charging load displays enormous unpredictability, it is hard to keep charging conduct to low load periods, prompting more prominent framework top distinction. This would at last outcome in lower usage productivity of appropriation organize hardware. Force nature of the circulation arrangement is influenced. The charger utilizes a few force electronic transformation gadgets as converters which instigate sounds in the source side current. Further the other challenges are as follows (Ahmad et al. 2017; Khan et al. 2018):

Availability of EV: With the constraint that EV can deliver energy only when parked, availability of EV is a major challenge. The small market share of EVs further compounds this problem.

Synchronization: The time and place based uncertain availability of EVs makes synchronizing a large no of EVs for charging/discharging very challenging.

User reluctance: For significant peak load support, a substantial no of EVs are required as single EV has very low power capability compared to required peak demand. Acceptance of large no of users to allow access to their EV battery power to someone else is a big challenge. Fast degradation of battery due to frequent charge/discharge cycles needs to be offset by incentives provided to participate in peak load shaving.

Infrastructure: Due to limited usage of EVs, the present parking infrastructure lacks support for V2G technologies. Implementing infrastructure for integrating EVs into grid is a challenge for highly urbanized space-constraint areas.

Control Complexity: With the wide deployment of EVs in future, controlling bidirectional power between EV and grid with numerous constraints e.g., battery characteristics, weather, user preference etc. is going to be very complex. The system operation control will be further complexed with implementation of V2G features e.g., peak load shaving, demand response.

In general, EV charging requires a long charging process. Up to this point, because of arrangement and cash requirements, the charging stations, charging heaps and other charging foundation are not generally conveyed. The previously mentioned reasons make it plausible that EV clients will be compelled to stop and pause, which brings about holding up tension.

What's more, EV clients exchange off between the rest of the battery vitality, the area dispersion of charging offices, and their itinerary items, which effectively brings about range nervousness.

Hence, more specialists and EV administrators are directing their concentration toward battery trading. Battery trading can give another completely energized battery, which doesn't require draining the vitality of the old battery (Khan et al. 2017). This chapter is based on the authors extended work published in (Ahmad et al. 2020)

5. DIFFERENT BATTERY SWAPPING TECHNIQUES

The different swapping techniques are discriminated on the basis of the position of the battery in the vehicle and the point of application of the robotic arm and are given as follows:

Sideways Swapping: This is generally utilized if there should arise an occurrence of vans and different vehicles where the sideways position is the most advantageous.

Back Swapping: This is used for the vehicles in which the battery is set from in reverse. For the most part on account of vehicles with huge boot space.

Base Swapping: This is utilized for vehicles whose battery is put at the base of the vehicle. The trading station is worked in the way with the end goal that the vehicle is put on a raised stage and the batteries are traded from the base utilizing an automated arm and different extras for the most part, put underneath the ground level.

Top Swapping: This is most usually utilized for the electric transports wherein the batteries are set at the top and upon the appearance of the transport, the housetop opens and the trading is finished by methods for the mechanical arm(Ahmad et al. 2020).

6 FUNCTIONING OF THE SWAPPING STATION

So as to set up and effectively execute the trading strategy for transports, vans and different vehicles, broad arranging should be finished covering all the essential necessities going from the accessibility of the batteries, chargers to information stockpiling and the board through cloud and correspondence between the parts to guarantee between operability. The BSS for the vehicles will work effectively gave there is nonstop correspondence between the various parts of the framework viz. the keen vehicle, trading station and the data framework. The data framework will be utilized to speak with the vehicle just as with the station. The vehicle will speak with the data framework through the WAVE correspondence while the station will use the nearby web for the correspondence. At the point when the vehicle's battery is drained, the data framework will get the notice from the vehicle, mentioning a battery trading administration. The data framework will advise the station about the area regarding the vehicle, it's normal appearance time and its recognizable pieces of proof with the goal that it readies the battery when the vehicle arrives at the station. When the vehicle arrives at the station, the driver swipes his participation card and the data framework finds all the significant information related to this card. This information incorporates data concerning the vehicle, battery, vehicle's trading history, exchanges finished, and other appropriate data. Working staff will confirm this information and direct the client towards the trading region where the necessary battery exchanging will occur by methods for an automated arm. All the information is being supported upon distributed storage from where it will be available to the specialists just as clients to guarantee total straightforwardness inside the methods. When the trading is done, the vehicle's battery will be checked promptly for its condition of charge, condition of wellbeing, remaining charge, battery age, battery life, number of charging and releasing cycles experienced and as needs are the proprietor will be educated about the charges acquired and the assessed time by which he can get his completely energized battery. The mechanical arms will successively put the battery inside the charging racks in which each space has the capacity to charge the battery. The electric transport battery pack has a battery management system (BMS) that screens wellbeing, voltage, and temperature of the individual cells for charge security and adjusting. The HV battery connector will be appended physically and just when the connector and the bay have been fixed the charging starts. The shut contact between the connector and the bay will be shielded by methods for a hook which will permit the current to stream once the contract is finished. Additionally, the trading station can perform a bi-directional force stream with the network.

Burden bend is improved by the way toward charging in off pinnacle hours and infusion of intensity into the matrix during top hours by the put-away batteries. This procedure in the end expands the general effectiveness of intensity frameworks. So as to create income from its charging/providing designs, the trading station will flexibly when the interest is high (significant expense) and obtain when the interest is (low value), this planning of intensity will likewise be a generous factor while considering the monetary attainability of the trading stations.

Dispersion transformer, AC-DC converters, battery chargers, vehicle batteries, automated arms, charging racks, upkeep framework, control framework are the central segments of a BSS alongside other hardware fundamental for trading and charging of batteries. There exist two potential areas for the converters in the BSS. In the first, we have a focal AC-DC converter associated legitimately to the conveyance transformer, giving DC capacity to all the chargers. In the subsequent setup, we have AC-DC converters set locally at every one of the chargers giving DC power. Both the designs are doable however for the most part, the last one is favored in light of the fact that it gives the benefit of operability regardless of whether any of the converters aren't inactivity state. Likewise, the brought together AC-DC converter will a huge one as far as size, weight, and cost which may raise issues during establishment.

As appeared in Figure 1, any force dissemination substation (1) can likewise be checked by said BSS control room (2). The said control room (2) can control or potentially manage as well as screen the general working of the said BSS. The said charging racks (3) outfitted with any moderate as well as quick charging techniques able EVSEs combined with the said battery racks (4). The said trading track (5) can encourage any coordinations and additionally wellbeing check or potentially manual as well as programmed unscrewing and additionally screwing the battery of any xEV. The said trading path (6) can be any fixed and additionally moving path from passage point to leave the purpose of the said BSS. The said trading robot (7) can encourage any xEV released battery evacuation as well as getting any xEV perfect charged battery from said battery rack as well as establishment of any battery pack on any xEV through any battery trading strategy on the said trading path (6). The said battery/charger support room (8) can encourage any upkeep or potentially fix as and when required for any battery as well as the charger. The said BSS support room (9) can encourage any upkeep and additionally housekeeping as and when required for the perfect as well as ideal activity of the said BSS (Ahmad et al. 2020).

The swapping station works on the first-in-first-out standard for example the vehicle showing up first will be given the need for their trading administrations when contrasted with the individuals who show up later. So as to go to the interest of the considerable number of vehicles moving toward the trading stations, it is fundamental that the batteries be charged following trading so they are accessible to another showing up clients. This incorporated activity of stations will improve the vehicle for its course, permitting the vehicle to amplify its excursion with insignificant slack. Battery charging production lines and BSS fills in as a base for the present EV battery swapping frameworks (Ahmad et al. 2020).

Figure 1. A typical BSS processing unit

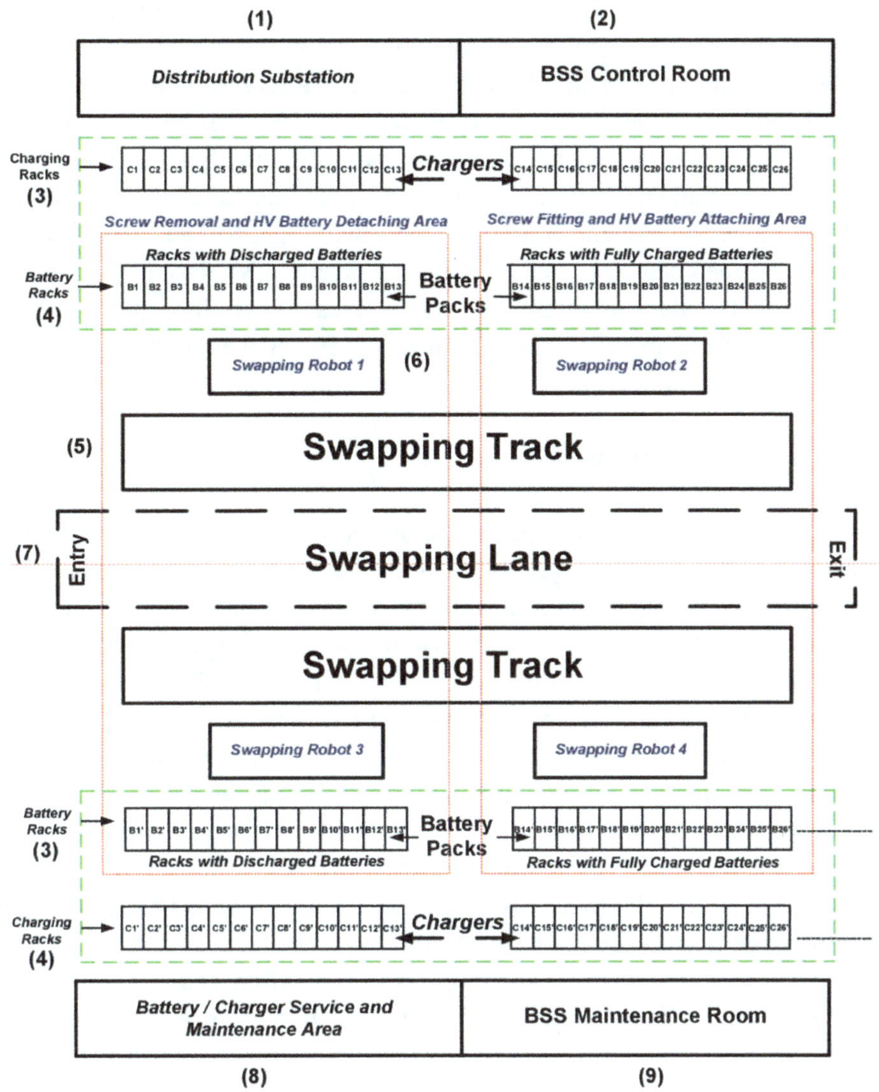

A coordinations framework is commonly utilized for the transportation of an enormous number of midway charged batteries. Battery coordinations procedure, BSS arranging, and development technique, and battery charging system for the BSSs are getting a great deal of consideration for investigation. Every one of these methodologies doesn't expect to fulfill the objective of "get vitality recharging whenever and anyplace rather they are planned to grow the inclusion and administration of a battery trading framework. Because of different limitations like land area, restricted accessibility of BSSs, swarm at stations, and different variables, there is a need to structure a progressively versatile and viable EV battery trading engineering. The detached mode doesn't understand the goal of "get vitality renewal whenever and anyplace." EV clients must head to a battery BSS for battery trading. Because of the undeniable limitations of area and number of the current battery BB, there may even now be a queueing and holding up marvel (Ahmad et al. 2020).

However, inactive battery swapping mode, the portability of the battery trading van evacuates the imperatives of area and amount of EV battery trading stations. Progressing to dynamic battery trading mode from the existing battery trading mode is the potential method to take care of this issue. Late writing additionally talks about the improvement of a novel quick EV battery trading gadget. The adaptable plan of this gadget can change over a van into a versatile EV BSS when introduced. This gadget exploits its segments like a situating pin, situating opening, situating track, situating space and PLC control framework for conveying exact situating and helpful trading. The procedure just takes a couple of moments because of its equal expulsion and establishment strategy for battery trading. Numerically it is just 3 mins in trial arrangements. A programmed battery trade is finished by driving EVs into the foreordained situation in battery trading van through collapsing incline steel. This fixes the area limitations and thus expands the amount of EV BSSs and add adaptability to the framework (Ahmad et al. 2020).

7. BSSS FOR PUBLIC TRANSPORT

One of the most vital and generally utilized open transportation vehicles is a downtown explorer transport. Because of their adaptability, monetary possibility, and capacity to work over every single imaginable course with fewer systems for upkeeps/administrations, they are utilized over practically all the courses inside a city. The trademark highlights of the electric transport are expressed as follows: right off the bat, the transports run over predefined courses with a predefined plan. Besides, the transports work all through the administration hours and stop just when any glitch happens. Thirdly, the transports have little league breaks between their excursions, which is at one time a solitary outing is finished transport is promptly arranged for the following outing with insignificant stoppage time. This is done to expand the arrangement of the armada transports without stopping them at the warehouses for extensive stretches. These highlights can convince just as discourage the idea of zap of the transports. The benefits of utilizing electric transport can be that the transport utility can be advanced to fulfill the timetable with to a limited degree with insignificant mischief to the earth.

One grave issue that would require consideration would be of charging the batteries once they are drained. This is the place the battery trading procedure turns into an imperative prerequisite of the transportation framework to maintain a strategic distance from the slacks in plan and streamline the utility of the transports (Mahoor et al. 2017; Ahmad et al. 2020). The merits of swapping technique for electric bus are as follows:

- The Battery Swapping procedure is a quick and effective strategy when contrasted with the diverse accessible vehicle charging plans, along these lines permitting the buyer to proceed with the drive absent a lot of postponement.
- The Battery Swapping strategy diminishes the client holding up time just as delays the battery life (better battery science) when contrasted with those which experience the quick charging plan (50 kW and that's just the beginning) as the battery is charged at the ideal voltage for a long term.
- Bidirectional power stream is joined in BSSs for association with the lattice. The heap bend is improved by the way toward charging in off-top hours and infusion of intensity into the matrix during top hours by the put away batteries. This procedure inevitably builds the general effectiveness of intensity frameworks. This is likewise increasingly practical for station proprietors when they have enough charged batteries for supporting their needs.

- The Battery Swapping procedure for the open travel electric transports would permit the improvement of these vehicles to fulfill their calendar to a limited degree with insignificant damage to nature. The battery pack of an operational electric transport has a limit running from 320kWh up to 590 kWh. Indeed, even the far reaching organization of quick DC charging (rating up to 240 kW, 200A, 480V) would expend adequate time that would in the long run disturb the ordinary activity of the transportation framework. This is the place the battery trading procedure turns into a key necessity of the transportation framework to maintain a strategic distance from the slacks in timetable and improve the utility of the transports.
- The work of the Battery Swapping strategy will be a financially savvy measure for the transport proprietors just as the station proprietors in view of commonly useful cost decreases emerging from utilities by energizing batteries on occasion of low burden and from improved efficiencies.
- It is, in this way, the goal of this examination is to give the nitty gritty commonly useful, condition agreeable, best in class and monetarily reasonable framework of the BSSs for the rock solid EVs.

8 BSS BENEFITS

A. Customer Perspective

The battery is accepted as the costlier piece of the client's view for having an EV. The talked about and anticipated BSS plan will permit clients to rent the battery rather than a protuberance venture which was evaluated between $12000 to $14000 for substitution of 24kWh in 2012. Long charging occasions, up-degree of family establishments to high power chargers, and the set number of open charging stations are some other concerning parts for EV proprietors. Level 1 chargers of a run of the mill rating between 1-6 kW are utilized for direct association in family units which takes 15 h for totally charging a 24-kWh battery. Be that as it may, up-gradation options for a less energizing time of to 7h are accessible as Level II chargers with a force rating of 3.3 kW. Be that as it may, the up-gradation cost is additional speculation and is evaluated as $849 in 2013. Regardless of whether proprietors will consent to contribute this sum then likewise they won't have the option to utilize it completely because of the restrictions in family unit current gracefully (Sarker et al. 2015).

The generally little size of the batteries additionally acts as a requirement for EV clients and it results principally to a restricted range. Simple access of proprietors to BSSs can wipe out these issues. In any case the conventional strategies are not practical as far as an overwhelming framework when contrasted with the proposed versatile BSSs. (Sarker et al. 2015; Ahmad et al. 2019a).

B. Power System Benefits

With the quickly developing EV advertise the requirement for quick chargers can turn into the prime worry of shoppers. So as to keep with the progressions in innovation they will be required to move up to the most recent accessible charger to fulfill their day by day vitality needs. In this angle, BSSs are clearly prevalent with the conventional charging foundations as the overhaul is just required at the BSS site though in the customary structure the update is to be made on the two sides i.e., at the family unit level and furthermore at the charging station. Moreover, as the framework is a pack of batteries, it can infuse power once again into the lattice, smoothing the net every day request bend and can offer types of assistance to the framework in general. This very element of the proposed BSS separated from being a capacity gadget offers extra advantages like recurrence guideline, load following, and willful hold arrangements (Sarker et al. 2015; Ahmad et al. 2019a).

9. BSS CHALLENGES

How about we talk about battery swapping, the fundamental advantage related with the trading model is speed. The entire activity could take under five minutes, basically a similar measure of time numerous individuals spend filling their gas tanks at stations today. Another in addition to isn't leaving your vehicle or manage conceivably tangled or messy lines. You could sit in your vehicle while the activity happens. In any case, the image is more unpredictable than this, and underneath the difficulties related with BSS has been picturized in Figure 2. Further, it has been clarified in detail (Ahmad et al. 2020).

Figure 2. Challenges of BSS

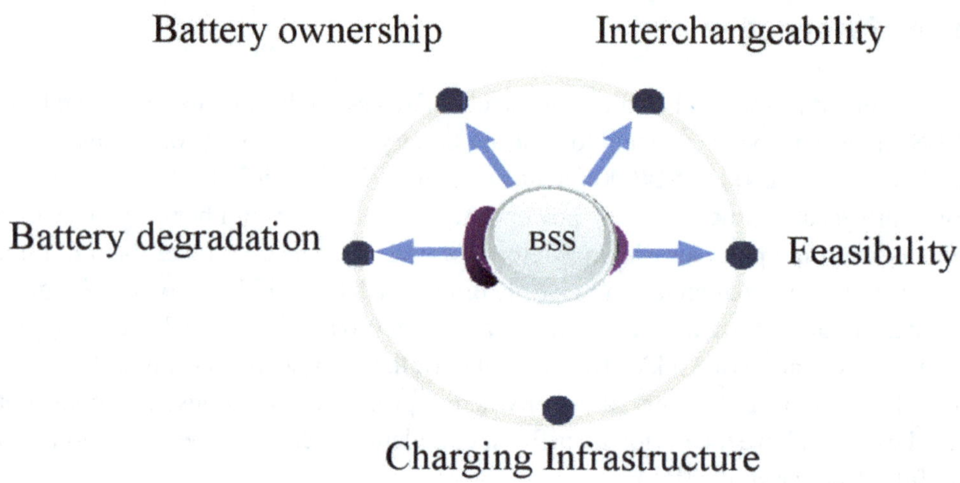

A. Interchangeability

Brand Compatibility and Cross-Platform highlights guarantee the accomplishment of innovation and make it a prevailing pattern in the market. The accessibility of comparative tradable battery packs of various producers are the main any expectation of a battery trading framework to stand firm as an essential alternative. This straightforward arrangement is reliant on the producer's assent. Be that as it may, this may confine the development, uniqueness, and adaptability of items planned solely by any maker. Besides, it will likewise confine item improvement as cell makers, dwelling at the highest point of the item chain, should configuration normalized cells so as to give comparable cells to next level assembling element, just through which further normalization can be delivered. One can contend that there may be diverse force sections for the battery pack, however this would then make an issue with gracefully and request and furthermore, conceivable similarity issues with vehicles not intended for sequential fueled battery packs.

B. Feasibility

Battery configuration is as of now acting like the most noteworthy obstacle of battery trading innovation. The structure of battery should consolidate heartiness in expelling and re-introducing it from the vehicle. In India, just hardly any vehicles are by and by offering such battery pack structures. For instance, Hero Maxiallows released battery to be traded effectively with charged battery and the released battery can be put to charging independently (Asaad et al. 2017a; Ahmad et al. 2018, 2019b).

C. Infrastructure

The foundation required for the battery packs is monstrous and considerably more unpredictable and costly than charging. BSSs in examination with charging stations apply a similar extent of interest with the prime necessity of charging the battery packs. The sole contrast is that the interest of the previous can be controlled (even in a pragmatic situation, the battery packs ought to consistently be accessible for sequential approaching clients and thus should offer to charge time in like manner). Further, to satisfy the day by day necessity of any station, the stacked quantum of charged battery packs ought to consistently surpass by a specific level of everyday request. Along these lines, the attainable procedure is to keep one battery pack inside the vehicle and another at a trading station i.e., two battery packs for each vehicle. Significantly in the wake of sifting through this it despite everything appears to be a removed future, as the financial matters of an across the nation battery trading framework isn't talking in favor when contrasted and an across the nation vehicle charging framework (Ahmad et al. 2019b).

D. Battery Degradation

The greatest battery charge extend is adversely influenced by the debasement of its presentation after some time. Thus the new battery packs will be increasingly preferred more in clients over the choice of other generally old battery packs as they will offer low vitality stockpiling because of corruption after some time and this will reflect in the mileage of EVs. Thus the substance of clients will lie just in new battery packs, which extensively, will lessen the working pattern of any battery pack (Sarker et al. 2015).

E. Battery Ownership

The vehicle proprietor possesses the battery: For this situation the proprietor should buy an extra battery pack which can be utilized as a substitution battery when the battery in the vehicle gets released. Which is less inclined to occur! The responsibility for the battery pack will consistently be invalid for a vehicle proprietor. This offers a few points of interest. Since the vehicle proprietor doesn't possess it he doesn't need to pay for it, which makes EV less expensive. Despite what might be expected, he is never recovering the traded battery and needs to pay an extra rent sum other than the expense of vitality. This can be charged on each trading or some other commonly concurred recurrence, a state on month. The later strategy for rent mount is exorbitant considering the least sum is comprehensive of least two battery packs and administration charge of trading station (while just a single battery is purchased if the vehicle is charged and has one claimed battery). The rent on-each-trade choice is considerably costlier (as trading stations will be a raised degree of speculation chances) and will twist clients for less use of trading stations to stay away from these extreme charges, particularly when a charging choice is accessible (Ahmad et al. 2020).

10 CONCLUSION

BSS systems are emerging as a promising option in contrast to the conventional battery charging station approach since they give a more extensive arrangement of business chances to the committed partners. This work has introduced the multidimensional parts of a perfect BSS sending. The total framework, strategies and advantages of BSS as for customer, power framework administrator and aggregator has been introduced. Further, the key difficulties related to the BSS, for example, Interchangeability, practicality, framework, battery corruption and battery possession are examined in detail.

REFERENCES

Ahmad, F., Alam, M., Alsaidan, I., & Shariff, S. M. (2020). *Battery Swapping Station for Electric Vehicles: Opportunities and Challenges*. IET Smart Grid. doi:10.1049/iet-stg.2019.0059

Ahmad, F., & Alam, M. S. (2017). Feasibility study, design and implementation of smart polygeneration microgrid at AMU. *Sustainable Cities and Society*, *35*, 309–322. doi:10.1016/j.scs.2017.08.007

Ahmad, F., & Alam, M. S. (2018a). Economic and Ecological Aspects for Microgrids Deployment in India. *Sustainable Cities and Society*, *37*, 407–419. doi:10.1016/j.scs.2017.11.027

Ahmad, F., & Alam, M. S. (2018b). Optimal Sizing and Analysis of Solar PV, Wind, and Energy Storage Hybrid System for Campus Microgrid. *Smart Sci*, *6*(2), 150–157. doi:10.1080/23080477.2017.1417005

Ahmad, F., & Alam, M. S. (2019). Assessment of power exchange based electricity market in India. *Energy Strateg Rev*, *23*, 163–177. doi:10.1016/j.esr.2018.12.012

Ahmad, F., Alam, M. S., & Asaad, M. (2017). Developments in xEVs charging infrastructure and energy management system for smart microgrids including xEVs. *Sustainable Cities and Society*, *35*, 552–564. doi:10.1016/j.scs.2017.09.008

Ahmad, F., Alam, M. S., & Shahidehpour, M. (2018). Optimal placement of electric, hybrid and plug-in hybrid electric vehicles (xEVs) in Indian power market. 2017 Saudi Arabia Smart Grid Conference, 1–7.

Ahmad, F., Alam, M.S., & Shariff, S.M. (2019a). A Cost-efficient Energy Management System for Battery Swapping Station. *IEEE Syst J*, 1–10. doi:10.1109/JSYST.2018.2890569

Ahmad, F., Member, S., & Alam, M. S. (2019b). A Cost-Efficient Approach to EV Charging Station Integrated Community Microgrid : A Case Study of Indian Power Market. *IEEE Trans Transp Electrif*, *5*(1), 200–214. doi:10.1109/TTE.2019.2893766

Amjad, M., Ahmad, A., Husain, M., & Umer, T. (2018). A review of EVs charging : From the perspective of energy optimization, optimization approaches, and charging techniques. *Transportation Research*, *62*(Part D), 386–417. doi:10.1016/j.trd.2018.03.006

Amrr, S. M., Alam, M. S., Asghar, M. S. J., & Ahmad, F. (2018). Low cost residential microgrid system based home to grid (H2G) back up power management. *Sustainable Cities and Society*, *36*, 204–214. doi:10.1016/j.scs.2017.10.016

Asaad, M., Ahmad, F., Alam, M. S., & Rafat, Y. (2017a). IoT Enabled Monitoring of an Optimized Electric Vehicle's Battery System. *Mobile Networks and Applications*, *23*(4), 994–1005. doi:10.100711036-017-0957-z

Asaad, M., Ahmad, F., & Saad Alam, M. (2017b). *IoT Enabled Electric Vehicle's Battery Monitoring System. EAI*. SGIOT. doi:10.4108/eai.7-8-2017.152984

Grigsby, L.L. (2017). *Power system stability and control*. Academic Press.

Kalkhambkar, V., Kumar, R., & Bhakar, R. (2016). Energy loss minimization through peak shaving using energy storage. *Perspectives on Science*, *8*, 162–165. doi:10.1016/j.pisc.2016.04.022

Khan, S., Ahmad, A., Ahmad, F., Shafaati Shemami, M., Saad Alam, M., & Khateeb, S. (2017). A Comprehensive Review on Solar Powered Electric Vehicle Charging System. *Smart Sci*, *6*(1), 54–79. doi:10.1080/23080477.2017.1419054

Khan, W., Ahmad, A., Ahmad, F., & Saad Alam, M. (2018). A Comprehensive Review of Fast Charging Infrastructure for Electric Vehicles. *Smart Sci*, *6*, 1–15. doi:10.1080/23080477.2017.1419054

Mahoor, M., Hosseini, Z. S., & Khodaei, A. D. K. (2017). Electric Vehicle Battery Swapping Station. *CIGRE Grid of the Future Symposium*, 1–5.

Malinowski, J., & Kaderly, K. (2004). Peak shaving - A method to reduce utility costs. *IEEE Region 5 Conference: Annual Technical and Leadership Workshop*.

Sarker, M. R., Pandžić, H., & Ortega-Vazquez, M. A. (2015). Optimal operation and services scheduling for an electric vehicle battery swapping station. *IEEE Transactions on Power Systems*, *30*(2), 901–910. doi:10.1109/TPWRS.2014.2331560

Shariff, S.M., Alam, M.S., & Ahmad, F. (2019). System Design and Realization of a Solar-Powered Electric Vehicle Charging Station. *IEEE Syst J*, 1–12. doi:10.1109/JSYST.2019.2931880

Singh, K., Amir, M., & Member, S. (2020). *An Integral Tilt Derivative Control Strategy for Frequency Control in Multimicrogrid System*. Academic Press.

Chapter 11
Drivers of EV Charging Infrastructure Entrepreneurship in India

Mohd Yasir Arafat
Aligarh Muslim University, India

Imran Saleem
Aligarh Muslim University, India

Thoudam Prabha Devi
University of Buraimi, Oman

ABSTRACT

The existing research advocating entrepreneurship as an important way to increase the uptake of electric vehicles (EVs) in developing countries and EV charging business is also playing a crucial role in increasing the adoption of EVs. EV charging is important for EV adoption, and entrepreneurship is also important for EV adoption; therefore, it is important that we must understand what mobilizes or prevents EV charging entrepreneurship. This chapter aimed at explaining drivers of EV charging entrepreneurship. A survey of 121 potential entrepreneurs shows that personal attitude, self-confidence, and opportunity perceptions shape the decision to engage in EV charging entrepreneurship. Policy measures to boost EV charging entrepreneurship have been suggested.

INTRODUCTION

The market share of Electric vehicles (EVs) has grown tremendously in India. In recent times, EVs have been more popular than other types of products, in part because of action by the Indian government to build out openly accessible charging infrastructure and give some tax exemption for electric vehicles (EVs). The Ministry of Heavy Industry is more interested in exploring ways to fostering new business creation that will expand private sector commercialization of EV charging services.

DOI: 10.4018/978-1-7998-6858-3.ch011

Even though the EV charging industry has huge market potential, EV charging business in Indian and elsewhere have not attracted significant investment because very few entrepreneurs have entered into the EV charging business. To explore how new start-ups could be boosted for the EV charging business, this study aimed at capturing drivers of new venture creation in the context of EV charging business. So that effective policies and programs can be developed.

Although, entrepreneurship is one of the important solutions to Electric Vehicle technology challenges in developing economies (Rady, 1992; Balachandra, 2011). However, apart from this realization, and what are the factors that promote or inhibit entrepreneurship in the EV industry especially, EV charging business, is missing from existing literature. Besides, three important points from the extant literature show the need for this research. First, extant research on EV entrepreneurship has given attention to the business model only, for example, customer engagement strategies (Loock, 2012), product and services (Causey et al., 2009), channels of distribution (Richter, 2012, 2013). Previous research deals with the success and performance of the business firm created that is more of managerial competence start-up process. This research rather, focuses firm creation process.

Second, most of the studies have focused on the challenges and problems being faced by the existing entrepreneurs of the EV charging business. This research focuses on the perceived challenges or barriers to starting a business. In this way, this research enhances our understanding of the nascent entrepreneurship.

Third, to encourage EV charging entrepreneurship, it is imperative to understand the perspective of the entrepreneur; what motivates or discourages entrepreneurs to start and run business in developing economies. Thus, in this way, this study will contribute a better understanding to drivers of EV charging entrepreneurship in developing countries. As the mindset of the people towards entrepreneurship in developing countries is different from their counterparts in the developed world; hence, by interviewing would-be entrepreneurs and by compiling the antecedents of their decision to become EV charging entrepreneurs, this paper discusses the potential policy measures to boost EV charging entrepreneurship.

This study examines the factors which may motivate or prevent starting an EV charging business. For this purpose, we are using a cognitive framework to explain the entrepreneurial phenomenon. Since the cognitive theoretical framework to study entrepreneurial activity has received much attention from entrepreneurship researchers and many studies have proved its theoretical soundness and robustness, hence, it is better to use this framework. Numerous factors are studied under the cognitive framework; we have chosen those which suit the objective and the respondents of this study.

This manuscript begins with brief descriptions of types of EVs and charging technology and business options available. The next section discusses the theoretical framework and hypotheses to be tested. In the next section, we present the methodology used to complete the study. The final section discusses the major findings and policy implications to boost EV charging entrepreneurship.

TYPES OF EVS AND CHARGING TECHNOLOGIES

Three types of EVs are prevailing in the market: (1) **Battery Electric Vehicle (BEV):** this type of EV is supported by battery pack only, (2) **Extended Range Electric Vehicle (EREV):** It is a combination of battery-electric vehicle and IC engine is also known range extender, (3) **Plug-in Hybrid Vehicle (PHEV):** This type of EVs are a combination of both EV and conventional with the addition of rechargeable battery system.

Similar to EV types, three types of charging technologies are also available. An EV can be recharged at three levels in decreasing order: DC fast charging, AC Level 2, and AC Level 1. AC Level charging has a very low cost of installation.

TYPES OF EV CHARGING ENTERPRISES

EV Mobility Operator (EMO)

All those enterprises which are running their EV charging point as core business activity or part of a major project.

Charging Point Network Operator (CPNO)

This type of business is operating charging point network and also provides maintenance service. More importantly, it looks after the network, therefore, it works for electric vehicle drivers. Sometimes, charging point network operators white label their services for charging brands or other enterprises.

Charging Point Services Operator (CPSO)

Charging point services operator is not commonly used but expected to grow in the future. This type of charging business refers to white label services.

Electric Vehicle Equipment Supplier (EVES)

Sometimes manufacturers of EV chargers are called as electric vehicle equipment supplier. Their core business is to manufacture EV charger, charging point, connectors, cables, etc., and supply them.

Charging Point Management System (CPMS)

This is a software management business that controls charging points. EV mobility operators, charging point network operators, and charging point services providers rely on CPMS software.

In this study, we define EV charging entrepreneurship as creating, sustaining and growth of new venture that is focused on developing, designing, manufacturing, and distributing EV charging service and products which includes all facets of EV charging value chain and also planning, installation, financing, consulting, installing maintenance and end of life management and disposal. Thus, this definition of EV charging entrepreneur implies an individual who starts or owns an EV charging business.

Conceptual Framework and Hypothesis

New entrepreneurial ventures transpire as a result of well calculated, thought out, and planned behavior. Therefore, entrepreneurs' new venture creations are considered illustrations of planned behavior. Previous research conducted proposes intentions are the single best forecaster of planned behavior (Learned, 1992). The electric vehicle industry provides immense opportunities for potential entrepreneurs. Entrepreneurship and the electric vehicles industry are complex and multifaceted; however, captivating when combined in research.

The business creation trend is analyzed by a variety of disciplines (psychology, management, economics, and sociology, etc.), and the substantial amount of models taken up, shows that none prevails and most of them bear some drawbacks. The recent trends in entrepreneurship research showed that researchers have shifted their focus towards the cognitive aspects of entrepreneurship. Thus, this research considers cognitive aspects of new venture creation, with the aim to describe the determinants of people's inclination to start a business in the Electric Vehicle (EV) industry or EV charging business. The independent variables affecting the EV charging entrepreneurship propensity are modeled containing factors describing perceptions of the individuals about the environment, fear of failure entrepreneurial self-efficacy, social norms, and whether the entrepreneur perceives EV industry as a viable opportunity. The hypothesized model is shown in Figure 1.

Figure 1. Hypothesized EV Charging Entrepreneurship Model

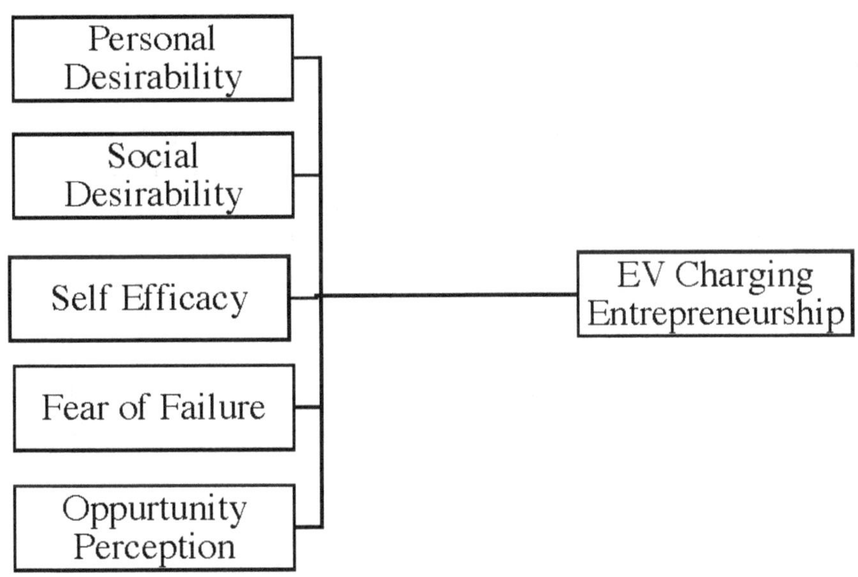

Personal Desirability or Professional Attraction

Personal desirability refers to the attractiveness of new business creation for an individual. If an individual has a positive attitude toward creating a new business or feel an attractive and desirable career, he is likely to create a new venture because any (planned) behavior is determined by attitude toward the behavior (Ajzen 1991). It has also been supported by previous studies that personal attraction has a positive impact on venturing decision (Ferreira et al. 2012; Liñan, Urbano, and Guerrero 2011; Arafat and Saleem 2017). Thus, the above discussion brings us to formulate following hypothesis:

Hypothesis 1: Personal (desirability) attraction towards entrepreneurship has a positive effect on those who want to start an EV charging business.

Social Desirability or Norms

Ajzen (1991) has argued that, intention to perform a particular activity is affected by the prevailing social norms or desirability. Social norms are the perceived societal force to carry out or not to carry out an activity. These are normative beliefs that an individual holds about the approval or disapproval of an activity or behavior by referent others (Ajzen 1991). The main referent others are members of family, colleagues, friends, and significant others. Social norms are also defined as social desirability (Krueger and Brazeal 1995; Guerrero et al 2008). Kolvereid and Isaksen (2006) conducted a study on the Norwegian context and established that social norms are having a positive influence start-up propensity or intention and many others (Roy, Akhtar & Das 2017; Karimi, et al., 2017) have also found the same results. In this way, the literature highlights a positive link between social desirability or norms and start-up. We, therefore, formulate the following hypothesis:

Hypothesis 2: Social desirability or norms have a positive impact on entrepreneurial propensity of creating EV charging firm.

Self-Efficacy or Entrepreneurial Ability

The planned behavior theory postulates, if an individual believes in his entrepreneurial ability; knowledge, and skills, he would perceive starting a business as a feasible option, and this perceived feasibility will shape the intention to become entrepreneur. The construct feasibility perception is also related Bandura's (1977) concept of self-efficacy. The concept of self-efficacy also similar to the concept of perceived behavioral control (Ajzen 1991). According to Armitage and Conner (2001), self-efficacy is "confidence in one's own ability to carry out a particular behavior" and perceived behavioral control is "perceived ease or difficulty of performing behavior". In the words of Ajzen, "Self-efficacy beliefs can influence the choice of activities, preparation for an activity, effort expended during the performance, as well as thought patterns and emotional reactions" (p. 184). Applying to the context of charging business, the following hypothesis is formulated:

Hypothesis 3: Self-efficacy or entrepreneurial ability has a positive influence on the propensity to start a new business in the EV industry.

Fear of Failure or Risk Perception

Fear of failure is also an important influencer of entrepreneurship or may prevent people from starting their EV charging business. The risk perception or fear of failure is a multifaceted, yet understudied construct (Cacciotti and Hayton, 2015). According to Arenius and Minniti (2005) risk perception or fear of failure is an expected risk of facing failure when undertaking an entrepreneurial activity. Many scholars (Arafat and Saleem, 2017; Ramos-Rodríguez, Medina-Garrido, and Ruiz-Navarro 2012; Fernández, Liñán, and Santos 2009) have found that fear of failure is negatively related to entrepreneurship. Therefore, people who have fear of failure their likelihood of starting a business, especially, EV charging business lower. This discussion leads to propose the hypothesis as following:

Hypothesis 4: Fear of failure or risk perception has a negative impact on the propensity to start an EV charging business.

Perception of Economic or Entrepreneurial Opportunity

According to the planned behavior theory, the action or behavior of people is affected by the attitude they have. Also, Ajzen (1991, p.188) defines attitude as "the degree to which a person has a favorable or unfavorable evaluation or appraisal of the behavior in question". When people see favorable entrepreneurial opportunities, they assess their ability to make use of these opportunities. This nurture a positive attitude towards the start of a business, consequently, affects entrepreneurial decision and action. Similarly, we contend that the perception of business opportunity in EV Charging business will affect positively the decision to engage in the EV charging business. Therefore, the following hypothesis is developed:

Hypothesis 5: Perception of entrepreneurial opportunities is positively related with the propensity to start a new EV charging business.

METHODOLOGY

In this study, a sample of students is used who are likely to choose their career as an entrepreneur just after their education is completed. We are using a convenient sampling which also commonly used in similar studies (Ferreira et al. 2012; Liñán and Chen 2009; Krueger, Reilly, and Carsrud 2000; Fayolle and Gailly 2015). For the data collection, we visited a central university of India which represents students from all geographical areas. We distributed questionnaires among the students of Electrical Engineering background because they are much aware of the EVs and its charging system. The response rate was 60%. Questionnaires that were not in complete form or inconsistent were rejected. Only 121 questionnaires were usable.

Logistic Regression

This research, employ the binomial logistic regression techniques to examine the effect of a series of predictor variables on the dependent variable. Due to the following reasons, we are employing this technique:

1. It doesn't consider the distribution of data.
2. The dependent variable is binary.
3. All the independent variables are also binary.

FINDINGS AND DISCUSSIONS

This section of this manuscript is divided into three sections the sections include; descriptive statistics, statistical tests, and a summary of the results. The descriptive statistics recognize the dependent and independent variables, the statistical testing established whether or not the suggested hypotheses were accepted or rejected, and finally a summary of the findings.

Although there was no control over how many males and females would respond, of 121 valid responses a total of 33 female students responded in comparison to 88 male students. Therefore, 27.3% of the respondents were females and 72.7% were males. This provided an impartial split in terms of the distribution of men and women in the electric vehicle industry. Although there were a larger percentage of male participants, this may be attributed to females' career choice, as the data was collected from the engineering students, in India the ratio of female students in engineering and technology is lower.

Descriptive Statistics

Descriptive statistics demonstrate that 34 percent of the respondents are willing to start their electric vehicle (EV) charging business, 62 percent have a positive attitude (desirous) towards starting their own business in the EV industry, 54 percent considered new start-up in EV industry as a socially desirable career, only 30 percent respondents are confident (or they feel that they have the ability to start-up) in their ability to start a business in the EV industry, 61 percent respondents perceive risk or think the risk perception or fear of failure may refrain them from starting their own business, and 61 percent perceive good entrepreneurial opportunities for new businesses of electric vehicle charging.

The result reveals that females have, on average, lower level of attraction towards new business start-ups in the EV charging industry. In case of, social desirability towards entrepreneurship in EV charging, males and females both have the same attitude. Self-efficacy or entrepreneurial ability to run a business of electric vehicle charging among males is two times more than that of females. In case of fear of failure or risk perception, female respondents are more risk-averse than their male counterparts. However, results highlight that women respondents see fewer opportunities than male students in the EV industry.

Regression Analysis

We ran logistic regression analysis to test the hypotheses developed in the previous section.

Table 1. Personal Desirability or Professional Attraction

	B	S.E.	Wald	df	Sig.	Exp(B)
Personal desirability	2.027	.527	14.794	1	.000	7.590
Constant	-2.079	.474	19.218	1	.000	.125

The result shows that the regression coefficient of Desirability has a positive sign. Moreover, the odds ratio of this variable is 7.59, which indicates students who are desirous of starting a business their likelihood to EV Charging business is eight times than the rest of the students. This finding is in congruence with previous research of Arafat, Saleem, & Dwivedi (2020), Arafat et al. (2020), Khan et al (2020).

Table 2. Social Desirability or Norms

	B	S.E.	Wald	df	Sig.	Exp(B)
Social desirability	.613	.393	2.435	1	.119	1.846
Constant	-.981	.303	10.495	1	.001	.375

The result indicates that social desirability has no significant impact on EV Charging Business creation propensity. This result also confirms the previous findings of Linan and Chen (2009).

Table 3. Self-Efficacy or Entrepreneurial Ability

	B	S.E.	Wald	df	Sig.	Exp(B)
Self efficacy	.696	.407	2.923	1	.087	2.006
Constant	-.859	.239	12.947	1	.000	.424

The result demonstrates that the individuals' entrepreneurial ability or confidence in their skills and knowledge have a significant and positive relationship with the propensity of becoming entrepreneur in the EV charging industry. The result shows the odds ratio for this variable is 2.006, which means individuals who feel that they have the capability of starting a business or entrepreneurial self-efficacy their likelihood of starting EV charging business is two times higher than of others. This result also confirms the previous findings of Linan and Chen (2009).

Table 4. Fear of Failure or Risk Perception

	B	S.E.	Wald	Df	Sig.	Exp(B)
Fear of failure	-.012	.282	.002	1	.965	.988
Constant	-.623	.273	5.206	1	.023	.536

The result indicates that fear of failure has no significant effect on entrepreneurial propensity to start a new business. This reveals that fear of failure is not a detractor to EV charging business. This result also confirms the previous findings of Arafat et. (2020).

Table 5. Perception of Economic or Entrepreneurial Opportunities

	B	S.E.	Wald	Df	Sig.	Exp(B)
Opportunity perception	1.482	.642	5.326	1	.021	4.400
Constant	-1.482	.495	8.943	1	.003	.227

The result shows that the regression coefficient of perception of economic opportunities has a positive sign. The odds value for this factor is 4.400, which shows people who see better economic opportunities in starting a business in the EV industry are four times more likely to do so than the rest of the students. This result also confirms the previous findings of (Khan, Arafat, Raushan & Saleem).

Based on the above results, the factors which have a significant influence on the propensity to start a new EV charging business are personal desirability, self-efficacy, and perception of entrepreneurial opportunities.

Implications

The results indicate that the social desirability, the coefficient of the perceived social acceptability variable in the model, is non-significant. Therefore, according to the results, there does not exist significant link between social desirability and the probability of being an entrepreneur in the EV industry. This may be due to less awareness of EV technology in society. Also, fear of failure does not prevent individuals to start an EV charging business. The result also indicates that self-efficacy has a positive and significant effect on start-up propensity which means starting an EV charging business is not under volitional control. This finding suggests people in India face less munificent environment for start –up in EV industry. According to Kolvereid and Isaksen (2006), if self-efficacy exerts strong effect on start-up propensity or behavior it means the national environment is less favorable for entrepreneurs.

Based on these results, the policymakers should

- frame policies and programs which focus on increasing personal attraction towards EV industry.
- increase awareness of EV technology in society through education, media etc.
- promote success stories of EV entrepreneurship around the globe.
- create and promote various schemes and policies focused on ease of doing business in EV industry.
- concentrate efforts to enable potential entrepreneurs to develop the skills essential for starting an EV charging business.

CONCLUSION

Research about entrepreneurship is a quite new and emerging field. Same can be said about studies in the EV charging business industry, which is unchartered territory. For understanding the dynamics of individual's intention of starting a business in the EV charging industry, this work is a first step. To study the influence of different cognitive perceptions on the propensity to be an EV charging entrepreneur, various cognitive variables have been examined and evaluated in this work. For this purpose, a survey of potential individuals likely to start new venture in EV industry was conducted. A sizeable sample of students of Electrical Engineering at Aligarh Muslim University was surveyed. We employed binomial logistic regression technique to test the hypothesized relationship. The findings suggest that the individuals who have attraction towards sustainable transportation, have confidence in their own knowledge and skills and perceive better economic opportunities in starting a business in electric mobility are more likely to become an EV entrepreneur.

REFERENCES

Ajzen, I. (1991). The theory of planned behavior. *Organizational Behavior and Human Decision Processes*, *50*(2), 179–211. doi:10.1016/0749-5978(91)90020-T

Arafat, M. Y., Khan, A. M., Saleem, I., Khan, N. A., & Khan, M. M. (2020). Intellectual and cognitive aspects of women entrepreneurs in India. *International Journal of Knowledge Management Studies*, *11*(3), 278–297. doi:10.1504/IJKMS.2020.109092

Arafat, M. Y., & Saleem, I. (2017). Examining start-up intention of Indians through cognitive approach: A study using GEM data. *Journal of Global Entrepreneurship Research*, *7*(1), 1–11. doi:10.118640497-017-0073-3

Arafat, M. Y., Saleem, I., & Dwivedi, A. K. (2020). Understanding entrepreneurial intention among Indian youth aspiring for self-employment. *International Journal of Knowledge and Learning*, *13*(3), 185–200. doi:10.1504/IJKL.2020.109882

Arafat, M. Y., Saleem, I., Dwivedi, A. K., & Khan, A. (2020). Determinants of agricultural entrepreneurship: A GEM data based study. *The International Entrepreneurship and Management Journal*, *16*(1), 345–370. doi:10.100711365-018-0536-1

Arenius, P., & Minniti, M. (2005). Perceptual variables and nascent entrepreneurship. *Small Business Economics*, *24*(3), 233–247. doi:10.100711187-005-1984-x

Armitage, C. J., & Conner, M. (2001). Efficacy of the theory of planned behaviour: A meta-analytic review. *British Journal of Social Psychology*, *40*(4), 471–499. doi:10.1348/014466601164939 PMID:11795063

Balachandra, P. (2011). Modern energy access to all in rural India: An integrated implementation strategy. *Energy Policy*, *39*(12), 7803–7814. doi:10.1016/j.enpol.2011.09.026

Bandura, A. (1977). Self-efficacy: Toward a unifying theory of behavioral change. *Psychological Review*, *84*(2), 191–215. doi:10.1037/0033-295X.84.2.191 PMID:847061

Bosma, N., Acs, Z. J., Autio, E., Coduras, A., & Levie, J. (2008). *Global entrepreneurship monitor executive report*. Babson Park.

Cacciotti, G., & Hayton, J. C. (2015). Fear and entrepreneurship: A review and research agenda. *International Journal of Management Reviews*, *17*(2), 165–190. doi:10.1111/ijmr.12052

Change, I. C. (2014). *Mitigation of climate change*. Contribution of Working Group III to the Fifth Assessment Report of the Intergovernmental Panel on Climate Change, 1454.

Delhi, S. I. N. (2015). Fame-India Scheme—Putting E-Mobility on Road. *Auto Tech Review*, *5*(4), 22–27.

Ehsani, M., Gao, Y., Longo, S., & Ebrahimi, K. (2018). *Modern electric, hybrid electric, and fuel cell vehicles*. CRC Press.

Fayolle, A., & Gailly, B. (2015). The impact of entrepreneurship education on entrepreneurial attitudes and intention: Hysteresis and persistence. *Journal of Small Business Management*, *53*(1), 75–93. doi:10.1111/jsbm.12065

Fernández, J., Liñán, F., & Santos, F. J. (2009). Cognitive aspects of potential entrepreneurs in Southern and Northern Europe: An analysis using GEM-data. *Revista de Economía Mundial*, (23), 151–178.

Ferreira, J. J., Raposo, M. L., Rodrigues, R. G., Dinis, A., & do Paço, A. (2012). A model of entrepreneurial intention: An application of the psychological and behavioral approaches. *Journal of Small Business and Enterprise Development.*

Gabriel, C. A., & Kirkwood, J. (2016). Business models for model businesses: Lessons from renewable energy entrepreneurs in developing countries. *Energy Policy*, *95*, 336–349. doi:10.1016/j.enpol.2016.05.006

Gulati, V. (2012). *National electric mobility mission plan 2020.* Department of Heavy Industry, Ministry of Heavy Industries & Public Enterprises, Government of India.

Gupta, M., Rafat, Y., & Alam, M. S. (2017). *Well to wheel cum tailpipe emission analysis: ICE vs xEV* (No. 2017-26-0107). SAE Technical Paper.

Høyer, K. G. (2008). The history of alternative fuels in transportation: The case of electric and hybrid cars. *Utilities Policy*, *16*(2), 63–71. doi:10.1016/j.jup.2007.11.001

India's Current Account Deficit to Touch $30 Billion in 2018. (2017). *The Economic Times.*

Khan, A. M., Arafat, M. Y., Raushan, M. A., Khan, M. S., Dwivedi, A. K., Khan, M. M., & Saleem, I. (2019). Role of intellectual capital in augmenting the start-up intentions of Indians—An analysis using GEM data. *Journal of Global Entrepreneurship Research*, *9*(1), 1–14.

Khan, A. M., Arafat, M. Y., Raushan, M. A., & Saleem, I. (2020). Examining the relevance of intellectual capital in improving the entrepreneurial propensity among Indians. *International Journal of Knowledge Management*, *16*(1), 123–140.

Kolvereid, L., & Isaksen, E. (2006). New business start-up and subsequent entry into self-employment. *Journal of Business Venturing*, *21*(6), 866–885.

Krueger, N. F., Reilly, M. D., & Carsrud, A. L. (2000). Entrepreneurial intentions: A competing models approach. *Journal of Business Venturing*, *15*(5/6), 411–432.

Liñán, F., & Chen, Y. W. (2009). Development and cross–cultural application of a specific instrument to measure entrepreneurial intentions. *Entrepreneurship Theory and Practice*, *33*(3), 593–617.

Liñán, F., Urbano, D., & Guerrero, M. (2011). Regional variations in entrepreneurial cognitions: Start-up intentions of university students in Spain. *Entrepreneurship and Regional Development*, *23*(3-4), 187–215.

Loock, M. (2012). Going beyond best technology and lowest price: On renewable energy investors' preference for service-driven business models. *Energy Policy*, *40*, 21–27.

Matulka, R. (2014). The history of the electric car. US Department of Energy.

Rady, H. M. (1992). Renewable energy in rural areas of developing countries: Some recommendations for a sustainable strategy. *Energy Policy*, *20*(6), 581–588.

Ramos-Rodríguez, A. R., Medina-Garrido, J. A., & Ruiz-Navarro, J. (2012). Determinants of hotels and restaurants entrepreneurship: A study using GEM data. *International Journal of Hospitality Management, 31*(2), 579–587.

Richter, M. (2012). Utilities' business models for renewable energy: A review. *Renewable & Sustainable Energy Reviews, 16*(5), 2483–2493.

Richter, M. (2013). Business model innovation for sustainable energy: German utilities and renewable energy. *Energy Policy, 62*, 1226–1237.

Shapero, A., & Sokol, L. (1982). *The social dimensions of entrepreneurship*. University of Illinois at Urbana-Champaign's Academy for Entrepreneurial Leadership Historical Research Reference in Entrepreneurship.

Society of Manufacturers of Electric Vehicles. (n.d.). *EV Industry*. Author.

Chapter 12
Cryptocurrency-Based Payment Framework for EV Charging

Wasif Karim
Aligarh Muslim University, India

Mohammad Saad Alam
Aligarh Muslim University, India

Yasser Rafat
CARET, Aligarh Muslim University, India

ABSTRACT

The world is transforming from paper or plastic currency towards digital currency. Everyone is trying to benefit from the decentralized system, the anonymity, and lower transaction cost offered by cryptocurrencies. One of the hurdles in the ascension of electric vehicle is its charging and the transaction for charging. In this chapter, a cryptocurrency-based payment framework is formulated for electric vehicle charging to eliminate the third-party involvement as in the plastic currency system. In this payment framework, smart contracts are deployed which would ensure the conditions to be satisfied, like the amount sent by the consumer. Smart contract includes the transfer function that would transfer the amount of ether and set charging time according to the charging rate, amount of ether sent, and power delivered. Overview of the platform prototype of payment is presented to demonstrate the implementation of the framework in the real world. At last, the authors look at the security issues, challenges ahead, and also suggested some of the improvements that could be further done.

I. INTRODUCTION

The world is looking forward to a transaction mode without involvement of any third party. Cryptocurrency, being a decentralized system (Nakamoto, 2008), is paving this way. It provides security with the advancing blockchain technology and the distributed ledger system. In the sequential chain of blocks, it stores the complete history of transactions. This provides transparency to the system (Bunjaku et

DOI: 10.4018/978-1-7998-6858-3.ch012

al., 2017). With the unique address for each transaction, complete anonymity is relished by companies (Bunjaku et al., 2017), (Lee, 2019).

These advancing features are magnetizing the users (Mahmood, 2021) and the world is moving towards the transformation from the centralized banking system to the decentralized cryptocurrency. At present, the total market value of all kinds of cryptocurrency has crossed USD 200 billion (Xu et al., 2019).

This chapter is aimed at designing a cryptocurrency-based payment framework for charging electric vehicles.

The automated payment for the charging of electric vehicles at charging stations will lead to speedy transactions. The provision of anonymity (Lee, 2019), (Bunjaku et al., 2017) for its users also attracts users, leading to the prosperity of the industry of clean and fuel-independent electric vehicles.

This payment system would be using smart contract (Loi et al., 2016) for the conditions to be fulfilled before the transaction occurs. These conditions would include whether the sender has that amount of digital currency in his account, what amount of charging is going to be done, and may include conditions for different pricing for different time slots.

This smart contract is dealt with solidity (Wöhrer & Zdun, 2018) programming language that enables the user to execute a payable function which would transfer the preset amount from the customer's digital account to the company's account only if all the conditions provided in that function is being fulfilled and there is no any issue.

The charging rate can be done in two ways. In one case, the power for charging being fixed and the amount of ether being paid depends only on the time for which charging is being done. In the second case, the pricing may be according to the energy consumption, for which we have to do interfacing with something which can provide us the energy being consumed.

If the smart contract is linked with some online stuff, the conversion rate of different currencies of the world and the digital currencies is to be made available for customers in a public function of the contract.

For the pricing to be done on the basis of peak hour and early hour, real world time is to be synchronized with the smart contract and pricing for different time slots is to be done.

With this system of transaction for the charging of electric vehicles, we will be enjoying less transaction cost (Huang et al., 2021), elimination of the risk of charge-backs, no third-party involvement and quick transaction. This results in satisfaction for the electric vehicle users. It will lead to getting rid of the dependence on the exhaustible and anti-environment fuel-based vehicles.

II. BACKGROUND

A. Blockchain

Blockchain is an information chain that is time-stamped and linked in a systematic order.

Two main features for which this technology is being adopted or trying to be adopted are:

(1) Once some information is recorded inside of this blockchain, it becomes almost impossible to do any kind of alteration.
(2) It is a distributed ledger that is completely open to all the members of that network.

Working of Blockchain:

To understand the working of this technology, we have to get what this block contains. There are broadly four types of information which are contained in each block:

(1) Data
(2) Hash of the Block
(3) Hash of the previous Block
(4) Nonce

Data stored in the block depends on the type of blockchain and its use. If a transaction blockchain is used, then the data could be the information of the transaction like the address of sender and receiver, the amount of transaction, etc.

Hash is something like the identity of that block. It is generated at the time of creation of the block by hashing functions like SHA256, which generates a unique 256 bits hash and whenever there is some alteration in the data of the block, then the hash changes, which reflects that something has been altered. This hash provides security against any tampering in the block. Someone who calculates the hash and adds the block to the blockchain is called a Miner and he/she is rewarded for that job.

Hash of the previous block provides a way of systematic linking, which puts the thing in order. In the code, this is compared with the hash of the block prior to this block and if it is not the same, it gives error (Savjee, n.d.).

Now, to understand this, let us assume that there is a blockchain of Ethereum transactions which contains five blocks. Let someone on the network tries to cheat by tampering with the block numbered 3 which contains the info of 5 ether sent by someone to him. Now he tampers the amount from 5 to 50 ether. With this tampering, the data inside the block 3 will change and as we know the hash is calculated taking data as input. This leads to changing the hash of the block 3. The "hash of the previous block" stored in the block 4 is the initially-calculated hash i.e., hash of block 3 before the tampering. Upon comparing, the hash of block 3 which is now different from the "hash of the previous block" stored in block 4, gives error. This makes the block 4 invalid and further comparison makes the entire block following block 3 invalid.

A question here may rise, if we recalculate the hash of every block following block 3, then there wouldn't be any error. But in reality, the length of the blockchain is quite long. Even though, with fast computers, this calculation of hash for all the following blocks could be done significantly fast. To avoid this, there is a second layer of protection in this technology, which is 'Proof of Work', which limits the block creation speed significantly (How Blockchain Transactions Work, n.d.). For example, in bitcoin, creating a block takes 10 minutes.

To understand the concept of Proof of Work, let's take a simple example. If we add a condition that the block having the hash started with 4 zeros is only acceptable. Now for fulfilling this condition, the miners have to generate hash, then check whether it starts with 4 zeroes, if not, change the nonce (Nonce is a random number or string given to a block.) to some other random number and recalculate the hash and process continues until it starts with 4 zeroes. This whole process consumes a substantial time.

Apart from this, there is further a third protection layer, which is the blockchain being a distributed ledger. The blockchain is open to all the members of that network. Also, whenever a miner mines a block, it is not simply added to the blockchain. The block having the generated hash and nonce is put to be verified by more than 50% of the participants, and then it is added to the blockchain (Savjee, n.d.).

So, for any tampering with a block, one has to breach these three layers of protection, which is almost impossible. This technology is also advancing with passing of the days, adding more protection against tampering. Inclusion of the Smart Contract in the blockchain is one of the advancements to this technology.

B. Smart Contract

Smart Contract is a piece of code usually written in Solidity programming language and executed by Ethereum (Antonopoulos & Wood, n.d.) – a public decentralized infrastructure for computation. It can be treated as a contract in digital form (YouTube, n.d.).

In Smart Contract, the programmer puts the conditions and makes it mandatory to be fulfilled to carry on the implication of that contract. If any of the conditions put by the deployer is not being satisfied by the user, he/she would not be able to go through the function provided by that smart contract like transaction or transfer of any asset or may be other functionalities.

Now let it be clearer by taking an example of a Smart Voting with a smart contract. The deployer of this contract has to write a piece of code which would include the functions of that voting. First, he/she has to authorize the address (Address is a 256 bits hexadecimal digits representing a contract or an account created by a human on the Ethereum network (Antonopoulos & Wood, n.d.).) of some of the users which he/she wants to have the voting right. A public function of the candidature showing names and relevant information is to be declared. There would be a public function that is used to receive the vote from the authorized addresses, like we would put a check that votes from only those addresses which are authorized will be valid, else it will show 'unauthorized error'. Solidity language provides a "require" function that enables this check. Also, there would be another "require" function to put restrictions on the address which has already voted so that a voter does not vote more than once. After the specified time period, the voting is to be stopped and the result is declared.

So, through this smart contract, we can have fair and smart voting.

These days, Smart Contracts are being used enormously for IoT based smart home and intelligent transportation (Wang et al., n.d.).

But besides these advantages, there are some challenges that need to be dealt with. One of the major issues is that of privacy of contract data as all the things like bytecode or crucial methods are publicly open. To tackle this privacy issue, Watanabe in his paper (Watanabe, 2015), presented an encryption-decryption system. Prior to deploying the smart contract, it needs to be encrypted and the members having these secret decryption keys will only have access to the smart contract.

Irreversibility of the blockchain is also an issue in case of smart contract, as if a smart contract has been deployed on the blockchain and there is some bug in the code of the contract, then it can't be edited to remove the bug (Wang et al., n.d.). So, the deployer has to be very sure that there is no error in the code of the contract.

C. Solidity

Solidity is a contract-oriented programming language. It is a high-level language specifically developed for writing and implementing Smart Contracts (Solidity, n.d.). The syntax of this language is having much similarity with that of JavaScript. Other programming languages like JavaScript, Python, etc. could also be used for Smart Contracts, but as Solidity is specific to this job, it is highly recommended.

Furthermore, it also supports inheritance allowing the smart contract to inherit all the properties of other smart contracts.

Visibility of the function or variable can be controlled by the deployer using the built-in keywords like private, internal or public and accessibility of the function with the keywords view, pure or payable takes into account the privacy. The transaction of ether (the digital currency on Ethereum network) can be processed simply with 'transfer' keyword and to know the amount of ether available in an account, there is 'balance' keyword (Dannen, n.d.).

The best part of this language is that there is a website "remix.ethereum.org", which can be used to write the contract, test it, debug it, and even it can be deployed on a local blockchain network. So, there is no trouble of "set up" like in other programming languages, just go to the website and write your code. It also provides access to a number of accounts having 100 fake ethers in each account, which allows users to test any transaction.

III. SMART CONTRACT FOR PAYMENT OF EV CHARGING

In this section, a smart contract for the transaction of ether for payment of EV Charging is executed and analyzed. This smart contract will accept the amount of ether from the address which is calling the function for payment. If the charger is supplying a constant power, the smart contract sets charging-time as per the amount of ether transferred and the power supplied. Another case may be there if the power supplied varies, which is discussed in a later section. This section illustrates the smart contract in step, starting from the simple payment and then adding more functions.

A. Simple Payment

Smart Contract for a Simple Payment

```
pragma solidity  ^0.6.6;
contract SimplePayment {
    function Pay() external payable { }
    function UpdateBalance() external view returns (uint) {
                return (address(this).balance);
    }
}
```

The above smart contract named 'SimplePayment' simply transfers ether from the address of the caller of the function named 'Pay' to the address of the smart contract. Every solidity program begins with the pragma statement mentioning the version in which the code is written. The keyword 'external' denotes the visibility of the function to all the addresses on the network. The keyword 'payable' in the 'Pay' function transfers ether set in the 'VALUE' field, if it is being called. In this function, the amount of ether is not fixed by the contract. The caller of the function can transfer any amount of ether. But the condition for fixed ether or a minimum value can be provided by the 'require' function. The function named 'UpdateBalance' verifies the transaction by checking the balance of the smart contract through the 'balance' keyword. This function uses the keyword 'view' which means the data of the blockchain can't be changed, but only be read and the 'returns' keyword means it will return the unsigned integer (uint) upon calling.

In Ethereum, whenever there is some addition or changes in the blockchain through mining, it is referred to as 'transaction'.

If the above contract is deployed in the online Remix IDE, it is mined and the transaction receipt is generated. Fig.1 shows the transaction receipt which includes the transaction status, address of contract, hash of the transaction, address from which the contract is deployed and other data.

This transfer of 20 ether is also verified in the Fig.2 by calling the 'UpdateBalance' function which gives $20*10^{18}$ wei (wei is a denomination of ether, 1 wei = 10^{-18} ether; In Remix IDE, all the transactions are done in wei.). The 'ACCOUNT' field in Fig.2 shows that the amount of ether deducted from the address is 20 ether (for the 'Pay' function) plus some fraction of a ether, that is for the transaction and execution cost, which is shown in Fig.3 in the form of 'gas', which is convertible to ether.

Figure 1. Transaction Receipt generated by Remix IDE

[vm] from: 0x5b3...eddc4 to: SimplePayment.(constructor) value: 0 wei data: 0x608...60033 logs:

status	0x1 Transaction mined and execution succeed
transaction hash	0x9ce32b2eb17f502ea836f18bfff17012cc656b14d9290240ecb91becd37c1d8c
contract address	0xd9145cce52d386f254917e481eb44e9943f39138
from	0x5b38da6a701c568545dcfcb03fcb875f56beddc4
to	SimplePayment.(constructor)
gas	3000000 gas
transaction cost	95957 gas
execution cost	31081 gas
hash	0x9ce32b2eb17f502ea836f18bfff17012cc656b14d9290240ecb91becd37c1d8c
input	0x608...60033
decoded input	{}
decoded output	-
logs	[]
value	0 wei

Figure 2. Deploy and Run Transactions in Remix IDE

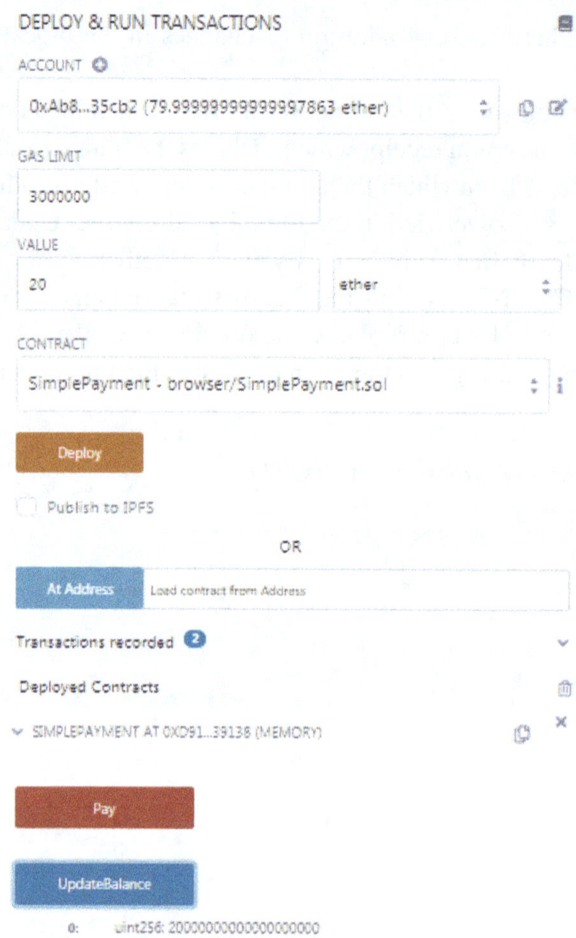

B. Smart Contract for Payment of Charging

This section illustrates the smart contract specifically for payment of EV charging. First, some of the state variables are to be declared. Charging rate is to be made public, it is declared as a public string type variable so as everyone has access to what amount is being charged per minute of charging. An integer type variable for charging rate is also declared, in units of szabo per minute (1 szabo = 10^{-6} ether), which could be used in further calculation. As mentioned earlier also, we are dealing with charging which is supplying constant power. So, this constant power supply is also declared as a public state variable and it is supposed to be 50 kW in this smart contract. Integer type public state variables for the amount sent to smart contract and the charging time (which would be calculated based on amount sent, power delivered and rate of charging fixed by the deployer of contract) are also declared.

For general instruction, a function named '_Instruction' is defined in the contract. This function returns some string type local variables which conveys the information like charging rate, which is set to 100 szabo (or 0.0001 ether) per minute for this contract. For clarity, the units of power supplied (kilowatt), charging time (second) and sent amount (szabo) are also mentioned in this function.

Figure 3. Transaction Receipt for Pay function called in Remix IDE

[vm] from: 0xab8...35cb2 to: SimplePayment.Pay() 0xd91...39138 value: 20000000000000000000 wei

status	0x1 Transaction mined and execution succeed
transaction hash	0xac96de459c9b25bb40a7c4d2c9854314e4f194934533d084ce4fd53bd45bc915
from	0xab8483f64d9c6d1ecf9b849ae677dd3315835cb2
to	SimplePayment.Pay() 0xd9145cce52d386f254917e481eb44e9943f39138
gas	3000000 gas
transaction cost	21370 gas
execution cost	98 gas
hash	0xac96de459c9b25bb40a7c4d2c9854314e4f194934533d084ce4fd53bd45bc915
input	0x537...a924c
decoded input	{}
decoded output	{}
logs	[]
value	20000000000000000000 wei

There should be some threshold for the charging, below which there may be inconvenience for the setup or may be uneconomical. For this purpose, the function also returns a string type variable for minimum time of charging, which is set to 10 minutes and this costs 1000 szabo or 0.001 ether.

Smart Contract for Payment of Charging

```
pragma solidity  ^0.6.6;
contract EVChargingPayment
{
     string public chargingRate = "100 szabo (= 0.0001 ether) per minute";
     uint chargingRate_in_szabo = 100;
//1 szabo = 10^-6 ether
     uint public powerSupplied = 50;
//if constant power of 50 kW is supplied
     uint public chargingTime;
     uint public sentAmount;
     function _Instruction() external view returns(string memory, string memo-
ry, string memory, string memory, string memory)
     {
```

```
            string memory chargingrate = " Charging Rate: 100 szabo (= 0.0001
ether) per minute";
            string memory powerUnit = "Power Supplied is in kilowatt.\n";
            string memory timeUnit = " Charging Time is in seconds. \n";
            string memory sentAmountUnit = "Sent Amount is in szabo (1 ether =
10^6 szabo). \n";
            string memory mininumTime = "There is a mininum charging time of 10
min (600 sec) which costs 0.001 ether
                                                            (equivalent to
approx 25 INR).\n";
            return (chargingrate, timeUnit, powerUnit, sentAmountUnit,  mininum-
Time);
      }
      function Pay() external payable
      {
            require (msg.value >=1e15 && msg.value <=1e18);
            sentAmount = msg.value/1e12;
//getting sent amount and converting from wei to szabo

            chargingTime =  (60*sentAmount)/(chargingRate_in_szabo);       //con-
verting it to seconds for more precision
      }
}
```

The function named 'Pay' is defined similar to that in the previous contract of simple payment. The 'require' keyword in this function is checking the condition within it to be fulfilled before the execution of the function. Two conditions are set in the 'require' function: one is to satisfy the minimum charging time (mentioned earlier) and another one is to limit the accidental excess transfer of ether. If the 'Pay' function is called with a value set to less than 10^{15} wei, say 10^{14} wei, the execution fails and the transaction gets reverted as shown in Fig.4.

Figure 4. Transaction receipt giving error in execution for transfer of 100 szabo or 1014 wei (which is less than minimum value, 1015 wei)

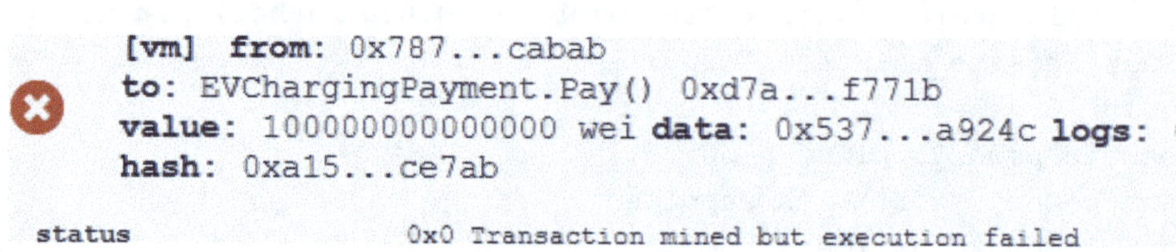

Fig.5 verifies the functions of the smart contract upon deploying and running in Remix IDE. It shows that the instructions are being displayed upon calling the public function named '_Instruction'. It shows the charging rate, units of power, charging time and amount sent to smart contract. It also conveys the details of minimum charging, what minimum amount is to be required and its equivalence to Indian Rupee. All the public variables are listed upon deploying the contract as in Fig.5. The function 'Pay' is called in Fig.5, with the 'VALUE' field set to 6000 szabo. The public variable 'sentAmount' verifies this. The charging time is calculated in the 'Pay' function with the given formula and it is displayed with the public variable 'chargingTime' in seconds, which is 3600 second or 1 hour in this case (for 6000 szabo). The formula used in this contract can be manipulated as the owner of the charging station wishes. The charging rate is also to be set by the owner.

If we want to do effective charging considering early hours for cheaper and peak hour for costlier, the smart contract is to be linked with real world time. The solidity language provides keywords 'now' or 'block.timestamp' which gives the number of seconds from the Unix epoch (which is 00:00:00 UTC, 1 Jan 1970). This could be utilized for time management in the contract. But security measures are to be taken while using these keywords as miners may leverage to some extent. With this time inclusion, the smart contract is made to do pricing accordingly. In peak hours, charging rate is manipulated to higher value for costlier charging while in early hours, it provides low charging rate leading to cheap charging.

The charger in our framework is supposed to be supplying constant power throughout. It may breach transparency to the framework as may be the charger is not supplying the mentioned power. To encounter this issue, a device measuring power supplied by a charger is to be interfaced with the smart contract so that the contract retrieves the power supplied continuously and does pricing accordingly.

IV. IMPLEMENTATION AND SECURITY ISSUES

A. Overview of Platform Prototype for Payment

This automated payment framework is to be employed on a platform for EV charging. The charging station employing this framework needs to have prompts issued through a screen like an ATM. The system is required to have the public blockchain running in it and the payment smart contract deployed in the blockchain. The customer must be the member of that network where the smart contract is deployed so that they have an address which is required for payment of charging. The customer has to set the 'ACCOUNT' field with his own address and call the '_Instruction' function to get the relevant information. Then setting the 'VALUE' field to an amount which customer wants to pay for charging, 'Pay' function is to be called. Based on charging rate, the charging time is calculated and returned by the contract. The charger is required to be interfaced to the smart contract. The charger receives the signal of charging time from the contract and the charger gets activated for the specified charging time and starts supplying power. After that many seconds, it gets deactivated automatically. The charging of electric vehicles through smart charger and smart contract is carried out in this way.

There would be some up-to-date exchange rate display for different paper currency (USD, INR, etc.) and ether (along with other units like szabo). This could be done in two ways, either deploying a second smart contract which imports the online converter, or simply fetching the exchange rates from an authenticated website and displaying them over the screen.

Figure 5. Running transaction for charging payment after deploying in Remix IDE

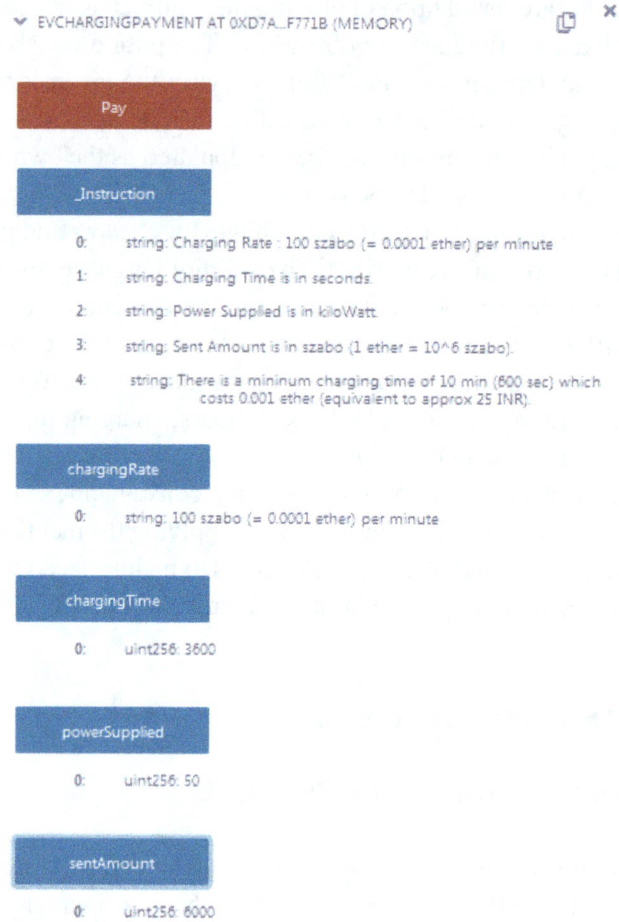

B. Security Vulnerability and Solution

Every emerging technology has to face some of the security issues in its initial stage of development. Blockchain being the decentralized ledger poses the issue of security as the attacker has access to most of the things. For execution, smart contracts are based on the blockchain technology. Any anonymous person can develop and deploy a vulnerable smart contract. Due to the feature of immutability (Huang et al., 2019), once any malicious smart contract has been deployed, it goes out of control. The immutability of the smart contract also leads to some issues in a legitimate smart contract. In this case of payment smart contract, it is to be deployed on the public blockchain only after all the variables and functions are set effectively. Once the smart contract is deployed, there will be no control over variables of charging rate and power supplied. So, the owner of the charging station has to consider the economic viability and other perspectives to set the values to variables before deploying on blockchain.

The vulnerability termed as "Locked Money" (Huang et al., 2019) is encountered when a transaction is being carried out and the field of receiver address is missed to be filled. As in most cases, the default address is "0x0", so the money sent gets locked in this address. To avoid this, the address field is to be checked prior to transfer. The "Reentrancy" is referred to as vulnerability due to interaction between two contracts, in which one contract may call back into another one and get back refunds. The well-known "TheDAO hack" had encountered this issue (Dika & Nowostawski, n.d.).

The website-based Remix IDE also provides the tool for debugging transactions to deal with the bug in the contract. Zeppelin Solutions has developed ZeppelinOS (Araoz, 2017), which is an open-source platform for testing and developing smart contracts and fixing security vulnerabilities. There are some other development tools like "HackThisContract'', "Security audits", "Oyente" (Dika & Nowostawski, n.d.), etc. for testing and developing smart contracts. A smart contract must be debugged and tested through the testing tools available before deploying it to public blockchain to avoid the security issues.

V. CONCLUSION AND FUTURE WORK

This chapter proposes the framework of blockchain-based ethereum payment for EV charging. The payment is done with the help of a smart contract which is executed and the charging time is allotted according to the amount of ether sent to the address of smart contract, power delivered and the charging rate. The contract also displays the required information of charging rate, power supplied and units. The blockchain based transaction and issue of charging time ensures transparency.

The code for the smart contract is formulated and for analysis, it is locally deployed in Remix IDE. The smart charger interfaces with this smart contract, get the signal of charging time and get activated only for the specified time. An overview of the platform prototype of payment is also presented to demonstrate the implementation of the framework in the real world. This is how EV charging through a decentralized system is employed and secured through cryptography. The last section highlights the security issues in the smart contract and how it could be avoided to some extent using the open-source tools for testing.

Some of the tests and adaptations are required for more improvement. Effective pricing for EV charging could be done by considering peak hour and early hour if real world time could be linked with the smart contract. Charging through a variable output power could also be done on interfacing a power measuring device to the smart contract.

ACKNOWLEDGMENT

We gratefully thank all the faculty members and other staff of Centre of Advanced Research in Electrified Transportation (CARET) and mentor Zaurez Ahmed for their immense support and favor in the whole process.

REFERENCES

Antonopoulos, A. M., & Wood, G. (n.d.). *Mastering Ethereum*. Available at https://www.docdroid.net/MecfDvO/masteringethereum-pdf#page=17

Araoz, M. (2017). *Introducing zeppelinos: The operating system for smart contract applications*. https://blog.zeppelin.solutions/introducingzeppelinos-the-operating-system-for-smart-contract-applications-82b042514aa8

Bunjaku, F., Gjorgieva-Trajkovska, O., & Miteva-Kacarski, E. (2017). Cryptocurrency - Advantages and Disadvantages. *Journal of Economics, 2*(1).

Dannen. (n.d.). *Introducing Ethereum and Solidity*. Academic Press.

Dika & Nowostawski. (n.d.). *Security Vulnerabilities in Ethereum Smart Contracts*. Academic Press.

How Blockchain Transactions Work (Adding Data to Blockchain). (n.d.). https://youtu.be/I_VBuzSV9vA

Huang, Li, Lai, Zhao, Wu, Li, Tong, & Lai. (2021). *A Novel Power Market Mechanism Based on Blockchain for Electric Vehicle Charging Stations*. doi:10.3390/electronics10030307

Huang, Y., Bian, Y., Li, R., Zhao, J., & Shi, P. (2019). Smart contract security: A software lifecycle perspective. *IEEE Access: Practical Innovations, Open Solutions, 7*, 150184–150202. doi:10.1109/ACCESS.2019.2946988

Lee. (2019). *Rise of Anonymous Cryptocurrencies: Brief Introduction*. IEEE.

Loi, L., Duc-Hiep, C., & Hrishi, O. (2016). Making Smart Contracts Smarter. *ACM SIGSAC Conference on Computer and Communications Security (CCS '16).*

Mahmood. (2021). *Blockchain Technology: Initiatives and Use Cases in the Industry*. doi:10.4018/978-1-7998-6650-3.ch009

Nakamoto, S. (2008). *Bitcoin: A peer-to-peer electronic cash system*. www.bitcoin.org/bitcoin.pdf

Savjee. (n.d.). https://www.savjee.be/videos/simply-explained/how-does-a-blockchain-work/

Solidity. (n.d.). https://solidity.readthedocs.io/en/v0.7.1/

Wang, Ouyang, Yuan, Ni, Han, & Wang. (n.d.). *Blockchain-Enabled Smart Contracts: Architecture, Applications, and Future Trends*. Academic Press.

Watanabe, H. (2015). Blockchain contract: A complete consensus using blockchain. *IEEE 4th Glob. Conf. Consum. Electron. (GCCE)*, 577–578. 10.1109/GCCE.2015.7398721

Wöhrer & Zdun. (2018). *Smart Contracts: Security patterns in the Ethereum ecosystem and Solidity*. IEEE.

Xu, L., Chen, L., Gao, Z., Carranco, L., Fan, X., Shah, N., Diallo, N., & Shi, W. (2019). *Supporting blockchain-based cryptocurrency mobile payment with smart devices*. IEEE.

YouTube. (n.d.). https://www.youtube.com/watch?v=ZE2HxTmxfrI

Chapter 13
Recent Advancements in PEM Fuel Cell Technologies for Electrified Transportation

Mohammad Arqam
Griffith University, Australia

Sameer Usmani
Australian National University, Australia

Sadegh Aberoumand
Griffith University, Australia

Vinay Kumar
Griffith University, Australia

ABSTRACT

The interest in developing clean and environmentally-friendly energy devices to be used on vehicles is intensifying because of emissions from conventional internal combustion engines considered as one of the significant contributors to the rapidly changing climate. Fuel cell energy devices, especially the proton exchange membrane (PEM) type, are the solid contender to replace the conventional vehicle propulsion technology in the transport sector. The PEM fuel cell technology needs a lot of efforts to overcome some existing problems such as durability, hydrogen storage, and cost for its successful worldwide commercialisation. This chapter deals with the durability, cost, and performance challenges related to the utilization of PEM fuel cell technology in electrified transportation. Recent advancements concerning the current challenges have been discussed. Moreover, issues of hydrogen storage and infrastructure are outlined.

DOI: 10.4018/978-1-7998-6858-3.ch013

1. INTRODUCTION

With the increase of global concerns on climate change, sustainability and development, the interest in using innovative, renewable energy technologies especially PEM fuel cell have been a focus of attention in recent years in the transport sector, due to their higher effeciency, high power to weight ratio, quick start-up time, environmentally friendly nature, potential to replace conventional engine technology and low Green house gas (GHG) emissions (A. Alaswad et al., 2015). A fuel cell is an electrochemical device that converts chemical energy (stored in hydrogen) into electrical energy. Unlike conventional batteries, PEM fuel cell uses external source to supply hydrogen as a reactant to the cell. Similarly, the electrodes in a PEM fuel cell do not participate in the reaction. The applications of PEM fuel cells include vehicles, sub-marines, spacecraft and electricity generators (A. Alaswad et al., 2015, 2016a). PEM fuel cell devices are especially reasonable for use in passenger vehicles, for example, light duty vehicles and heavy duty vehicles.

The transport sector have been cited as one of the major contributors of the GHG emissions into the environment for the past several years (A. Alaswad et al., 2016a). Researchers have been trying to develop fuel cell systems that may capable of using different fuels rather than hydrogen (Ou et al., 2012). Through this developement, the fossil fuel consumption could be significantly reduced in the transport sector and may pave the way for alternative technologies of the same nature for the better and sustainable future (Hao et al., 2010). Consequently,two different approaches have been found in the open literature to manage vehicle GHG emissions issues. The primary approach concerns with the type of fuel that can be applied either by upgrading quality of the conventional fuels or through using alternative fuels (such as bio-fuels), while the secondary approach deals with the vehicle emissions that must be in accordance with the emission standards/measures. These approaches might play a vital role in reducing fuel consumption and GHG emissions in the transport sector (Achour et al., 2011, 2012; Achour & Olabi, 2016; Saboohi & Farzaneh, 2009).

The concept of fuel cell was first introduced by Sir William Grove in 18th century and it was developed in mid of 19th century. Since then, special focus were laid upon development of fuel cell as the researchers/scientists were keen to explore strategies to power spacecrafts through using fuel cell technology (A. Alaswad et al., 2016a). Fuel cells are classified on the basis of electrolyte being utilised. Among all, the most prominent is the PEM fuel cell which uses solid polymer as electrolyte and permeable carbon electrodes (coated with platinum alloy catalyst). The PEM fuel cell uses clean and pure hydrogen supplied from reformers or capacity tanks. As far as the matter of PEM fuel cell working is concerned, the hydrogen is supplied to the anode (where electrons are separated from protons). The protons pass through the permeable membrane and strike the cathode, while the membrane does not allow the electrons to pass through it so that they can make their way through the outer circuit and produce electricity. The water is formed as a by-product on cathode where oxygen combines with electrons and protons. Pure oxygen can be supplied or it can be separated from the air, at the electrode during the process.

Due to rapid emphasis on clean energies technologies in transportation section, collosal upsurge was observed in 2012 when 45,700 fuel cells were shipped, which was twice that of the previous year. This pointed towards a fact that the demand of using fuel cell in the transport sector was observed by the industry. Moreover, the transport sector had witnessed major advancements when a company (Hyundai) manufactured I x 35 fuel cell electric vehicle (FCEV), and Toyota presented Mirai 2015 (Costamagna & Srinivasan, 2001; A. G. Olabi, 2013; Abdul Ghani Olabi, 2010; Wilberforce et al., 2016).

In this chapter, an overview of the recent advancements of the proton exchange membrane (PEM) fuel cell technology especially in the transport sector has been provided. The current challenges and issues related to the cost, durability and performance have been discussed. Hydrogen storage and hydrogen refuelling infrastructure as major concern are also overviewed.

2. PEM FUEL CELL COMPONENTS

In the light of the literature, PEM fuel cell has been broadly divided into three main components as shown in **Table 1** (Bıyıkoğlu, 2005; Carton & Olabi, 2017; Costamagna & Srinivasan, 2001; Hermann et al., 2005; James et al., 2010; Litster & McLean, 2004; Abdul Ghani Olabi, 2012; Tawfik et al., 2007). The main functions have been listed to illustrate their participation in the power generation mechanism.

Table 1. PEM fuel cell components

Components	Functions	References
Membrane electrode assembly (MEA)	• Main component of PEM fuel cell • Consists cathode/anode porous electrode, catalyst layer, gas diffusion layer (GDI) and proton conducting electrolyte	(Bıyıkoğlu, 2005; Costamagna & Srinivasan, 2001; Litster & McLean, 2004)
Anode/cathode current collector	• Often called bi-polar plates • Act as electron conductors • Circulating fuel inside the cell • Water management in the cell • Thermal management of the fuel cell • Graphite based composite materials used for bipolar plates • Metal plates can be used where, mechanical strength is of importance	(Hermann et al., 2005; James et al., 2010; Litster & McLean, 2004)
Auxillary components	• Transportation and compression of gasses • Thermal and water management	(Carton & Olabi, 2017; Hermann et al., 2005; Abdul Ghani Olabi, 2012; Tawfik et al., 2007)

3. MATERIALS FOR PEM FUEL CELL COMPONENTS

The sub-components of the MEA, particularly electrodes, are the main cost contributor to the fuel cell stack. Nafion membranes are used as electrolyte for low temperature (<80°C) applications, while H_3PO_4-pbi used for high temperature applications (110°C -180°C). As of now, electro-catalyst platinum and platinum alloys are the most useful materials for electrodes irrespective of their market cost. The popular choice for GDL is porous carbon fabric often sealed with a PTFE (Teflon) coating or carbon paper (Carton & Olabi, 2017; Cele et al., 2010; Khaligh & Li, 2010; Abdul Ghani Olabi, 2012; Tawfik et al., 2007).

The commonly used materials of PEM are nafion and polybenimidazole (PBI), materials such as sulphonated aromatic hydrocarbon polymer membrane, inorganic/organic composite membrane, polymer blend membrane and phosphoric acid doped nafion polybenimidazole (H_3PO_4/Nafion-PBI) are also developed. The platinum and its alloys are used for electrode catalyst layers. Other materials are also prepared by substituting platinum with other metals. The GDL is usually made up of porous carbon paper or carbon cloth teflon coating, however carbon fibre based gas diffusion materials have also been developed. Graphite is most commonly used for bi-polar plates, other choices may include metals (non coated), non metals (non porous), composites (polymer-carbon and polymer metals) (A. Alaswad et al., 2016a). The primary parts of the PEM fuel cell stack are shown in **Fig. 1**.

Figure 1. PEM fuel cell stack components

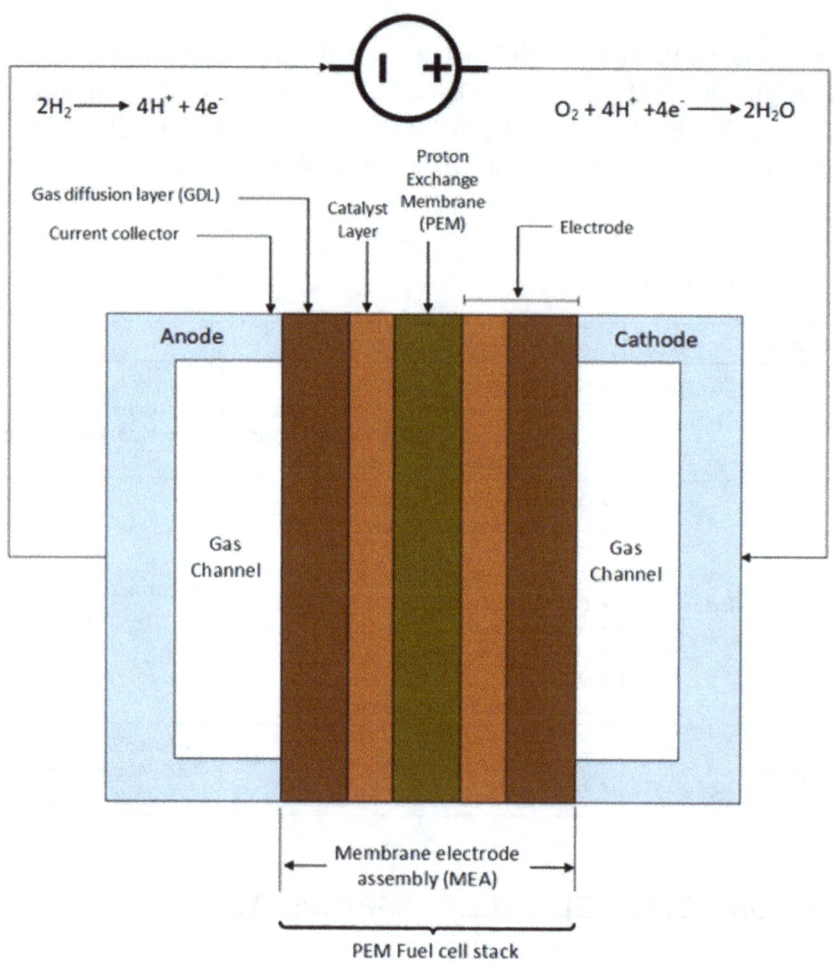

4. APPLICATIONS IN TRANSPORT SECTOR

The transport sector contributes 17% of the global GHG emissions, and it is considered as one of the sectors which can be improved by implementing clean energy technologies (Sharaf & Orhan, 2014). Thus, critical modifications are expected in this sector in order to achieve the target of the Kyoto Protocol and Paris Accord (Cacciola et al., 2001), which is also evident from responses, received from industries. The industries have shown interest by substantial investment in technologies with the potential to reduce GHG emissions along with high energy conversion efficiency.

Electric vehicles (EVs), driven and operated only by batteries or fuel cells are considered as a prime contender to reduce GHG emissions from the transportation sector. EVs are more efficient than conventional internal combustion engines because of their high conversion efficiency of electric motors. Following the trend, these virtues entice industries to replace conventional internal combustion engines with EVs [22,23]. The conversion efficiency to drive force of the electric motor is 90%, which is comparatively higher than internal combustion engines, which has 25% of efficiency, mainly because electric motors are in direct contact with the wheels. Minimal energy is lost in the process to run a vehicle, unlike the operation of internal combustion engines. Moreover, the regenerative braking mechanism (often refers to an energy recovery mechanism that helps slowing down a vehicle by converting its kinetic energy into a form that can be stored and utilised when needed) in EVs can restore approx. 50% of an electric vehicle's energy to the storage cells. These are the factors, which promote EVs as an eco-friendly vehicle over internal combustion engine drive vehicles (Elzen et al., 2004; Hofman et al., 2004).

Unlike EVs, combustion of hydrocarbon-based fuel to propel a conventional vehicle discharges GHG to the environment and other harmful gases through the exhaust system. Furthermore, there are other emissions related to the production of fossil-based fuels, whereas EVs emit zero GHG emissions during the operation. Concerning the source of power, the overall emission from EVs depends on how electrical energy is created, and there are options where considerably lower GHG emissions can be achieved with production of hydrogen fuel through renewable sources. Following a similar pattern of incremental improvements to IC engine vehicles in the last 100 years, advancements towards energy efficiency, overall weight reduction and reliability in EVs will make them more competitive and attractive, leading to the replacement of IC engine-driven vehicles.

Electric vehicles can be propelled either with fuel cells or batteries, but it has been found that vehicles equipped with fuel cells are better than batteries for a higher range (more than 160 km) in terms of overall cost, GHG emissions, refuelling time and well to wheel energy efficiency utilising natural gas or biomass as refuelling source (Pollet et al., 2012). Hybridization of fuel cell systems (FCS) with devices which have higher specific energy like batteries or ultra-capacitors has essential advantages over basic fuel cell systems, thrashing the generally moderate FCS transient response, reducing the warm-up time and enhancing the hydrogen economy of the FCS to achieve maximum power (Pollet et al., 2012). According to the report by Carbon trust (Pollet et al., 2012), fuel cells can power up to 491 million cars across the globe, and PEMFC technology can be a vital technology in transport applications because of its high power density, compactness, high energy conversion efficiency and light-weight. However, the high cost of this technology is the primary barrier in successful implementation at a commercial scale.

The energy units in transport applications can be categorized into light-duty vehicles (LDVs), auxiliary power units (APU), light-duty fuel cell electric vehicles (L-FCEV), heavy duty fuel cell electric vehicle (H-FCEV), marine propulsion and ariel vehicles. APU are responsible for power transmission to air conditioning and electronic system in the vehicle (*Development of Auxiliary Power Unit for Recreational Yachts | Www.Fch.Europa.Eu*, n.d.). LDV consist of a wheel chair, two-wheel vehicles such as bikes, scooters and electric bicycle instead of transport vehicles such as carts, forklifts, pallet trucks and tow trucks. Forklifts have adequately demonstrated the commercially viable fuel cells in the transportation sector and currently, approx. One thousand three hundred fuel cell-based forklifts are operating in the United States (US).

Many automobile manufacturers progressively advancing for the commercialization of light duty FCEVs such as Daimler AG, Mazda, Volvo, General Motors, Volkswagen, Toyota, Nissan, and Honda (Sharaf & Orhan, 2014). For instance, Honda displayed their first FCV *FCX model Clarity* in Los Angeles in 2007 and was made available to the customers by the mid of 2008. In fact this was a first fuel cell vehicle which was launched in US (Andújar & Segura, 2009). On the other hand, Japan claims there will be two million FCEVs in Japan as of 2025 (Pollet et al., 2012).

Buses, locomotives, trucks, vans, service fleet and other vehicles comes under the category of Heavy duty fuel cell electric vehicles (H-FCEVs) (Sharaf & Orhan, 2014). A number of Fuel cell Electric Buses (FCEBs) have been developed and operating in the United States and other part of the world during the past couple of years (Andújar & Segura, 2009). As per the statstics, 25 fuel cell buses in the United States and almost 30 in western europe were deployed in 2012. In fact, fuel cell buses gradually becoming a source of inspiration and a mode of demonstrating considerable research and development in the respective field to the heavy duty FCEVs industries. PEM fuel cells as power generators are widely accepted in transport sector considering the fact that they can make use of regenerative braking where the energy released during braking has been stored back and used when required. A matter of fact, the amateur phase of PEM fuel cell technologies and the absence of large scale manufacturing are still responsible for not making FCEBs economically viable and commercially available.

The expansion of Fuel cell applications in marine transportation has also been observed in recent years. It is highly likely that in the near future, PEM fuel cell will be acting as primary power source to the marine propulsion in conjunction with diesel engines since they have already been used as secondary power units. The potential market for the PEM fuel cell technology to be used for marine propulsion may include freight ships, yachts, ferries, and submarines (Leo et al., 2010). Fuel cells have enough potential to propel boats, ships and other marine transportation due to their higher efficiencies, easy operation, and low emmisions however some issues are also there such as lifetime, reliability, shock resistance, and response to the salty ocean water that need to be rectified before its successful commercialisation (Cacciola et al., 2001). Moreover, there are still some sort of unclear explanations about the volume and weight that has to be assigned to the fuel cell to be used on ships since the mode of use is different from the other transport sectors (Pollet et al., 2012).

The first hybrid (PEM fuel cell and battery) yatcht was developed in Germany in 2003 and the first business traveller ship was put in place in 2008. A number of Fuel cell driven/propelled submarines or the ones with fuel cell as an auxiliary device to assist with propulsion have also effectively put into services. Fuel cell technology with an Air independent propulsion (API) system is now considered as a potential technology to fuel the marine transportation in the countries like Spain, Portugal, Germany, Greece and Italy (Leo et al., 2010).

Likewise, the space industry also showed interest to opt for the fuel cell technology for the space propulsion. In 1960s, NASA uses Alkaline fuel cell (AFC) and PEM fuel cell (PEMFC) technologies for its space programs/propulsion. The one obvious benefit of using fuel cell in space applications is the only by-product remain during electrochemical reaction is water that is of a great significance. The market for the fuel cells to be used on commercial and military air vehicles seems a bit unrealistic due to the fact that it requires high energy density, reliability and durability especially for the development of Unmanned Aerial Vehicles (UAVs) which are still in the initial phase of development.

5. CHALLENGES, ISSUES AND DEVELOPMENTS

While effective utilization of this technology in the vehicle sector has occurred in USA and Europe, the fuel cell technology is still confronting various difficulties to its successful commercialization. The cost of fuel cell is noteworthy issue besides performance, durability and other critical issues that need to be rectified. The infrastructure and hydrogen storage is an underlying issue for the technology to promote commercialization of transportation sector.

5.1 The Cost

The fuel cell cost can be divided into three categories namely material cost, labour cost and capital cost (Marcinkoski et al., 2011; J. Wang, 2015). It has been observed that the capital and labour costs which include the design and fabrication costs of the componenets, can be minimised through large scale assembling. The cost of material include bipolar plates, membrane, catalysts, are basically reliant on the market and technological advancements as well(Odeh et al., 2013; Sun et al., 2011). According to Carbon Trust, the cost of fuel cell at commercial scale should be under $36/kW. Platinum (Pt), with over 30,000 tons of reserves and 250 tons of annual production mined in the United States, Russia and South Africa. Given its high esteem, the larger part of Pt used in FCEV can be recycled at the end of vehicle's life (Wilberforce et al., 2016).

The cost of FCEV can be minimised by controlling material costs (e.g Pt), improving durability and increasing power density. The Pt usage in electro-catalyst layer can certainly degrade the cost of the PEM fuel cell technology and it would make the technology feasible to manufacture at large scale (A. Alaswad et al., 2016a). The Pt cost can be reduced by the application of platinum alloy catalyst or using core shell catalyst. The alloy of platinum with cheap metals such as chromium (Cr) and ruthenium (Ru) is called as Pt-alloy catalysts (Gasteiger et al., 1993; Long et al., 2000) and if low-cost nanoscopic metal cores such as Cr and Ru are capped by a platinum shell, they are called as "core shell catalysts" (Nagabhushana et al., 2006). The Pt content in FCEV was reduced by increasing the specific power of the catalyst from 2.8 kW/g to 5.8 kW/g in 2008 to 2012 respectively. The recent data show that catalyst specific power tends to 8.0 kW/g. Very few organizations are currently working on overall cost reduction of FCEV by incorporating technologies to reduce Pt usage, simplify system designs and power densities.

Alternative way to reduce FCEV cost is by eliminating Pt from its alloys. The elimination of Pt open provision to utilise cheap metals such as iron (Fe) and cobalt (Co) as replacement in FCEV technology. A patent by ACAL Energy Ltd. as FlowCath fuel cell design, which uses a liquid polymer cathode solution in place of platinum based solid cathode in PEM fuel cells (Costamagna & Srinivasan, 2001; *Hydrogen Fuel Cell That's As Durable As A Conventional Engine*, n.d.). In contrast to other technologies, hydrogen used at the anode in the regular form, while proton and electron are absorbed into the proprietary redox catalyst solution, flowing from the stack to a regenerator vessel (as shown in **Fig 2**). The model presented in **Fig 2** is a basic design breakthrough that can possibly reduce Pt cost by 66% (*Hydrogen Fuel Cell That's As Durable As A Conventional Engine*, n.d.). The model uses comparatively smaller number of components in the system by neglecting water recovery and fuel humidification, enhancing the durability through cathode replacement, which usually shows performance degradation with time and that may shorten the product life. The anticipated cost of fuel cell by ACAL Energy is $36/kW (*Hydrogen Fuel Cell That's As Durable As A Conventional Engine*, n.d.).

Another patent filed by ITM Power, demonstrate the high power densities by ousting the perfluorosulfonic acide membranes. The high power densities of fuel cell meet the same power demand with smaller in size, reduced cost and lighter stack. The expected cost of fuel cell was estimated to be $35/kW ("Performance Update on ITM Power's Fuel Cell Membrane," 2013).

In one of the very important patents, fuel cell stack configuration was upgraded by a group of researchers from the University College London and Imperial College (Costamagna & Srinivasan, 2001). The "Flexi-Planar" design utilises a laminated, printed circuit boards (PCB) materials arranged in a fuel cell stack to add provision of internal water, fuel and air channels. The incorporation of PCBs eliminate many components and reduce the overall cost of the conventional fuel cell. The expected cost of fuel cell was \$26/kW with potential areas of cost reduction as stack assembly, sealing, fuel, water management and air.

Figure 2. FlowCath fuel cell design (Hydrogen Fuel Cell That's As Durable As A Conventional Engine, n.d.)

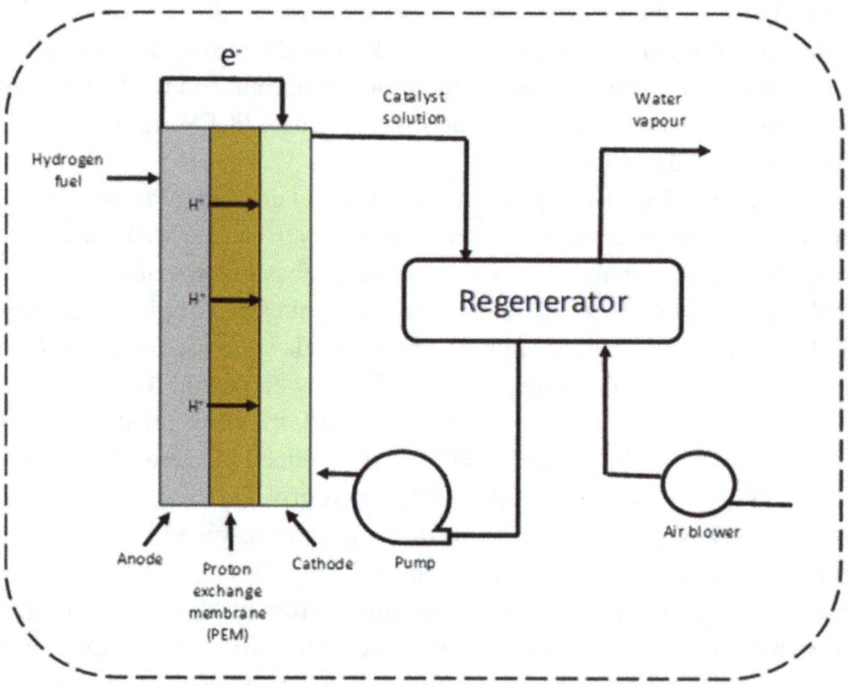

5.2 The Durability

The mismanagement of water and thermal issues in fuel cells cause degradation of accumulated materials and catalyst, leading to poor reliability and short term durability (Bıyıkoğlu, 2005; Choi et al., 2012). The degradation is a direct consequence of chemical reaction, fuel starvation, oxidant and corrosion of different fuel cell components resulting in flooding or dehydration. The flooding causes the corrosion of the electrodes and dehyderation harm the membrane, catalyst layers and gas dispersion media. The issue can be improved by modification in desing and materials or by controlling the flow rates, temperature and humidity.

The PEM fuel cell can be upgraded with long term durability by developing low cost, high corrosion resistance bi-polar plate. In general, graphites are used in a bi-polar plates because of its high surface conductivity, high corrosion resistance and moderately low surface contact resistance but it has its anomaly because the material is brittle and a high risk associated with dismantling/breaking during the operation, especially in transport applications (A. Alaswad et al., 2016a). In place of graphite, carbon composites have been recommended to produce economical and long term durable bi-polar plates (Baroutaji et al., n.d.; Carton & Olabi, 2017; H Wang, 2003). Another recommendation is, metals and their alloys can be used to make a thinner bi-polar plates for high power density and higher durability (Hermann et al., 2005). Metals such as Ti, Al, SS and Ni should be avoided, because they are most likely to corrode in PEM fuel cell environment (Pozio et al., 2003). The corrosion causes an increase in electrical resistance, which decreases the power output and efficiency of the PEM fuel cells (Antunes et al., 2010; Hermann et al., 2005; Mehta & Cooper, 2003). In order to avoid this issue, metal based and carbon based corrosion resistant coating can be used on metals (Joseph et al., 2005). Carbon based coatings include conductive polymers such as polypyrrole (PPY), polyaniline (PANI), graphite and composite coverings viz. Ag-PTFE, TiC-ETFE (Fu et al., 2008; Hentall et al., 1999; Mawdsley et al., 2013; Show, 2007), while metal based coatings involve metal nitrides, metal carbides and nobel metals (Cho et al., 2005; Hung et al., 2009; Ren et al., 2012; Heli Wang & Turner, 2007; Yang et al., 2010; Zhang et al., 2011) and oxides such as Ruthenium oxide (RuO_2) and fluorine doped Tin (II) oxide (SnO_2: F) (Heli Wang & Turner, 2007; Weakley, 2012).

The carbon corrosion in PEM fuel cell is considered as source of major degradation in catalyst layer. Platinum nanoparticles that may sustain/survive on carbon black (Pt/C) are considered as most encouraging/useful electro catalyst for PEM fuel cell. It may be because of adequate access to electrons, gasses and protons by the platinum nanoparticles in catalyst layer. The Pt nanoparticles accelerate the electrochemical degradation and isolate from the carbon when working under high current condition. It has been observed that blend of graphene and Pt/C could be effective under high current density conditions by transporting electrons (Choi et al., 2012). Alternatively, graphene has shown better conductivity and low electrical resistance under high current density conditions which, present that graphene can give comparatively higher durability than carbon black with its one of a kind graphitized basal plane (Owejan et al., 2008).

Excessive hydration is another factor which causes reliability issue. It causes uncertain voltage at low current density, voltage loss at high current density, inconcisisten start-up at cold conditions and enhance carbon corrosion in the catalyst layer (Lee et al., 2015). It has been concluded from literature that the design membrane and its material determination must be balanced for hydration, especially in transport applications where fuel cell is subjected to large load variations. Substitute approach is to construct the nafion membranes for long term durability of PEM fuel cell. On the other hand, sPP-b-PAES (6.5K) -3.0 membrane has been proposed by researchers (Ball & Weeda, 2015), providing stable dimensional fluctuations, better proton conductivity, better overall performace and high durability as compared to nafion membranes.

ACAL Energy Ltd. in June 2013 conducted a long term durability test and reported that the PEM fuel cell can run upto 10,000 hours (equivalent to 300,000 miles) without any indication of degradation (Choi et al., 2012). On comparing with a light weight diesel engine, it is concluded that the PEM fuel cell could be able to perform as good as diesel engines under the same operation conditions. The test surpassed the US department of Energy (DOE) milestone of 5000 h runtime (equaivalent to 150,000 miles) with only 10% degradation rate. The fuel cell design/technology by ACAL Energy Ltd. does not rely on platinum (as the catalyst) to sustain/carry the reaction in the fuel cell. The fluid behaves both as catalyst and a coolant for the cell.

In The Fuel Cell and Hydrogen 2 Joint Undertaking (FCH2JU) yearly plan of 2016 (Pottmaier et al., 2015), the primary focus was given to the development of new manufacturing technologies for the PEM fuel cell to be used in transport sector. In has observed that regardless of the level of maturity, PEM fuel cell technology still confronting some issues/challenges, for example, production cost, efficiency and manufacturability. FCH2JU proposed and established a project that may look after the processes involved in PEM fuel cell stack manufacturing and components production. The target was to increase the production rate from 100 stacks per year to 50,000 as of 2020, with a commulagtive power production rate of around 5 MW/year. Another target was to design compact, high performance and Balance-of-Plant (BOP) PEM fuel cell systems that may include an intercooler, humidifier, compressor, turbine and valves. The projected outcomes include the power generation cost must lie below 107 $/kW for the production rate of 50,000 units of fuel cell per year and to limit the maximum power degradation rate by 10% even after 6000 hours of running passeneger vehicle.

5.3 The Hydrogen Infrastructure

Since hydrogen is considered to be the lightest chemical element among other elements/fuels, often comes under the category of least storage density. Hydrogen refuelling infrastructures have been a matter of serious consideration in recent years because of the cost of establishement. The Fuel cell vehicle market is solely governed by the hydrogen refuelling infrsatructures (Muellerlanger et al., 2007). The widespread existing gasoline delivery systems from the refineries may not be a good fit in the case of hydrogen. There is a need to design and develop an efficient, cost effective hydrogen refuelling infrastructure to produce, dispense and transport hydrogen fuel to end users (*FCT - Fuel Cell Applications - Fuel and Infrastructure*, n.d.). Researchers and industry experts around the globe have been trying and paying serious attention towards hydrogen compression, supply chains, optimum designs to reduce sowmehow the total cost involved in building hydrogen refuelling infrastructures. A typical hydrogen infrastructure includes hydrogen supply, hydrogen storage and a dispensing unit is depicted in **Fig. 3**.

Figure 3. Hydrogen infrastructure (FCT - Fuel Cell Applications - Fuel and Infrastructure, n.d.)

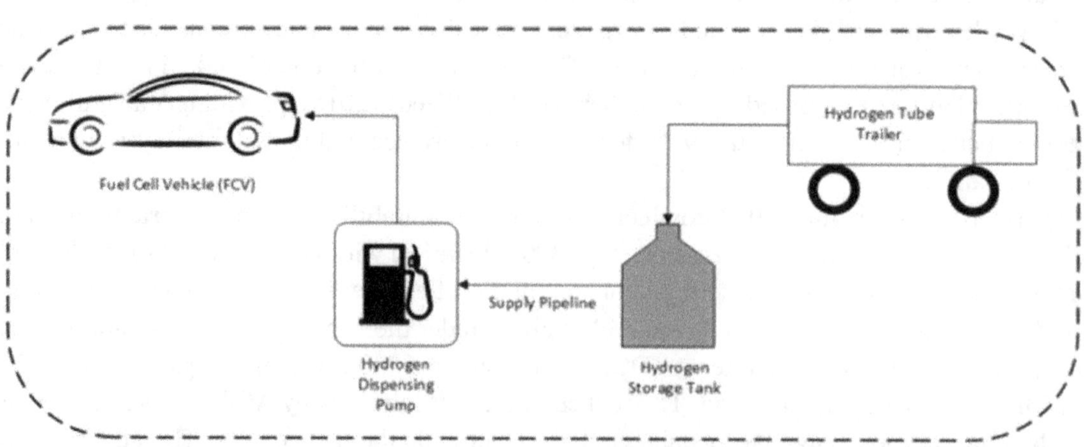

Two major approachs of hydrogen delivery have been cited in the open literature viz. direct supply or onsite production (A. Alaswad et al., 2016b). Hydrogen can be delivered either in a liquefied form or compressed gas. The compressed gas (hydrogen) can only be supplied in a small quantity while the liquefied hydrogen can be delivered in large volumes. Onsite hydrogen production can be accomplished/done through using natural gas reformation where methane splits into molecules of hydrogen and oxygen. Electrolysis of water molecules is another way of producing hydrogen where electric current used to split water into molecules of hydrogen and oxygen. However, reformation of natural gas found to be the most common method used for hydrogen production. Biomass and solar energy are found to have enough potential to act as alternatives for the onsite hydrogen production in the years to come (Abed Alaswad et al., 2016a).

Hydrogen storage is governed/determined by the phases of the gas. A cryogenic storage vessel is required if the hydrogen is supplied in liquefied form to keep the temperature in the desired liquid range. A capacity tank is normally released off on the site when the compressed natural gas form hydrogen is delivered and afterward displaced. In the cases where hydrogen produced nearby, it is then transferred to a compressed storage vessel from the reformer through a compressor. Hydrogen liquefaction is an intensive energy process that requires a huge plant and can only be possible at locations where a plenty of hydrogen is produced. As a safety measure, hydrogen must be delivered to the vehicle storage tank at an appropriate pressure and temperture and in a controlled way through the hydrogen refueller (A. Alaswad et al., 2016a; Singh et al., 2015). The hydrogen dispensing is same as that of refilling petrol or diesel with the aid of nozzle and hoses connected to the vehicle fuel tank. The world has witnessed the issues related to the establishment and cost of the oil and gas infrasturctures which took a long time to put in place during the initial few decades- thanks to the politico-financial strategies and subsidies that helped carrying oil and gas infrastructures to its present form. Likewise, there is a need to put serious efforts to replicate the same for the hydrogen to be successful in the coming years. The governments around the world start to think carefully and must adjust/redefine their energy needs/requirements for the sustainable future as far as the expansion/use of hydrogen as an alternative fuel is concerned (Abed Alaswad et al., 2016b).

FCH2JU in conjunction with many European fuel cell industry partners established a number of projects such as Hydrogen Mobility Europe (H2ME) to overcome/handle the issue of hydrogen refuelling infrastructures throughout the Europe. The aim was to expand the number of Hydrogen Refuelling Infrstructures (HRIS) in the continent itself (Emilio, 2020). As of now, there are 73 HRIS spreaded between Germany (18), Scandinavia (12), France (8), and UK (6) with the introduction of 29 new HRIS in 2019.

5.4 The Hydrogen Storage

The amount of hydrogen in the vehicle fuel tank will decide the cruising/speed range of the vehicle. On comparing with gasoline, it has only one tenth of energy for the same volume. Therefore, it is essential/mandatory to enhance the efficiency and storage capacity of the hydrogen onboard to obtain/attain better performance same as that of gasoline engines. Fuel Cell technology based/driven vehicles have already been developed that can go up to 300 miles without refuelling, but the hydrogen storage system is heavy and expensive to store sufficient amount of hydrogen needed to propel the vehicle (Ogden & Yang, 2005; Singh et al., 2015). Out of three effective means/ways for hydrogen storage on the vehicles, one is to store hydrogen in a high pressure tank in the gaseous form as shown in **Fig. 4**. This is entirely different from compressed natural gas (CNG) vehicles, usually store gas at 20-25 MPa while the hydrogen vehicle uses 35-70 MPa.

Figure 4. Block diagram of hydrogen storage tank in the vehicle (Emilio, 2020).

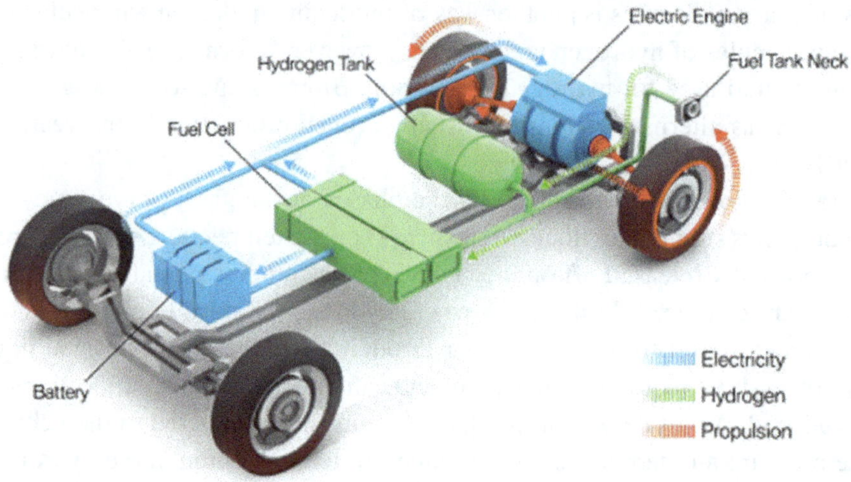

To extend the vehicle speed/cruising range, there is an urge to develop tanks that can carry more hydrogen (up to 70 MPa or probably more). However, these systems might be huge and cost ineffective. Since hydrogen diffuses more easily and effectively through different ordinary materials often used for high pressure vessels, and also through tiny gaps (sufficient to seal other gasses), special focus has been given to find a set of materials that can withstand high pressures without compromising structural rigidity (Zhevago, 2016). The currently developed high pressure tanks made up of premium carbon reinforcement (PCR) which can withstand 150 MPa pressure to meet the structural requirement issued for safety (Zhevago, 2016). This is close enough to the US DOE's aim of 40g/l and 5.5% of the system volumetric efficiency/capacity and systems gravimetric capacity. Neverthless, it is still difficult to meet the ultimate target of DOE (70g/l volumetric capacity and 7.5% gravimetric capacity) by using this technology (Zhevago, 2016).

Another approach to store hydrogen is to cool the gas down to 20 K at the atmospheric pressure. By doing so, hydrogen can be stored in a larger amount on board yet it requires a thermally insulated double wall tank structure that helps keep the temperature low (Muellerlanger et al., 2007). One more interesting and effective approach is to store hydrogen in a solid form. Different alloys such as metal hydrides can absorb hydrogen in higher densities and they are smaller and lighter in weights compared to that of storage tanks. The hydrogen can be discharged/extracted from these metals with the help of water (Stolten & Grube, 2010). A considerable efforts are required to build these materials to store and deliver hydrogen as and when occasioned. Metal Organic Frameworks (MOFs) are cited as strong contender for hydrogen storage applications due to their rich chemistry, tendency to adjust the surface roughness and pore sizes, vast physical variety, large surface area and porosity. Moreover, MOFs may store more hydrogen compared to other storage technologies, as shown in **Fig. 5**. Because of their higher mass density, MOFs are not yet complying with the current target of DOE in spite of the rapid growth in recent years (Hirscher et al., 2010; A.G. Olabi, 2014).

Figure 5. The amount of hydrogen per litre vs storage technology

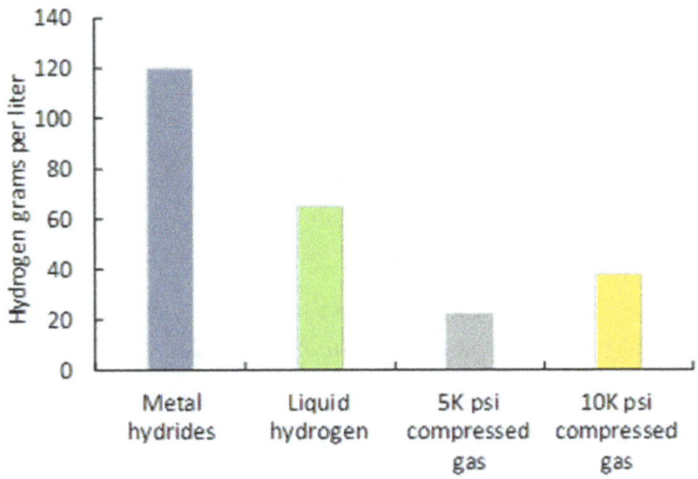

6. CONCLUSION

Implementation of clean and efficient energy technlogies in transportation sector has emphasised the mass production and adoption of fuel cell technology worldwide. Proton exchange membrane (PEM) fuel cell device offers more advantages over the conventional internal combustion engines through emission reduction and higher/improved efficiency. To meet the current requirements for the transport sector, there is urgent need to overcome several challenges including poor long-term durability, high cost, hydrogen storage and refuelling infrastructure associated with the fuel cell technology and its widespread commercialisation. The long term mechanical strength/durability cited as a key performance indicator for the successful commercialisation of fuel cell technology/devices in the transport sector. Degradation of the main components over time such as flow plates and membrane is the main cause/reason behind the low long-term durability of the fuel cell devices. Hence, introducing/developing solid/strong materials for the PEM fuel cell is of utmost importance. The high cost of power generation from fuel cell devices is another strong issue that should be rectified for the successful commercialisation of fuel cell technology/vehicles. Finding a platinum-free catalyst and bringing down the platinum content are the key areas to lower the cost of power generation without compromising the performance. This review chapter has highlighted the primary challenges the fuel cell technology is experiencing and current developments towards reaching the targets and breaking market barriers for the successful commercialisation of the technology.

REFERENCES

Achour, H., Carton, J. G., & Olabi, A. G. (2011). Estimating vehicle emissions from road transport, case study: Dublin City. *Applied Energy*, *88*(5), 1957–1964. doi:10.1016/j.apenergy.2010.12.032

Achour, H., Marashly, A., & Olabi, A.-G. (2012). Assessing energy consumption of the transport sector in Aleppo, Syria. *Sustainable Manufacturing and Renewable Energy*, *1*(3–4), 87–102.

Achour, H., & Olabi, A. G. (2016). Driving cycle developments and their impacts on energy consumption of transportation. *Journal of Cleaner Production*, *112*, 1778–1788. doi:10.1016/j.jclepro.2015.08.007

Alaswad, A., Baroutaji, A., Achour, H., Carton, J., Al Makky, A., & Olabi, A. G. (2016a). Developments in fuel cell technologies in the transport sector. *International Journal of Hydrogen Energy*, *41*(37), 16499–16508. doi:10.1016/j.ijhydene.2016.03.164

Alaswad, A., Baroutaji, A., Achour, H., Carton, J., Al Makky, A., & Olabi, A. G. (2016a). Developments in fuel cell technologies in the transport sector. *International Journal of Hydrogen Energy*, *41*(37), 16499–16508. doi:10.1016/j.ijhydene.2016.03.164

Alaswad, A., Baroutaji, A., Achour, H., Carton, J., Al Makky, A., & Olabi, A. G. (2016b). Developments in fuel cell technologies in the transport sector. *International Journal of Hydrogen Energy*, *41*(37), 16499–16508. doi:10.1016/j.ijhydene.2016.03.164

Alaswad, A., Dassisti, M., Prescott, T., & Olabi, A. G. (2015). Technologies and developments of third generation biofuel production. *Renewable & Sustainable Energy Reviews*, *51*, 1446–1460. doi:10.1016/j.rser.2015.07.058

Andújar, J. M., & Segura, F. (2009). Fuel cells: History and updating. A walk along two centuries. *Renewable & Sustainable Energy Reviews*, *13*(9), 2309–2322. doi:10.1016/j.rser.2009.03.015

Antunes, R. A., Oliveira, M. C. L., Ett, G., & Ett, V. (2010). Corrosion of metal bipolar plates for PEM fuel cells: A review. *International Journal of Hydrogen Energy*, *35*(8), 3632–3647. doi:10.1016/j.ijhydene.2010.01.059

Ball, M., & Weeda, M. (2015). The hydrogen economy – Vision or reality? *International Journal of Hydrogen Energy*, *40*(25), 7903–7919. doi:10.1016/j.ijhydene.2015.04.032

Baroutaji, A., Carton, J. G., Stokes, J., & Olabi, A. G. (n.d.). *Design and Development of Proton Exchange Membrane Fuel Cell using Open Pore Cellular Foam as Flow Plate Material*. Academic Press.

Bıyıkoğlu, A. (2005). *RETRACTED: Review of proton exchange membrane fuel cell models*. Elsevier.

Cacciola, G., Antonucci, V., & Freni, S. (2001). Technology up date and new strategies on fuel cells. *Journal of Power Sources*, *100*(1–2), 67–79. doi:10.1016/S0378-7753(01)00884-9

Carton, J. G., & Olabi, A.-G. (2017). Three-dimensional proton exchange membrane fuel cell model: Comparison of double channel and open pore cellular foam flow plates. *Energy*, *136*, 185–195. doi:10.1016/j.energy.2016.02.010

Cele, N. P., Sinha-Ray, S., Munda, J., & Jimoh, A. A. (2010). The state of the art proton exchange membrane fuel cells for clean energy. *2010 IEEE International Energy Conference*, 865–869. 10.1109/ENERGYCON.2010.5771804

Cho, E. A., Jeon, U.-S., Hong, S.-A., Oh, I.-H., & Kang, S.-G. (2005). Performance of a 1kW-class PEMFC stack using TiN-coated 316 stainless steel bipolar plates. *Journal of Power Sources*, *142*(1–2), 177–183. doi:10.1016/j.jpowsour.2004.10.032

Choi, H.-J., Jung, S.-M., Seo, J.-M., Chang, D. W., Dai, L., & Baek, J.-B. (2012). Graphene for energy conversion and storage in fuel cells and supercapacitors. *Nano Energy*, *1*(4), 534–551. doi:10.1016/j.nanoen.2012.05.001

Costamagna, P., & Srinivasan, S. (2001). Quantum jumps in the PEMFC science and technology from the 1960s to the year 2000: Part II. Engineering, technology development and application aspects. *Journal of Power Sources*, *102*(1–2), 253–269. doi:10.1016/S0378-7753(01)00808-4

Development of Auxiliary Power Unit for Recreational yachts. (n.d.). Retrieved September 7, 2020, from https://www.fch.europa.eu/project/development-auxiliary-power-unit-recreational-yachts

Elzen, B., Geels, F. W., Hofman, P. S., & Green, K. (2004). Socio-technical scenarios as a tool for transition policy: An example from the traffic and transport domain. *System Innovation and the Transition to Sustainability: Theory. Evidence & Policy*, 251–281.

Emilio, M. D. P. (2020, March 12). EETimes—EC to Bet on Hydrogen Fuel Cell Vehicles. *EETimes*. https://www.eetimes.com/ec-to-bet-on-hydrogen-fuel-cell-vehicles/

FCT - Fuel Cell Applications—Fuel and Infrastructure. (n.d.). Retrieved September 7, 2020, from http://www.fuelcelltoday.com/applications/fuel-and-infrastructure

Fu, Y., Hou, M., Xu, H., Hou, Z., Ming, P., Shao, Z., & Yi, B. (2008). Ag–polytetrafluoroethylene composite coating on stainless steel as bipolar plate of proton exchange membrane fuel cell. *Journal of Power Sources*, *182*(2), 580–584. doi:10.1016/j.jpowsour.2008.04.051

Gasteiger, H. A., Markovic, N., Ross, P. N. Jr, & Cairns, E. J. (1993). Methanol electrooxidation on well-characterized platinum-ruthenium bulk alloys. *Journal of Physical Chemistry*, *97*(46), 12020–12029. doi:10.1021/j100148a030

Hao, H., Wang, H., Song, L., Li, X., & Ouyang, M. (2010). Energy consumption and GHG emissions of GTL fuel by LCA: Results from eight demonstration transit buses in Beijing. *Applied Energy*, *87*(10), 3212–3217. doi:10.1016/j.apenergy.2010.03.029

Hentall, P. L., Lakeman, J. B., Mepsted, G. O., Adcock, P. L., & Moore, J. M. (1999). New materials for polymer electrolyte membrane fuel cell current collectors. *Journal of Power Sources*, *80*(1–2), 235–241. doi:10.1016/S0378-7753(98)00264-X

Hermann, A., Chaudhuri, T., & Spagnol, P. (2005). Bipolar plates for PEM fuel cells: A review. *International Journal of Hydrogen Energy*, *30*(12), 1297–1302. doi:10.1016/j.ijhydene.2005.04.016

Hirscher, M., Panella, B., & Schmitz, B. (2010). Metal-organic frameworks for hydrogen storage. *Microporous and Mesoporous Materials*, *129*(3), 335–339. doi:10.1016/j.micromeso.2009.06.005 PMID:21226135

Hofman, P. S., Elzen, B. E., & Geels, F. W. (2004). Sociotechnical scenarios as a new policy tool to explore system innovations: Co-evolution of technology and society in the Netherland's electricity domain. *Innovation*, *6*(2), 344–360. doi:10.5172/impp.2004.6.2.344

Hung, Y., Tawfik, H., & Mahajan, D. (2009). Durability and characterization studies of polymer electrolyte membrane fuel cell's coated aluminum bipolar plates and membrane electrode assembly. *Journal of Power Sources*, *186*(1), 123–127. doi:10.1016/j.jpowsour.2008.09.079

Hydrogen Fuel Cell That's As Durable As A Conventional Engine. (n.d.). Retrieved September 7, 2020, from http://www.acalenergy.co.uk/news/release/acal-energy-system-breaks-the-10000-hour-endurance-barrier/en/

James, B. D., Kalinoski, J. A., & Baum, K. N. (2010). *Mass Production Cost Estimation For Direct H2 PEM Fuel Cell Systesm for Automotive Applications. 2010 Update*. Directed Technologies. doi:10.2172/1218888

Joseph, S., Mcclure, J., Chianelli, R., Pich, P., & Sebastian, P. (2005). Conducting polymer-coated stainless steel bipolar plates for proton exchange membrane fuel cells (PEMFC). *International Journal of Hydrogen Energy*, *30*(12), 1339–1344. doi:10.1016/j.ijhydene.2005.04.011

Khaligh, A., & Li, Z. (2010). Battery, ultracapacitor, fuel cell, and hybrid energy storage systems for electric, hybrid electric, fuel cell, and plug-in hybrid electric vehicles: State of the art. *IEEE Transactions on Vehicular Technology*, *59*(6), 2806–2814. doi:10.1109/TVT.2010.2047877

Lee, J. Y., Yu, D. M., Kim, T.-H., Yoon, S. J., & Hong, Y. T. (2015). Multi-block copolymers based on poly(p-phenylene)s with excellent durability and fuel cell performance. *Journal of Membrane Science*, *492*, 209–219. doi:10.1016/j.memsci.2015.04.013

Leo, T. J., Durango, J. A., & Navarro, E. (2010). Exergy analysis of PEM fuel cells for marine applications. *Energy*, *35*(2), 1164–1171. doi:10.1016/j.energy.2009.06.010

Litster, S., & McLean, G. (2004). PEM fuel cell electrodes. *Journal of Power Sources*, *130*(1–2), 61–76. doi:10.1016/j.jpowsour.2003.12.055

Long, J. W., Stroud, R. M., Swider-Lyons, K. E., & Rolison, D. R. (2000). How To Make Electrocatalysts More Active for Direct Methanol OxidationAvoid PtRu Bimetallic Alloys! *The Journal of Physical Chemistry B*, *104*(42), 9772–9776. doi:10.1021/jp001954e

Marcinkoski, J., James, B. D., Kalinoski, J. A., Podolski, W., Benjamin, T., & Kopasz, J. (2011). Manufacturing process assumptions used in fuel cell system cost analyses. *Journal of Power Sources*, *196*(12), 5282–5292. doi:10.1016/j.jpowsour.2011.02.035

Mawdsley, J. R., Carter, J. D., Wang, X., Niyogi, S., Fan, C. Q., Koc, R., & Osterhout, G. (2013). Composite-coated aluminum bipolar plates for PEM fuel cells. *Journal of Power Sources*, *231*, 106–112. doi:10.1016/j.jpowsour.2012.12.074

Mehta, V., & Cooper, J. S. (2003). Review and analysis of PEM fuel cell design and manufacturing. *Journal of Power Sources*, *114*(1), 32–53. doi:10.1016/S0378-7753(02)00542-6

Muellerlanger, F., Tzimas, E., Kaltschmitt, M., & Peteves, S. (2007). Techno-economic assessment of hydrogen production processes for the hydrogen economy for the short and medium term. *International Journal of Hydrogen Energy*, *32*(16), 3797–3810. doi:10.1016/j.ijhydene.2007.05.027

Nagabhushana, K. S., Weidenthaler, C., Hocevar, S., Strmcnik, D., Gaberšček, M., Antozzi, A. L., & Martelli, G. N. (2006). *Preparation, characterization and properties of Pt-CuCo-reduced and Pt-on-Cu skin type bimetallic carbon-supported (Vulcan XC72) electrocatalysts*. /paper/Preparation%2C-characterization-and-properties-of-and-Nagabhushana-Weidenthaler/65e5dd1b1d89bf0f657ff1e8007e0b74d5640168

Odeh, A. O., Osifo, P., & Noemagus, H. (2013). Chitosan: A low cost material for the production of membrane for use in PEMFC-A review. *Energy Sources. Part A, Recovery, Utilization, and Environmental Effects*, *35*(2), 152–163. doi:10.1080/15567036.2010.523760

Ogden, J., & Yang, C. (2005). *Implementing a Hydrogen Energy Infrastructure: Storage Options and System Design*. https://escholarship.org/uc/item/4xb940vg

Olabi, A. G. (2010). The 3rd international conference on sustainable energy and environmental protection SEEP 2009-Guest Editor's Introduction. *Energy*, *35*(12), 4508–4509. doi:10.1016/j.energy.2010.09.053

Olabi, A. G. (2012). *Developments in sustainable energy and environmental protection*. Elsevier. doi:10.1016/j.energy.2011.12.037

Olabi, A. G. (2013). *State of the art on renewable and sustainable energy*. Elsevier. doi:10.1016/j.energy.2013.10.013

Olabi, A. G. (2014). 100% sustainable energy. *Energy*, *77*, 1–5. doi:10.1016/j.energy.2014.10.083

Ou, X., Yan, X., Zhang, X., & Liu, Z. (2012). Life-cycle analysis on energy consumption and GHG emission intensities of alternative vehicle fuels in China. *Applied Energy*, *90*(1), 218–224. doi:10.1016/j.apenergy.2011.03.032

Owejan, J. P., Gagliardo, J. J., Sergi, J. M., & Trabold, T. A. (2008). Two-Phase Flow Considerations in PEMFC Design and Operation. *ASME 2008 6th International Conference on Nanochannels, Microchannels, and Minichannels*, 1733–1740. 10.1115/ICNMM2008-62037

Performance update on ITM Power's fuel cell membrane. (2013). *Fuel Cells Bulletin*, *11*(1), 11. doi:10.1016/S1464-2859(13)70026-9

Pollet, B. G., Staffell, I., & Shang, J. L. (2012). Current status of hybrid, battery and fuel cell electric vehicles: From electrochemistry to market prospects. *Electrochimica Acta*, *84*, 235–249. doi:10.1016/j.electacta.2012.03.172

Pottmaier, D., & Baricco, M. (2015). Materials for hydrogen storage and the Na-Mg-B-H system. *AIMS Energy, 3*(1), 75–100. doi:10.3934/energy.2015.1.75

Pozio, A., Silva, R. F., De Francesco, M., & Giorgi, L. (2003). Nafion degradation in PEFCs from end plate iron contamination. *Electrochimica Acta*, *48*(11), 1543–1549. doi:10.1016/S0013-4686(03)00026-4

Ren, Z., Zhang, D., & Wang, Z. (2012). Stacks with TiN/titanium as the bipolar plate for PEMFCs. *Energy*, *48*(1), 577–581. doi:10.1016/j.energy.2012.10.020

Saboohi, Y., & Farzaneh, H. (2009). Model for developing an eco-driving strategy of a passenger vehicle based on the least fuel consumption. *Applied Energy*, *86*(10), 1925–1932. doi:10.1016/j.apenergy.2008.12.017

Sharaf, O. Z., & Orhan, M. F. (2014). An overview of fuel cell technology: Fundamentals and applications. *Renewable & Sustainable Energy Reviews*, *32*, 810–853. doi:10.1016/j.rser.2014.01.012

Show, Y. (2007). Electrically conductive amorphous carbon coating on metal bipolar plates for PEFC. *Surface and Coatings Technology*, *202*(4–7), 1252–1255. doi:10.1016/j.surfcoat.2007.07.065

Singh, S., Jain, S., Ps, V., Tiwari, A. K., Nouni, M. R., Pandey, J. K., & Goel, S. (2015). Hydrogen: A sustainable fuel for future of the transport sector. *Renewable & Sustainable Energy Reviews*, *51*, 623–633. doi:10.1016/j.rser.2015.06.040

Stolten, D., & Grube, T. (Eds.). (2010). *18th World Hydrogen Energy Conference 2010 - WHEC 2010: Proceedings. 4: Parallel sessions book: storage systems, policy perspectives, initiatives and cooperations*. Forschungszentrum, IEF-3.

Sun, Y., Delucchi, M., & Ogden, J. (2011). The impact of widespread deployment of fuel cell vehicles on platinum demand and price. *International Journal of Hydrogen Energy*, *36*(17), 11116–11127. doi:10.1016/j.ijhydene.2011.05.157

Tawfik, H., Hung, Y., & Mahajan, D. (2007). Metal bipolar plates for PEM fuel cell—A review. *Journal of Power Sources*, *163*(2), 755–767. doi:10.1016/j.jpowsour.2006.09.088

Wang, H. (2003). Stainless steel as bipolar plate material for polymer electrolyte membrane fuel cells. *Journal of Power Sources*, *115*(2), 243–251. doi:10.1016/S0378-7753(03)00023-5

Wang, H., & Turner, J. A. (2007). SnO2:F coated ferritic stainless steels for PEM fuel cell bipolar plates. *Journal of Power Sources*, *170*(2), 387–394. doi:10.1016/j.jpowsour.2007.04.028

Wang, J. (2015). Barriers of scaling-up fuel cells: Cost, durability and reliability. *Energy*, *80*, 509–521. doi:10.1016/j.energy.2014.12.007

Weakley, S. A. (2012). *Pathways to Commercial Success: Technologies and Products Supported by the Fuel Cell Technologies Program* (PNNL-21804, 1057844; p. PNNL-21804, 1057844). doi:10.2172/1057844

Wilberforce, T., Alaswad, A., Palumbo, A., Dassisti, M., & Olabi, A.-G. (2016). Advances in stationary and portable fuel cell applications. *International Journal of Hydrogen Energy*, *41*(37), 16509–16522. doi:10.1016/j.ijhydene.2016.02.057

Yang, L., Yu, H., Jiang, L., Zhu, L., Jian, X., & Wang, Z. (2010). Improved anticorrosion properties and electrical conductivity of 316L stainless steel as bipolar plate for proton exchange membrane fuel cell by lower temperature chromizing treatment. *Journal of Power Sources*, *195*(9), 2810–2814. doi:10.1016/j.jpowsour.2009.11.018

Zhang, D., Duan, L., Guo, L., Wang, Z., Zhao, J., Tuan, W.-H., & Niihara, K. (2011). TiN-coated titanium as the bipolar plate for PEMFC by multi-arc ion plating. *International Journal of Hydrogen Energy*, *36*(15), 9155–9161. doi:10.1016/j.ijhydene.2011.04.123

Zhevago, N. K. (2016). Other methods for the physical storage of hydrogen. In *Compendium of Hydrogen Energy* (pp. 189–218). Elsevier. doi:10.1016/B978-1-78242-362-1.00008-0

Chapter 14
V2G Pilot Projects:
Review and Lessons Learnt

Sanchari Deb
School of Engineering, University of Warwick, Coventry, UK

Essam A. Al Ammar,
King Saud University, Saudi Arabia

Hasan AlRajhi
Umm AlQura University, Saudi Arabia

Ibrahin Alsaidan
Qassim University, Saudi Arabia

Samir Shariff
Taibah University, Saudi Arabia

ABSTRACT

Electric vehicles (EVs) are not only a viable energy efficient mode of transport, but they have considerable capacity of providing flexible and quick responding storage alternative based on vehicle-to-grid (V2G) scheme. V2G technology facilitates bidirectional flow of energy to and from the vehicle by a power converter. However, there is skepticism regarding the economic profitability of the V2G scheme. Despite the aforementioned challenges, the V2G technology is explored in matured markets. A number of V2G pilot projects across the world have investigated different aspects of V2G integration such as technological readiness, economic feasibility, social benefits, and challenges of V2G. This work aims to review the existing pilot projects on V2G functionality.

DOI: 10.4018/978-1-7998-6858-3.ch014

1. INTRODUCTION

Energy crisis, global warming, and environmental degradation are the crucial problems of 21st century. Electric Vehicles (EVs) are a clean mode of transport and an environment friendly alternative to tackle the problem of poor air quality. Moreover, low or negligible maintenance and operating cost are the additional benefits associated with EVs (Deb et al. 2018a; Teixeira & Sodre 2016). Motivated by the aforementioned advantages many countries are promoting the use of EVs for public as well as private transport. However, the transition from Internal Combustion Engine (ICE) driven vehicles to EVs is not completely barrier free. Lack of public charging stations, improper siting of charging stations, uncoordinated activities in establishing charging infrastructure are some of the key factors inhibiting the large scale deployment of EVs (Deb et al. 2018a; Amini et al. 2017; Ahmad et al. 2017). The increased load due to uncoordinated EV charging in the charging stations may hamper trouble free operation of the power system, thereby causing voltage instability, harmonics, power losses, as well as degradation of the reliability indices (Deb et al. 2018b; Deb et al. 2017a; Deb et al. 2017b; Deb et al. 2019). The non-uniform distribution of load in a particular location during a particular time is detrimental to the power utility. However, in recent years Vehicle Grid Integration (VGI) has become an attractive choice to manage the load by either returning electricity to the grid or by controlling the charging rate of the EVs. Moreover, the utilities can benefit from delayed or controlled charging (Sachan et al. 2020; Sachan et al. 2018). VGI facilitates plug-in EVs (PEVs) to act as generation or demand assets for grid operators. VGI can be unidirectional (V1G) or bidirectional (V2G) (Tan et al. 2016) and can help in frequency regulation, provide flexibility (Lezama et al. 2020) as well as reliability enhancement of the power system (Liu et al. 2018; Wu & Sioshansi 2019). V2G technology is explored worldwide and researchers are working to deal with the challenges associated with V2G technology. A number of V2G pilot projects across the world have investigated different aspects of V2G integration such as technological readiness, economic feasibility, social benefits, and challenges of V2G. This work aims to review the existing pilot projects on V2G functionality qualitatively as well as quantitatively and systematically capture the key lessons learnt from them.

2. V2G TECHNOLOGY

V2G technology facilitates bidirectional flow of energy to and from the vehicle by a power converter (Tomic& Kempton 2007). This novel technology is developed at University of Delaware by the research group of Professor Kempton (Kempton et al. 2013; Shinzaki et al. 2015). The architecture used for V2G technology at University of Delaware is as shown in Fig. 1.

3. V2G STANDARDS

The V2G standards proposed by different international organizations are listed in Table 1. From Table 1, it is observed that ISO has proposed several standards on different aspects of V2G such as communication and security

Figure 1. Architecture of V2G technology used at University of Delaware

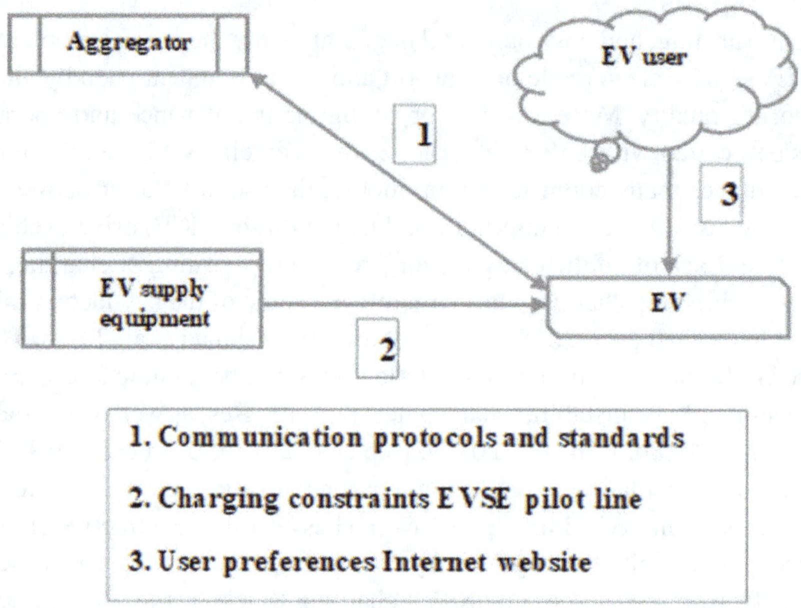

Table 1. V2G Standards (Das et al. 2020); (Kester et al. 2019)

Organization	Designation	Details
ISO	ISO 15118-1	Road vehicles- V2G communication interface-Part 1: general information and use case definition
	ISO 15118-1	Road vehicles- V2G communication interface-Part 2: network and application protocol requirements
	ISO 15118-3	Road vehicles- V2G communication interface-Part 3: physical and data link layer requirements
	ISO 15118-4	Road vehicles- V2G communication interface-Part 4: network and application protocol conformance test

4. POLICIES AND REGULATIONS FOR V2G

The concept of V2G is new globally. It would be useful to understand how policies and regulations have evolved in different parts of the world to support V2G. This section systematically captures the policies and regulations in major EV markets. The existing EV policies of some selected countries that support V2G are reported in Table 2. There is no clear policy on VGI in most of the countries. From Table 2, it is observed that there exist a number of policies for increasing EV adoption and sustainable development of EV charging infrastructure. EV adoption and development of sustainable charging infrastructure are prerequisite for successful implementation of V2G. It is expected that the policies listed in Table 2 will help in EV adoption thereby help in execution of V2G in near future. However, favorable policies for EV adoption are not going to be sufficient to enable implementation of V2G. From Table 2, it is observed that USA has strong policies on EV adoption as well as charging infrastructure development as compared to other countries.

Table 2. Global Policies indirectly supporting V2G[xi]

Country	Policy	Description	Timeline
USA (California)	EVSE policy for multi-unit dwelling	Encourages establishment of EVSE in designated parking or common space	2012 (updated)
	State Agency EVSE Installation	Encourages state agencies to identify locations for placement of EVSE and accommodate future EVSE demand Locations recommended are state employee parking facilities, new as well as existing agency buildings	2012
	EVSE Open Access Requirements	EVSE service providers may not charge a subscription fee or require membership for use of their public charging stations	2013
	The California Building Standards Commission	Establishes building standards for EVSE installation in parking lots and multi-unit dwellings	2013
	EVSE Policies for Residential and Commercial renters	Encourages installation of EVSE at residential locations	2014
	PEV Charging Electricity Exemption	Electricity used to charge PEVs at state owned parking facility is exempted from California law	2014
	EVSE Local Permitting Policies	A city or country must adopt an ordinance to streamline permitting process for EVSE	2015
	PEV Parking Space Regulation	Establishes special parking spots for EV users	2017
Germany	Joint Agency for Electric Mobility	Co-ordinate all Federal governments e mobility activities	2010
	Electric Mobility Act	Provide preferential treatment to EVs and charging stations	2014
Japan	Next Generation Vehicle Charging Infrastructure Deployment Promotion Project	Strategic and quick deployment of charging infrastructure Encourage municipalities and expressways operating organizations to issue charger deployment plans	2013
	Extension support by Tokyo	Introduction of additional fiscal incentives for charger deployment	2018
China	Notice of subsidy for charging infrastructure deployment	Subsidies for charging station deployment	2014
	Made in China 2025	Guidelines for development of NEV value chain	2015
	Guiding opinions on Accelerating on Construction of EV charging station	Construction and planning of charging stations	2015

5. CASE STUDIES

5.1. Electrification of a Porto Santo Island with Renewable

The Portuguese island of Porto Santo was primarily powered by diesel generators at high cost due to fuel importation. The local government wanted to utilize abundant sun and wind resources available in that island. However, intermittency issues hindered the progress.

The utility named EEM in collaboration with Renault and the Mobility House introduced VGI with special focus on V2G. Moreover, they developed the infrastructure and ecosystem of V1G EVs, V2G EVs, and stationary second-life Renault vehicle batteries in order to interface with the island's renewable.

The Mobility House built upon its energy Marketplace, as well as its smart charging platform, Charge Pilot, to optimize the flow of electricity to and from batteries, both stationary and in EVs.

The results indicated that CO2 emissions and costs have declined, and grid reliability and wind and solar utilization has increased.

5.2. EV benefits to Stadium

The stadium named Johan Cruijff ArenA in Amsterdam, is the home to the Ajax soccer team. It has managed charging and V2G technologies that supplement the stadium's solar and battery storage system. During non-event days, the system generates revenue by ancillary services such as frequency response. During events such as Champions League soccer matches and Ed Sheeran concerts, bidirectional EVs and stationary storage stand ready to shave expensive load peaks and provide backup power if necessary.

6. REVIEW OF SELECTED INTERNATIONAL PILOT PROJECTS ON V2G TECHNOLOGY

A comprehensive description of selected V2G pilot projects is presented in this section thereby elaborating the technical, economic as well as policy aspects of V2G scheme.

6.1. University of Delaware V2G Development and Demonstration

The V2G technology was discovered by the research group of Prof. Willett Kempton in 1990. The University of Delaware has already developed a set of communication or interacting technologies, policies, and market strategies to achieve V2G technology. Dr. Willett Kempton, Dr. Suresh Advani, and Dr. Ajay Prasad are the researchers at the University of Delaware conducting research on the V2G technology. The research works conducted at the University of Delaware on V2G scheme are summarized as in Table 4 in the Appendix.

6.2. US Air Force V2G Pilot Project

The US Department of Defense aimed at introduction of V2G services with the motive to decrease the net operating cost of EV fleet. This project was jointly funded by US Department of Defense, California Energy Commission, and Environmental Security Technology Certification Program. Lawrence Berkeley National Laboratory, Bosch Software Innovations, and Honeywell Inc. were partners in this project. The final project goal is to analyze the potential of EVs to support critical infrastructures during emergency especially in terms of frequency regulation (Murnay 2013). This project currently deploys 13 Nissan Leafs, 4 Ford C-Max and 2 Chevrolet Volt. A cost benefit analysis of the project is as shown in Table 3. From the findings of the project, it is clear that the regulation revenue from V2G is promising. However, the project also concluded that battery degradation, uncertainty in schedules are some of the challenges associated with V2G schemes.

Table 3. Cost Benefit Analysis of US Air Force V2G Pilot Project

Parameter	Value
Energy cost ($)	13.11
Revenue up($)	84.06
Revenue down($)	24.55
Daily energy cost ($)	0.72/ EV
Daily regulation revenue ($)	6.03/ EV
Net Benefit ($)	111.40 /EV/ month

6.3. Parker Project

Parker project is a Danish demonstration project focusing on V2G technology[v]. The project builds on the findings of EDISON (Binding et al. 2010) and Nikola project (Anderson et al. 2014) that analyzed the capability of EVs in supporting power system. The Parker project analyzed the grid applications, grid readiness, as well as scalability and replicability of V2G scheme[v]. Nissan, Mitsubishi Corporation, Mitsubishi Motors Corporation, Nuvve, Frederiksberg Forsyning A/S, Insero A/S, Enel X and the Electrical Engineering Department of Denmark Technical University were partners in this project. The main findings of the Parker project are[v]:

1. EVs models of PSA, Mitsubishi, and Nissan together with DC V2G chargers (Enel X) can support V2G and are ready to provide advanced frequency regulation services to the grid.
2. A field-test in Copenhagen concluded that it is possible to commercialize this technology though the provision of Frequency Containment Reserve (FCR).
3. Further steps must be taken for universal support of V2G and VGI services across all EV brands, standards, and markets.

6.4. Redispatch V2G

The Redispatch V2G is a joint project of European Transmission System Operator, Tenne T, the energy service provider, Mobility House, and automotive manufacturer, Nissan. The project aimed to make use of the potential of EV batteries for storing locally produced electricity and feeding it back to the grid during emergency[vi]. During the project phase, Nissan EVs were being used as mobile energy storage systems in the TenneT control area in Northern and Southern Germany for reduction of local overloads. The load and energy management software developed by The Mobility House enabled automatic control of the EV charging and discharging process. The battery cells in EVs provided opportunity for optimizing the grid. A decentralized storage option for storing excess renewable energies was created. An intelligent load control system supporting the grid stabilization and providing cost savings for end users was also developed.

6.5. INVENT Project

The INVENT project was initiated by Nuevve in 2017 to investigate whether EVs can be turned into VPP thereby feeding energy back to the grid via charger[vii]. Nuevve deployed its V2G enabled charging platforms on 50 new UC San Diego EV chargers. As part of the demonstration project, Nuvve also plans to use UC San Diego's solar forecasting technology to inform its electric vehicle charging schedule. This technology includes weather models, machine learning tools, and sky imagers – fish-eye lenses that capture a 360-degree view of the horizon and generate a 3-D model of the clouds they observe. The devices are connected to a sophisticated forecasting system that uses what it observes to predict solar power output in 15-minute increments. Thus, we can conclude that this project analyzed the technical aspects of V2G scheme rigorously. However, the commercial and policy aspects of V2G were neglected in this project.

6.6. ACES Project

The ACES project aimed to analyze technical and economic benefits that can be achieved by EV integration in Bornholm. The research undertaken as a part of this project was supported by real EV usage pattern, grid data, as well as field testing [viii]. The research simulated a full scale penetration scenario of EVs in Bornholm in order to access the impact of new aggregating EVs on the energy system. A small pilot project involving 50 publicly and privately owned Nissan vehicles and V2G chargers to effectively balance the power system was also a part of this project.

6.7. Grid Motion Project

The Grid Motion Project is a collaborative project involving different partners such as Group PSA, Direct Energie, Nuvve, Enel, Proxiserve, and Denmark Technical University. **Grid Motion project aimed to evaluate possible savings achieved by EV users through the implementation of smart charging and discharging strategies for EVs.** The project was carried out with two complementary types of users:

1. 50 Peugeot iOn, Partner Electric, Citroën C-ZERO or Berlingo owners tested "smart" unidirectional charging, in line with their mobility needs, when electricity prices are generally lower, such as night-time in France
2. A fleet of 15 B2B EV Peugeot iOn or Citroën C-ZERO vehicles with Enel bidirectional charging stations provided grid balancing services through short charging and discharging cycles, again carried out in line with mobility needs.

6.8. Massachusetts Electric School Bus Pilot Project

The Massachusetts Department of Energy Resources (DOER) initiated a pilot project to test the feasibility of using electrical school bus for proving V2G services. The electric school bus pilot project was funded through the Regional Greenhouse Gas Initiative (RGGI) with roughly $2 million. And, the project was administered by the Massachusetts State Department of Energy Resources. The project was planned to support four school districts with electric school buses, charging equipment as well and technology and systems needed to support V2B/V2G demonstrations[x]. The project concluded that electric school buses are still an emerging technology and requires more testing and validation [x]

6.9. V2H Pilot Projects

Stakeholders have expressed interest in whether EVs can serve as a backup power source during outage or demand response (DR) events. This value proposition could be achieved through a vehicle-to-home (V2H) configuration (Lazzeroni et al. 2019). V2H would island a home during an outage to provide on-site resiliency or for residential DR participation (providing load drop by islanding the house). Notably, V2H is distinct from, and less complex than, a vehicle-to-grid (V2G) configuration which requires different equipment set up and integrates an EV into the electric distribution system to provide grid support services through bidirectional power flow, rather than the unidirectional power flow of a V2H configuration. A V2H system is composed by at least one EV, a bi-directional charger, home loads, small-scale distributed generation, a smart meter, home grid itself and a home energy management system (HEMS). Researchers across the world are exploring various aspects of V2H technology. Globally, pioneering V2H projects have demonstrated cutting-edge insights on different aspects of V2H technology through learning by doing. A list of major V2H pilot projects and their findings is presented in Table 5 in the Appendix.

7. FUTURE RESEARCH DIRECTION

India is lagging behind compared to other countries in terms of deployment of V2G scheme. Researchers must come forward and actively engage in research that is essential for successful deployment of V2G scheme. A list of topics where rigorous research effort is required in developing countries are identified as follows:

1. Techno-Economic analysis of V2G scheme in the Developing countries market
2. Feasibility analysis of using EVs as VPP by V2G functionality
3. Development of framework for utilizing EVs as VPP in the regulatory context
4. Development of incentive mechanism to promote V2G scheme

For countries with established V2G technology the research must focus on addressing the gaps of smart charging such as:

1. Limited SCM Energy Management System (EMS) offerings with little/no connectivity to utility control systems
2. Planning tools use deterministic load profiles, even though EV loads are highly stochastic
3. EV infrastructure planning tools and EMS are not integrated - deviations could compromise reliable grid operation
4. SCM systems have limited cybersecurity protocols

8. CONCLUSION

EVs have the potential of reducing CO_2 emissions of the transport sector considerably. Apart from being an environment friendly mode of transport, EVs also have the potential to provide grid services at the time of emergency by V2G functionality. This work discusses comprehensively the V2G scheme with special focus on the internationally executed pilot projects on V2G scheme. The global status of V2G scheme is also reported and it is observed that Europe is leading in deployment of V2G scheme. This work also discusses some selected pilot projects on V2G functionality qualitatively as well as quantitatively and systematically capture the key lessons learnt from them. Moreover, the usefulness of V2G scheme in Indian context is analyzed and whether the lessons learnt from the international pilot projects are useful for India is discussed lucidly. The key research areas where India must focus for successful deployment of V2G scheme are also identified and reported. Thus, this work will serve as a guide to India for promoting V2G scheme.

REFERENCES

Ahmad, F., & Alam, M. S. (2017). Feasibility study, design and implementation of smart polygeneration microgrid at AMU. *Sustainable Cities and Society*, *35*, 309–322. doi:10.1016/j.scs.2017.08.007

Ahmad, F., Alam, M. S., & Asaad, M. (2017). Developments in xEVs charging infrastructure and energy management system for smart microgrids including xEVs. *Sustainable Cities and Society*, *35*, 552–564. doi:10.1016/j.scs.2017.09.008

Al-Awami, A. T., Amleh, N. A., & Muqbel, A. M. (2017). Optimal demand response bidding and pricing mechanism with fuzzy optimization: Application for a virtual power plant. *IEEE Transactions on Industry Applications*, *53*(5), 5051–5061. doi:10.1109/TIA.2017.2723338

Amini, M. H., Moghaddam, M. P., & Karabasoglu, O. (2017). Simultaneous allocation of electric vehicles' parking lots and distributed renewable resources in smart power distribution networks. *Sustainable Cities and Society*, *28*, 332–342. doi:10.1016/j.scs.2016.10.006

Amrr, S. M., Alam, M. S., Asghar, M. J., & Ahmad, F. (2018). Low cost residential microgrid system based home to grid (H2G) back up power management. *Sustainable Cities and Society*, *36*, 204–214. doi:10.1016/j.scs.2017.10.016

Andersen, P. B., Marinelli, M., Olesen, O. J., Andersen, C. A., Poilasne, G., Christensen, B., & Alm, O. (2014, October). The Nikola project Intelligent electric vehicle integration. In IEEE PES Innovative Smart Grid Technologies, Europe (pp. 1-6). IEEE. doi:10.1109/ISGTEurope.2014.7028765

Arslan, O., & Karasan, O. E. (2013). Cost and emission impacts of virtual power plant formation in plug-in hybrid electric vehicle penetrated networks. *Energy*, *60*, 116–124. doi:10.1016/j.energy.2013.08.039

Binding, C., Gantenbein, D., Jansen, B., Sundström, O., Andersen, P. B., Marra, F., & Træholt, C. (2010, July). Electric vehicle fleet integration in the Danish EDISON project-a virtual power plant on the island of Bornholm. In *IEEE PES general meeting* (pp. 1–8). IEEE. doi:10.1109/PES.2010.5589605

Calearo, L., Thingvad, A., Ipsen, H. H., & Marinelli, M. (2019). *Economic Value and User Remuneration for EV Based Distribution Grid Services. In 2019 IEEE PES Innovative Smart Grid Technologies Europe*. IEEE.

Das, H. S., Rahman, M. M., Li, S., & Tan, C. W. (2020). Electric vehicles standards, charging infrastructure, and impact on grid integration: A technological review. *Renewable & Sustainable Energy Reviews*, *120*, 109618. doi:10.1016/j.rser.2019.109618

Deb, S., Kalita, K., & Mahanta, P. (2017a, December). Review of impact of electric vehicle charging station on the power grid. In *2017 International Conference on Technological Advancements in Power and Energy (TAP Energy)* (pp. 1-6). IEEE. 10.1109/TAPENERGY.2017.8397215

Deb, S., Kalita, K., & Mahanta, P. (2017b, December). Impact of electric vehicle charging stations on reliability of distribution network. In *2017 International Conference on Technological Advancements in Power and Energy (TAP Energy)* (pp. 1-6). IEEE. 10.1109/TAPENERGY.2017.8397272

Deb, S., Kalita, K., & Mahanta, P. (2019). Distribution Network planning considering the impact of Electric Vehicle charging station load. In *Smart Power Distribution Systems* (pp. 529–553). Academic Press. doi:10.1016/B978-0-12-812154-2.00022-5

Deb, S., Tammi, K., Kalita, K., & Mahanta, P. (2018a). Review of recent trends in charging infrastructure planning for electric vehicles. *Wiley Interdisciplinary Reviews. Energy and Environment*, *7*(6), e306. doi:10.1002/wene.306

Deb, S., Tammi, K., Kalita, K., & Mahanta, P. (2018b). Impact of electric vehicle charging station load on distribution network. *Energies*, *11*(1), 178. doi:10.3390/en11010178

Deivanayagam, R. (2017). Vehicle-to-grid Technology: Concept, Status, and Challenges. *The Journal of Undergraduate Research at the University of Illinois at Chicago*, *10*(1). Advance online publication. doi:10.5210/jur.v10i1.8013

Godarzi, A., Niaki, S. N., Ahmadkhanlou, F., & Iravani, R. (2011, January). Local and global optimization of exportable vehicle power based smart microgrid. In *ISGT 2011* (pp. 1–7). IEEE. doi:10.1109/ISGT.2011.5759144

Hassan, A. S., Marmaras, C. E., Xydas, E. S., Cipcigan, L. M., & Jenkins, N. (2013). *Integration of wind power using V2G as a flexible storage*. Academic Press.

Hosseini, S. S., Badri, A., & Parvania, M. (2012, September). The plug-in electric vehicles for power system applications: The vehicle to grid (V2G) concept. In *2012 IEEE International Energy Conference and Exhibition (ENERGYCON)* (pp. 1101-1106). IEEE. 10.1109/EnergyCon.2012.6347734

Ju, L., Li, H., Zhao, J., Chen, K., Tan, Q., & Tan, Z. (2016). Multi-objective stochastic scheduling optimization model for connecting a virtual power plant to wind-photovoltaic-electric vehicles considering uncertainties and demand response. *Energy Conversion and Management*, *128*, 160–177. doi:10.1016/j.enconman.2016.09.072

Kahlen, M., & Ketter, W. (2015, February). Aggregating electric cars to sustainable virtual power plants: The value of flexibility in future electricity markets. *Twenty-Ninth AAAI Conference on Artificial Intelligence*.

Kahlen, M. T., Ketter, W., & van Dalen, J. (2018). Electric vehicle virtual power plant dilemma: Grid balancing versus customer mobility. *Production and Operations Management*, *27*(11), 2054–2070. doi:10.1111/poms.12876

Kempton, W., Marra, F., Andersen, P. B., & Garcia-Valle, R. (2013). Business models and control and management architectures for EV electrical grid integration. In *Electric Vehicle Integration into Modern Power Networks* (pp. 87–105). Springer. doi:10.1007/978-1-4614-0134-6_4

Kempton, W., & Tomić, J. (2005). Vehicle-to-grid power implementation: From stabilizing the grid to supporting large-scale renewable energy. *Journal of Power Sources*, *144*(1), 280–294. doi:10.1016/j.jpowsour.2004.12.022

Kempton, W., & Tomić, J. (2005). Vehicle-to-grid power fundamentals: Calculating capacity and net revenue. *Journal of Power Sources*, *144*(1), 268–279. doi:10.1016/j.jpowsour.2004.12.025

Kester, J., Noel, L., Lin, X., de Rubens, G. Z., & Sovacool, B. K. (2019). The coproduction of electric mobility: Selectivity, conformity and fragmentation in the sociotechnical acceptance of vehicle-to-grid (V2G) standards. *Journal of Cleaner Production*, *207*, 400–410. doi:10.1016/j.jclepro.2018.10.018

Lazzeroni, P., Olivero, S., Repetto, M., Stirano, F., & Vallet, M. (2019). Optimal battery management for vehicle-to-home and vehicle-to-grid operations in a residential case study. *Energy*, *175*, 704–721. doi:10.1016/j.energy.2019.03.113

Letendre, S. E., Kempton, W., & Stavros, R. (2002). The V2G concept: A new model for power? *Public Utilities Fortnightly*, *140*(4), 16–16.

Lezama, F., Soares, J., Canizes, B., & Vale, Z. (2020). Flexibility management model of home appliances to support DSO requests in smart grids. *Sustainable Cities and Society*, *55*, 102048. doi:10.1016/j.scs.2020.102048

Liu, H., Wang, B., Wang, N., Wu, Q., Yang, Y., Wei, H., & Li, C. (2018). Enabling strategies of electric vehicles for under frequency load shedding. *Applied Energy*, *228*, 843–851. doi:10.1016/j.apenergy.2018.06.122

Liu, Z., Zheng, W., Qi, F., Wang, L., Zou, B., Wen, F., & Xue, Y. (2018). Optimal dispatch of a virtual power plant considering demand response and carbon trading. *Energies*, *11*(6), 1488. doi:10.3390/en11061488

Marnay, C. (2013). *Los Angeles air force base vehicle to grid pilot project*. Academic Press.

Musio, M., & Damiano, A. (2012, October). A virtual power plant management model based on electric vehicle charging infrastructure distribution. In 2012 3rd IEEE PES Innovative Smart Grid Technologies Europe (ISGT Europe) (pp. 1-7). IEEE. doi:10.1109/ISGTEurope.2012.6465684

Musio, M., Lombardi, P., & Damiano, A. (2010, September). Vehicles to grid (V2G) concept applied to a virtual power plant structure. In *The XIX International Conference on Electrical Machines-ICEM 2010* (pp. 1-6). IEEE. 10.1109/ICELMACH.2010.5608261

Mustafa, M. A., Zhang, N., Kalogridis, G., & Fan, Z. (2013, February). Smart electric vehicle charging: Security analysis. In 2013 IEEE PES Innovative Smart Grid Technologies Conference (ISGT) (pp. 1-6). IEEE.

Noel, L., de Rubens, G. Z., Kester, J., & Sovacool, B. K. (2019). Navigating expert skepticism and consumer distrust: Rethinking the barriers to vehicle-to-grid (V2G) in the Nordic region. *Transport Policy, 76*, 67–77. doi:10.1016/j.tranpol.2019.02.002

Noel, L., & McCormack, R. (2014). A cost benefit analysis of a V2G-capable electric school bus compared to a traditional diesel school bus. *Applied Energy, 126*, 246–255. doi:10.1016/j.apenergy.2014.04.009

Pandžić, H., Morales, J. M., Conejo, A. J., & Kuzle, I. (2013). Offering model for a virtual power plant based on stochastic programming. *Applied Energy, 105*, 282–292. doi:10.1016/j.apenergy.2012.12.077

Parsons, G. R., Hidrue, M. K., Kempton, W., & Gardner, M. P. (2014). Willingness to pay for vehicle-to-grid (V2G) electric vehicles and their contract terms. *Energy Economics, 42*, 313–324. doi:10.1016/j.eneco.2013.12.018

Pasetti, M., Rinaldi, S., & Manerba, D. (2018). A virtual power plant architecture for the demand-side management of smart prosumers. *Applied Sciences (Basel, Switzerland), 8*(3), 432. doi:10.3390/app8030432

Raab, A. F., Ferdowsi, M., Karfopoulos, E., Unda, I. G., Skarvelis-Kazakos, S., Papadopoulos, P., . . . Strunz, K. (2011, September). Virtual power plant control concepts with electric vehicles. In *2011 16th International Conference on Intelligent System Applications to Power Systems* (pp. 1-6). IEEE. 10.1109/ISAP.2011.6082214

Sachan, S., & Adnan, N. (2018). Stochastic charging of electric vehicles in smart power distribution grids. *Sustainable Cities and Society, 40*, 91–100. doi:10.1016/j.scs.2018.03.031

Sachan, S., Deb, S., & Singh, S. N. (2020). Different charging infrastructures along with smart charging strategies for electric vehicles. *Sustainable Cities and Society, 60*, 102238. doi:10.1016/j.scs.2020.102238

Shinzaki, S., Sadano, H., Maruyama, Y., & Kempton, W. (2015). *Deployment of vehicle-to-grid technology and related issues* (No. 2015-01-0306). SAE Technical Paper.

Singh, M., Kumar, P., & Kar, I. (2010, September). Analysis of vehicle to grid concept in Indian scenario. In *Proceedings of 14th International Power Electronics and Motion Control Conference EPE-PEMC 2010* (pp. T6-149). IEEE. 10.1109/EPEPEMC.2010.5606641

Taheri, H., Ghasemi, H., Rahimi-Kian, A., & Akbari, T. (2011, May). Modeling and optimized scheduling of virtual power plant. In *2011 19th Iranian Conference on Electrical Engineering* (pp. 1-6). IEEE.

Tan, K. M., Ramachandaramurthy, V. K., & Yong, J. Y. (2016). Integration of electric vehicles in smart grid: A review on vehicle to grid technologies and optimization techniques. *Renewable & Sustainable Energy Reviews, 53*, 720–732. doi:10.1016/j.rser.2015.09.012

Teixeira, A. C. R., & Sodré, J. R. (2016). Simulation of the impacts on carbon dioxide emissions from replacement of a conventional Brazilian taxi fleet by electric vehicles. *Energy, 115*, 1617–1622. doi:10.1016/j.energy.2016.07.095

Thingvad, A., Calearo, L., Andersen, P. B., Marinelli, M., Neaimeh, M., Suzuki, K., & Murai, K. (2019). *Value of V2G Frequency Regulation in Great Britain Considering Real Driving Data. In 2019 IEEE PES Innovative Smart Grid Technologies Europe*. IEEE.

Thingvad, A., & Marinelli, M. (2018). Influence of v2g frequency services and driving on electric vehicles battery degradation in the nordic countries. *31st International Electric Vehicles Symposium & Exhibition & International Electric Vehicle Technology Conference 2018*.

Thingvad, A., Ziras, C., Hu, J., & Marinelli, M. (2017, August). Assessing the energy content of system frequency and electric vehicle charging efficiency for ancillary service provision. In *2017 52nd International Universities Power Engineering Conference (UPEC)* (pp. 1-6). IEEE. 10.1109/UPEC.2017.8231947

Tomić, J., & Kempton, W. (2007). Using fleets of electric-drive vehicles for grid support. *Journal of Power Sources*, *168*(2), 459–468. doi:10.1016/j.jpowsour.2007.03.010

Tsoleridis, C., Chatzimisios, P., & Fouliras, P. (2016). Vehicle-to-Grid Networks: Issues and Challenges. In Smart grid: networking, data management, and business models (pp. 347-369). CRC Press.

Vandael, S., Holvoet, T., Deconinck, G., Kamboj, S., & Kempton, W. (2013, October). *A comparison of two GIV mechanisms for providing ancillary services at the University of Delaware. In 2013 IEEE international conference on smart grid communications (SmartGridComm)*. IEEE.

Vasirani, M., Kota, R., Cavalcante, R. L., Ossowski, S., & Jennings, N. R. (2012). Technical Report: Virtual Power Plants of Wind Power Generators and Electric Vehicles. Tech. rep.

Vasirani, M., Kota, R., Cavalcante, R. L., Ossowski, S., & Jennings, N. R. (2013). An agent-based approach to virtual power plants of wind power generators and electric vehicles. *IEEE Transactions on Smart Grid*, *4*(3), 1314–1322. doi:10.1109/TSG.2013.2259270

Wang, Q., Rongrong, L., & Rui, J. (2016). Decoupling and Decomposition Analysis of Carbon Emissions from Industry: A Case Study from China. *Sustainability*, *8*(10), 1059–1076. doi:10.3390u8101059

Wu, F., & Sioshansi, R. (2019). A stochastic operational model for controlling electric vehicle charging to provide frequency regulation. *Transportation Research Part D, Transport and Environment*, *67*, 475–490. doi:10.1016/j.trd.2018.12.005

Yang, J. J., Zhao, J. H., Wen, F. S., Xue, Y. S., Li, L., & Lv, H. H. (2014). Development of bidding strategies for virtual power plants considering uncertain outputs from plug-in electric vehicles and wind generators. *Dianli Xitong Zidonghua*, *38*(13), 92–102.

Zecchino, A., Knezović, K., & Marinelli, M. (2017, September). Identification of conflicts between transmission and distribution system operators when acquiring ancillary services from electric vehicles. In 2017 IEEE PES Innovative Smart Grid Technologies Conference Europe (ISGT-Europe) (pp. 1-6). IEEE. doi:10.1109/ISGTEurope.2017.8260127

Zecchino, A., & Marinelli, M. (2018). Analytical assessment of voltage support via reactive power from new electric vehicles supply equipment in radial distribution grids with voltage-dependent loads. *International Journal of Electrical Power & Energy Systems*, *97*, 17–27. doi:10.1016/j.ijepes.2017.10.034

Zecchino, A., Prostejovsky, A. M., Ziras, C., & Marinelli, M. (2019). Large-scale provision of frequency control via V2G: The Bornholm power system case. *Electric Power Systems Research*, *170*, 25–34. doi:10.1016/j.epsr.2018.12.027

Zecchino, A., Thingvad, A., Andersen, P. B., & Marinelli, M. (2019). Test and Modelling of Commercial V2G CHAdeMO Chargers to Assess the Suitability for Grid Services. *World Electric Vehicle Journal*, *10*(2), 21. doi:10.3390/wevj10020021

Ziras, C., Zecchino, A., & Marinelli, M. (2018, June). Response accuracy and tracking errors with decentralized control of commercial v2g chargers. In 2018 Power Systems Computation Conference (PSCC) (pp. 1-7). IEEE. doi:10.23919/PSCC.2018.8442488

APPENDIX

Table 4. Summary of research works on V2G scheme at University of Delaware

Author	Year	Domain	Summary	Findings
Parsons et al.	2014	V2G and power markets	Survey of more than 3,000 respondents regarding possible V2G contracts	• Drivers did not like fixed requirements like required plug-in time per month • They seemed prefer arrangements such as being paid more or less depending on the amount of time they stayed plugged in
Tomic& Kempton	2007	V2G and power markets	Examination of two actual electric vehicle fleets namely 100 Think City vehicle and 252 Toyota RAV4, their operating cycles, and the value of revenue from these vehicles if they were equipped for V2G power	With Level 2 charging infrastructure of 6.6 kW the annual net profit from Think city fleet is 7000 to 70,000 USD providing regulation down only From RAV4 fleet, the annual net profit is 24,000 to 260000 providing regulation up and down
Kempton &Tomic	2005	V2G and power markets	Analysis of V2G as storage for large-scale renewable electricity	V2G could stabilize large-scale (one-half of US electricity) wind power with 3% of the fleet dedicated to regulation for wind and 8–38% of the fleet providing operating reserves or storage for wind.
Vandael et al.	2013	V2G technology	Comparison between centralized and decentralized to control V2G scheme	Centralized mechanism is beneficial but involves computational burden
Noel &Cormack	2014	Policy	Cost benefit analysis of a V2G enabled Electric School Bus compared to traditional diesel school bus	The V2G-capable electric bus provides the school savings of $6,070 per seat in net present value
Kempton &Tomic	2005	V2G and power markets	Calculation of the capacity and net revenue from V2G scheme	Economic motivation for V2G scheme is compelling. Societal advantages of developing V2G include an additional revenue stream for cleaner vehicles, increased stability and reliability of the electric grid

Table 5. Requirements, Availability, and Estimated Cost of using electrical school buses for V2G technology[x]

Category	On Board system requirements	Availability	Estimated cost	Off Board system requirements	Availability	Estimated cost
Bus upgrades	Bidirectional charger on bus (hardware) and communication module (software	Available from Lion but not fully developed or certified	$12,000 approx	Combo plug	Available from Lion	$ 8000 per bus
Charging infrastructure	Capable of bidirectional flow	Level 2 EVSE capable of bidirectional with the necessary power is not yet available	$25,000 / unit	DC fast charger with inverter capable of bidirectional power flow	Available from Coritech	$60,000 / unit
Building controller	Communication device (including hardware and software)	EPRI has one available (not UL certified) - requires additional software	$25,000 / unit	Communication device (including hardware and software)	EPRI has one available (not UL certified, and Kisensum has integrated with Coritech	$25,000 / unit for EPRI $20,000 for Kisensum w/Coritech
Installation and Commissioning	----------------	----------------	$10,000 / site			$10,000 if 3-phase power, 480V is available otherwise $20,000 if 3-phase power needs to be run to bus

Table 6. Pilot Projects on V2H

Pilot Project	Year	Country	Findings
Nissan V2H Australia[Xiii]	2015	Australia	A V2H-equipped LEAF could be an alternative to a home battery system such as Tesla's Powerwall. It could save consumers money by storing solar energy for night time use, and/or by charging at off-peak rates overnight and discharging power during the day. It is estimated that the 40 kWh version of the LEAF offers up to 4 times the energy storage capacity of a typical home battery
INVADE [xiv]	2017-2020	Norway, Germany, Spain, Netherland, Bulgaria	Demonstrated how big data, machine learning and analytics parts of the INVADE platform can be integrated into the existing smart home solutions demonstrated in the Triangulum Smart City solutions in Stavanger with respect to V2H, batteries and boilers.
TM4, Hydro-Québec V2H Pilot [xv]	2012-2014	Canada	Tested the use of energy stored in the battery when an electric vehicle is parked at home.
Franklin, Tennessee Nissan Leaf Pilot [xvi]	2015	North America	Use of LEAF vehicles to assist in powering its headquarters facilities during peak electrical demand times, anticipating significant cost savings
Leaf to Home [xvii]	2010	Japan	Concluded that Leaf to Home system which could power a typical Japanese house during two days.
Toyota Smart Home [xviii]	2012	Japan	Developed a V2H system for the mutual sharing of power between EVs and homes. The V2H system was using Prius PHEVs, in 10 households

Chapter 15
Implementation and Demonstration of Electric Vehicle-to-Home (V2H) Application:
A Case Study

Mahdi Shafaati Shemami
CARET, Aligarh Muslim University, India

Marzieh Sefid
https://orcid.org/0000-0001-6142-5879
Aligarh Muslim University, India

ABSTRACT

This chapter emphasizes the utilization of the plug-in hybrid electric vehicle (PHEV) as a backup power source for residential loads in under-developing and developing countries. It works as a source of energy in residential micro-grid based on the condition of vehicle battery without harming its function as an EV (electric vehicle). The suggested V2H system uses solar PV power to charge vehicle battery; therefore, the entire system works as a residential nano-grid system. The EV is considered as a load of home when its batteries are charged by solar PV or grid. However, the main emphasis is given to use solar PV power to reduce charging from the grid. The key objectives of this work are to minimize the energy cost of a household by reducing the dependency of residential loads on the power grid to enhance the reliability of power supply to residential loads during load shedding and blackouts and to maximize the utilization of power produced by solar PV array mounted on the rooftop.

DOI: 10.4018/978-1-7998-6858-3.ch015

INTRODUCTION

Climate change and pollution of the environment are the two challenges that the world is facing today. Carbon dioxide (CO2) emission analysis of the world is shown in Figure 1 shows that transportation is a major contributor to increase CO2 emission (International Emissions Agency 2015). Especially in the U.S., personal vehicles (light-duty vehicles) represent 60% of the transportation emissions is shown in Figure 2 (Transportation Overview I Center for Climate and Energy Solutions n.d.).

Figure 1. World CO_2 emissions by sector in 2013 (International Emissions Agency 2015)

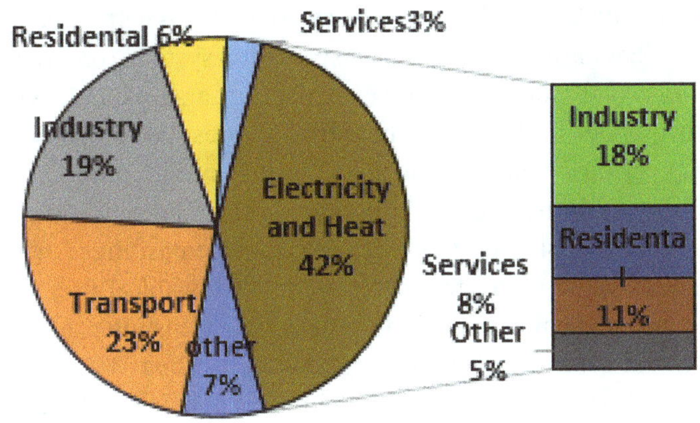

Figure 2. U.S. transportation sector CO_2 emission, by source (Transportation Overview Center for Climate and Energy Solutions n.d.)

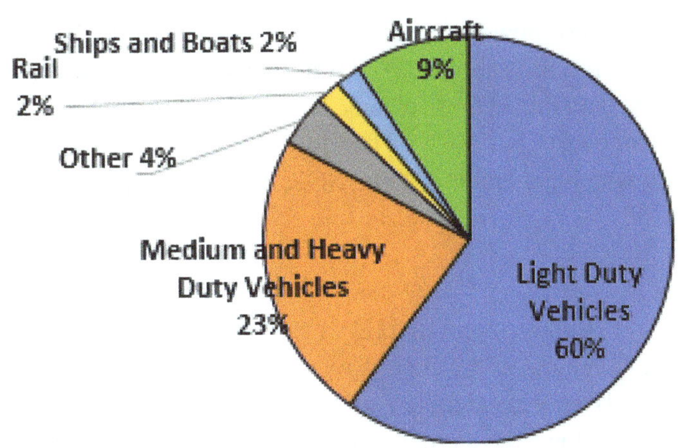

Emissions of private-sector vehicles (2 wheelers and 4 wheelers) can be decreased by the utilization of EVs and PHEV. The EVs/ PEVs / PHEVs (it can be called xEVs) have achieved considerable attention in the last years because of rising attention about the energy crisis, environmental pollution and rapid climate change. XEVs are vehicles with a large battery pack. Thus, the higher mileage and lower cost per unit (Rs/km) led to an increase of EV penetration in the vehicle market in recent years

(All-Electric Vehicles n.d.). It is estimated that the worldwide sales of EVs will reach a staggering 3.8 million by 2020 (Mehta et al. 2016).

In India over 85% of the houses of rural area in five states out of the six states of Jharkhand, Bihar, Madhya Pradesh, West Bengal, Uttar Pradesh, and Odisha had electrical energy less than 8 h. In some areas for less than 50 W loads also there is a power supply problem, in some areas electrical energy is not available for the entire day (Jhunjhunwala, Lolla, and Kaur 2016). Generally, in India, a Lead-acid battery-based UPS/inverter system utilized in the household as a backup power supply. The rated capacity normally used is generally 2 kWh to 5 kWh that regularly, one or two batteries in series or parallel connection are used for this case. The EVs battery in comparison UPS based system has a large capacity. The small EV has a battery pack in the range of 12–18 kWh capacity, the middle-sized EVs battery pack is in the range of 22–32 kWh, and the large sedan vehicle has 60–90 kWh battery package (University 2017).

India has the objective of having an all-electric transportation system to decrease fuel import charge by 2030 in the National Mission on Electric mobility (All electric cars by 2030? Nitin Gadkari inaugurates EV charging points at Niti Aayog | India News, The Indian Express n.d.). Presently, in India on the roads there are more than 18,000 electric vehicles (electromobilitysummit.com n.d.), this number of EV can be simply used through load shedding and power outage. India at present installed more than 3.45 GW solar PV system and the goal of installation of 100 GW has been arranged with an outlay of US$100 billion by the government of India (Ministry of New and Renewable Energy - Scheme / Documents n.d.). The manufacturing cost and payback period of the solar PV installation are falling day by day due to the fast growth in PV manufacturing technology (The Cost of Solar Will Drop Another 25% by 2022 n.d.). Consequently, solar PV panel's costs are becoming inexpensive and are utilized in many places. Using the solar PV in the residential application on the rooftop, in addition to reducing the electricity bill for owners of the house, on the generation side can save more energy by reducing energy loss in the transmission line. Transmission line losses in India as it is reported in (CEA 2017) are 23%. Therefore, the generation of 1 kW power by rooftop solar PV can save 1.23 kW power on the power plant side. Coal consumption for the generation of power in India is 545.9 MT (CEA 2017) and the typical efficiency of coal-fired power plants at present is about 33% (High Efficiency Low Emission Coal | HELE | WCA n.d.). For a 100 W bulb, based on the 33% efficiency, 876 kWh energy per year will consume and 6,150 kWh/ton is thermal energy of coal (CEA 2017). Therefore, 357 kg coal will be burned to provide enough energy for the 100 W bulb which generates 2.26 kg SO2, 840 kg CO2, and 2.31 kg NOx.

Background

Vehicle-to-Home (V2H) and Home-to-Vehicle (H2V) systems introducing the capabilities of vehicles for conserving energy in the vehicle and utilize it whenever it is necessary (Shafaati Shemami, Saad Alam, and Jamil Asghar 2018). In V2H and H2V system the bidirectional charger is not only for charging the EV battery, but it also allows the battery to feed power to the household loads to help the power grid. Usually, the average vehicle trip length from home-to-work is not long for a particular case, it is found to be 12 miles (19.4 km) (Morrow, Karner, and Francfort 2008). Thus, a substantial amount of unutilized energy of the battery can be utilized. The EV battery can provide energy at the peak demand period for the residential loads whenever the user reaches home and then the charging of the EV battery can be shifted to the off-peak-load period during night time. During the night, the EVs battery works as an energy storage system, further, solar energy can be utilized to increase energy production.

Nissan LEAF introduced the Control System to deliver electricity to homes through the EVs Power. The control system lets a Nissan EV provide the electricity stored in lithium-ion batteries package with residential home whenever EV is connected to the electricity distribution board. The inbuilt connector is CHAdeMO type which is a fast-charging protocol, it is available in global markets and guarantees a high level of adaptability, reliability, and stability (ENSTO 2016).

In another case, the basic V2H control system came to the marketplace by the Mitsubishi Company. The Outlander PHEV of Mitsubishi supply electrical power to residential loads and works as a source of energy by utilizing stored energy of the PHEV battery to run home appliances in a power shedding or blackouts or an emergency case (Press Release I News･Events I MITSUBISHI MOTORS n.d.).

Nichicon Corporation has introduced the V2H system in 2012 for Nissan LEAF and in 2013 developed it for business continuity plans (BCPs) and in 2014 offered a high-performance model (ZHTP1900R). The output power of the ZHTP1900R model input is 6 kW (less than 30 A) during charging and it can provide a maximum of 60A for two outlets (NICHICON CORPORATION I News Release I EV Power Station Connection Enabled For Electric Vehicles (EVs) from Mitsubishi Motors Corporation, thereby Expanding EV Applications n.d.).

DENSO Corporation in 2013 has started to demonstrate a V2H power supply system for Toyota Motors and Mitsubishi Motors EV. The V2H system is designed to work with DENSO's HEMS. It can charge EV for just 15 minutes with the DENSO system and the EV be able to move up to 20 km (DENSO Develops Vehicle-to-Home Power Supply System for Electric Vehicles I News I DENSO Global Website n.d.). Many companies developed their standard for charging like Nissan LEAF or Tesla so, there are three major charging levels: Level 1, Level 2, and level three which are shown in Table 1 separately (Group, Industry, and Unit 2019).

Table 1. Charging levels and mileage range

Type of Charging	Power Levels	Km of Range /hour of Charge	Charging Place
Level 1	110/120 VAC (15 to 20 Amps)	6.5-9.5 km/hour	Home or Workplace
Level 2 3.3 kW 6.6 kW 9.6 kW 19.2 kW	210/240 VAC 20 Amps 40 Amps 50 Amps 100 Amps	12.5-19.5 km/hour	Home, workplace, or public charging station
Level 3 Fast Charging	440 or 480 VAC	80 < 30 Min	Public charging station or commercial

Fast charging has five main standards which are CHAdeMO (introduced by Japanese Companies), U.S. standard which is CCS1, CCS2 introduced by European industry, the Chinese company introduced GB/T standard, and Tesla Supercharger established by Tesla Company (Agarwal et al. 2020).

This work proposes the use of EV as a backup power source for residential loads as a part of the residential micro-grid in underdeveloped countries and developing countries, without affecting operation EV as a vehicle. This work also presents the solar PV-based charging system for EV, therefore, the entire V2H and H2V system work as a nano-grid system. The EV battery can be charged through the power grid as a load of the home but the key contribution of this work is the use of solar PV panels as a power source to charge the EV to minimize the effect of EV charging stress on the power grid. Two charging methods of DC slow and fast charging are applied that in both methods constant voltage (CV) and constant current (CC) charging modes have been applied for better performance.

The main goal of this work is to reduce the energy cost of the home that interfaces with the rooftop solar PV source and EV battery. It also, reduce the dependency of home-to-grid and enhance the reliability of power supply to residential loads in case of load shedding and blackouts. Maximize the utilization of power produced by the solar PV panel mounted on the rooftop. Moreover, utilization of the DC bus for DC loads to reduce power loss in DC to AC and AC to DC conversion.

Load shedding, blackout, and the high price of electricity caused a lot of difficulty in the utilization of energy. It is a common problem in developing and undeveloped countries. Apart from the lower power consumption of each household (just to fulfill the basic need of energy), even that is not available reliably. Power cut/ load shedding is common for houses (Jhunjhunwala, Lolla, and Kaur 2016). Thus, the target is the optimization of power consumption for home by sharing or flattening the peak hour's energy consumption through the utilization of solar PV-based energy and EV and reducing the home dependency on the power grid.

In developing countries, a low power backup supply for emergency loads are extensively utilized in millions of household. The energy storage capacity of the lead-acid-based batteries which are normally used is very low (few kWh only depending upon the capacity of battery) (Shafaati Shemami, Saad Alam, and Asghar 2018). The capacity of the battery which is commonly utilized in a conventional Indian home is 12 V and 150 AH which have 2.16 kWh capacity. The efficiency of the lead-acid battery, in general, is 65% thus the effective energy which can be utilized is 1.44 kWh only. Thus, the backup period is very less. However, xEV has an enormous energy storage capacity (Shemami et al. 2018). This capacity of xEV can be easily utilized during the idle period of the vehicle as a standby power source for the household.

Design and Development of Vehicle-to-Home (V2H) Backup Power System

Background and Feasibility Study

In the previous section merit and the demerits of the state-of-the-art for the V2H systems have been discussed and the need for proposed work was described. The main goal of this work is to reduce the energy cost of the home that interfaces with the rooftop solar PV source and EV battery. It also, reduce the dependency of home-to-grid and enhance the reliability of power supply to residential loads in case of load shedding and blackouts. Maximize the utilization of power produced by the solar PV panel mounted on the rooftop. Moreover, utilization of the DC bus for DC loads to reduce power loss in DC to AC and AC to DC conversion. In this section, modeling, simulation and the results of the proposed system are described.

Proposed V2H System

In this work, the V2H system can work as a stand-alone system or grid-connected system. The designed V2H system as shown in Figure 3 has the advantage of both applications of stand-alone and grid-connected. Moreover, it is cheap as it involves only a standalone inverter which does not cost much. The utilization of the PV system that is integrated with the power grid supply provides cheap and reliable power sources for households.

The V2H system has a hierarchical control that the residential loads are divided into luxury loads (which consume more power), normal loads (that consume less power), and essential/emergency loads.

The washing machine, room heaters, air conditioners, etc. that are luxury loads take power directly from the utility network, these luxury loads remain switched off through load shedding or blackout. The fans, lights, fridge, room coolers are normal loads that generally used in typical Indian homes, receive power from the V2H system, which includes solar PV panel and EV as a standby power source. The emergency light and few fans categorized as emergency loads are connected to the battery-based storage, DC-to-AC inverter, or UPS which has low power storage just for a few hour back-ups.

The V2H system can be fitted for urban areas of developing countries that the power supply is not reliable and every day few hours of load shedding is usual. The utilization of solar PV energy help in decreasing grid stress and load demand. Therefore, the V2H system additionally works as a peak shaving power plant by disconnecting the residential loads from the power grid for the time of maximum solar insolation hours that the daily peak load hours are at the same time. However, when the stored energy of the EV battery in the night hours is exhausted if load shedding happening at this time, it causes a shutdown of domestic loads. Thus, it is necessary to have a standby power supply to provide minimum power for essential or emergency loads.

Figure 3. Detailed block diagram of power flow for the proposed system

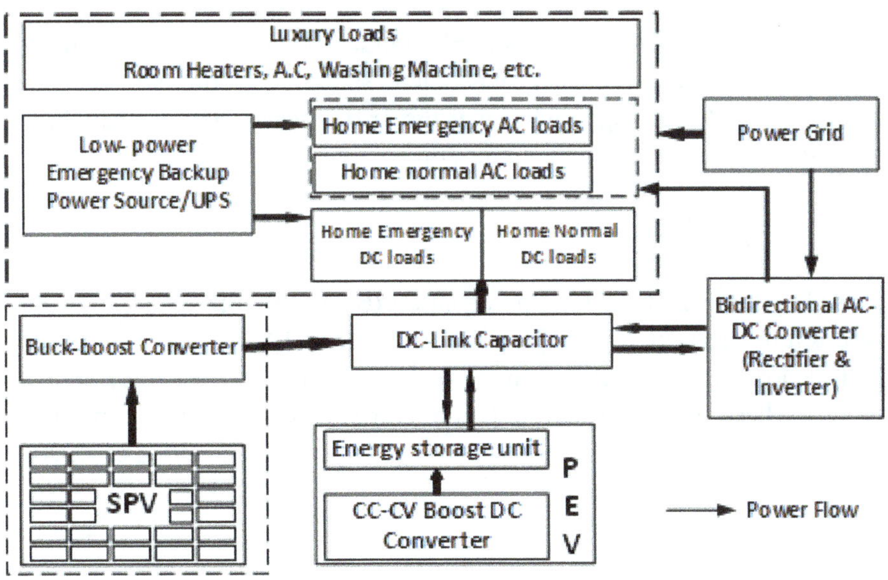

The proposed V2H system is developed to make a reliable and cheap power source for the residential home. Some constrain should be considered:

$$Power_{Generation} \geq Power_{Demand} \quad (1)$$

where

$$Power_{Generation} = Power_{PV}, Power_{Grid}, Power_{EV}, or\ Power_{Standby\ emergency} \tag{2}$$

In the case of economic intentions, the effort is to utilization and maximum harness solar energy, so, all possible solar PV panel energy should be utilized. Therefore, EV batteries can store surplus solar PV energy. Whenever solar insolation is high, the system statement is:

$$P_{PV} \geq P_{Load} + P_{EV\ Battery\ charging} \tag{3}$$

Also, during low insolation:

$$P_{EV\ Battery\ discharging} \geq P_{Load} - P_{PV} \tag{4}$$

In the same way, in the night whenever the EV battery discharged:

$$P_{Grid} \geq P_{Load} \tag{5}$$

When there is load shedding during the night:

$$P_{Standby\ essential/emergency} \geq P_{Essential/emergency\ load} \tag{6}$$

In reliability mode, the system provides a reliable standby power source, even in case of utilization of the solar PV power. This mode is necessary where load shedding is common and sometimes the load shedding is 10 to 12 hours. Accordingly, whenever the grid power supply is available and the SoC of the EV battery is low the suggested system expression is:

$$P_{Grid} = P_{Load} + P_{EV\ Battery\ charging} - P_{PV} \tag{30.7}$$

While load shedding happened it works on equation (6). In reliable mode, the emergency backup battery always keeps fully charged, either by power grid or solar PV panel power, to be able to provide energy for load during load shedding. The solar PV output power characteristics frequently repeat every 24 hours. Therefore, the battery level of charge should return to a similar state every 24 hours, the limitation on battery energy storage (BES) energy balance is stated in reference (Ha et al. 2017) as:

$$\begin{cases} \sum_{t=1}^{H_1} \eta_{BES} E_{BES}^{Ch}(t) = \sum_{t=1}^{H_2} E_{BES}^{dis}(t) \\ H_1 + H_2 = 24 \end{cases} \tag{8}$$

Where η_{BES} shows the efficiency of the battery energy storage (BES), and H1 is the total charging time and H2 is the total discharging time of the battery energy storage per day. The charge or discharge of the BES should not exceed the bearable margins to provide stable performance and extend the life of the battery, which is expressed (Ha et al. 2017), therefore:

$$E_{BES}^{min} \leq E_{BES}^{ch} \leq E_{BES}^{max} \tag{9}$$

$$E_{BES}^{min} \leq E_{BES}^{dis} \leq E_{BES}^{max} \tag{10}$$

where E_{BES}^{max} shows the maximum battery power capacity and E_{BES}^{min} shows the minimum power of the battery. Consequently, the above restriction should be considered as a critical limit for designing the V2H system. Based on the constraints and needs of the homeowner six modes of operation have been designed.

Modes of Operation for V2H

This work has six modes of operation which are based on power flow direction. Each mode of operation is investigated by drawing the block diagram and the direction of electrical flow in the following sections.

Mode I: Grid Supplying Power to Home

In this mode of operation, as shown in Figure 4 because of the low level of SoC of EV or unavailability of EV at home, this source of energy cannot be utilized. Moreover, solar PV panel source is also not available because of partial shedding on the power grid, cloudy weather conditions, or during night time. Hence, the power grid delivers power to the residential loads.

Mode II: Direct Grid Transfer – Grid Connected Charging

In this mode of operation, EV is available at home, but because of the low SoC of EV and need of vehicle, the charging of EV battery is necessary, therefore, power grid supplies power for both EV and residential loads. In this mode, power from solar PV power does not exist and EV gets charged through the fast charging method (43A). Figure 5 shows the direction of power flow from the grid to residential loads and EV.

Mode III: Solar PV Charging – Grid Connected Rectification

In this case, residential loads get power from the utility grid, and solar PV panel delivers power to charge EV battery. The slow charging method for the battery is provided in this case, due to the use of solar PV, which cannot generate high current for fast charging. Figure 6 shows the direction of power flow from the grid to home and Solar PV to EV.

Mode IV: Solar PV Connected to Home – EV Charging

In this mode IV of operation, the solar PV panel has sufficient power for both EV battery charging and residential loads. Solar PV generates 1300 W to feeds EV battery on slow charging mode (350W) and 820W for residential loads plus some energy which losses in the conversion, Figure 7 shows the direction of power flow from solar PV to EV and residential loads.

Figure 4. Block diagram of mode I and directions of power flow

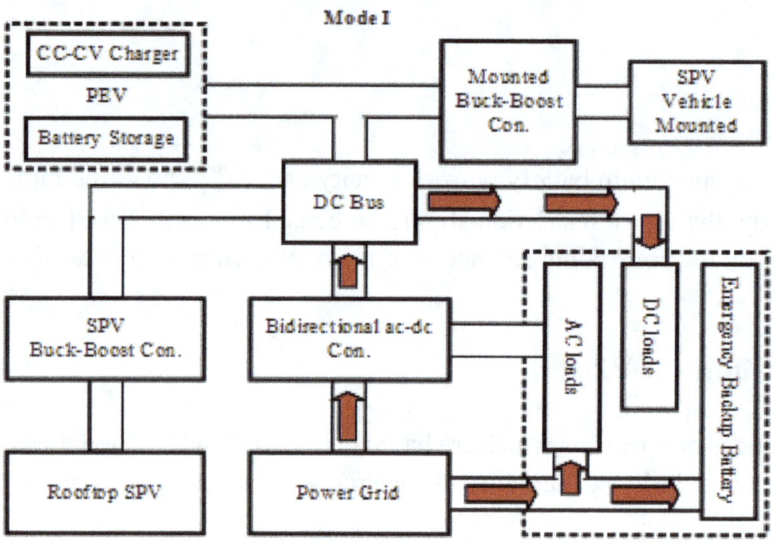

Figure 5. Block diagram of mode II and directions of power flow

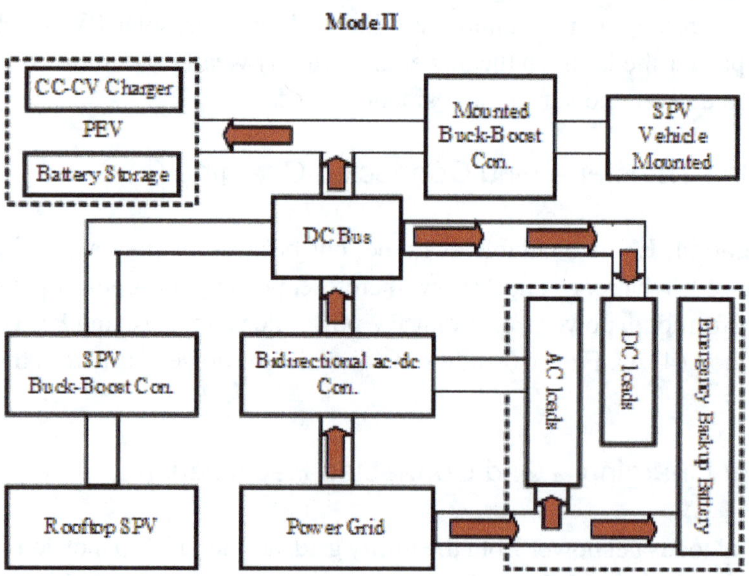

Mode V: EV Linked Power Source

In this mode as shown in Figure 8, both solar PV panel source and utility grid power are not presented because of partial load shedding or cloudy weather. Thus, EV feeds power to residential loads through DC link bus. Residential loads consist of residential AC loads and residential DC loads which are normal loads and luxury load are not connected to proposed V2H system.

Figure 6. Block diagram of mode III and directions of power flow

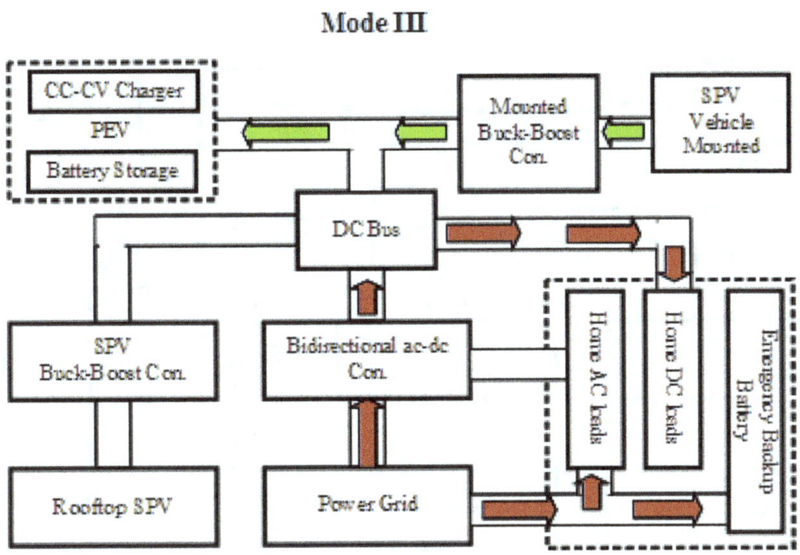

Figure 7. Block diagram of mode IV and directions of power flow

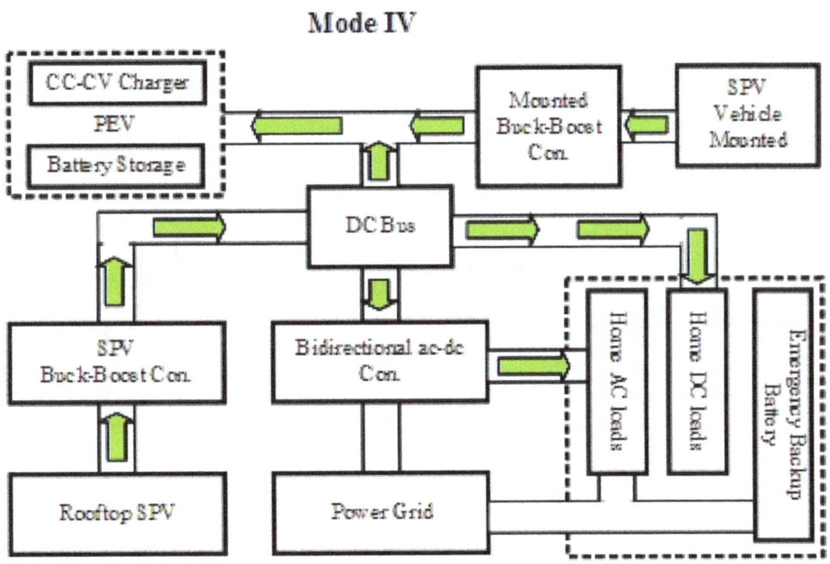

Mode VI: Backup Battery Source

In this mode of operation, EV, utility grid, and solar PV power sources are not available because of any reason, thus the backup battery source delivers the required power for the residential emergency AC and DC loads which emergency residential loads consist of both AC and DC emergency loads. Figure 9 shows the block diagram and the directions of power flow for this model.

Figure 8. Block diagram of mode V and directions of power flow

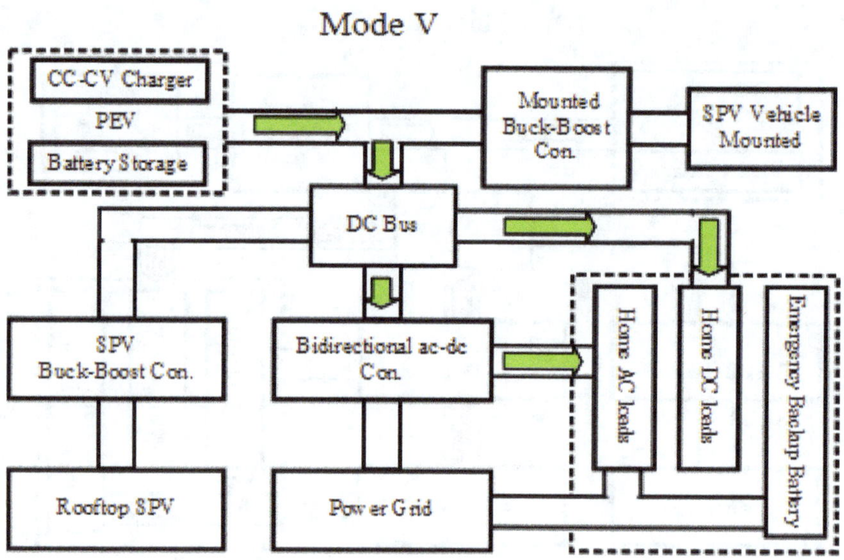

Figure 9. Block diagram of mode VI and directions of power flow

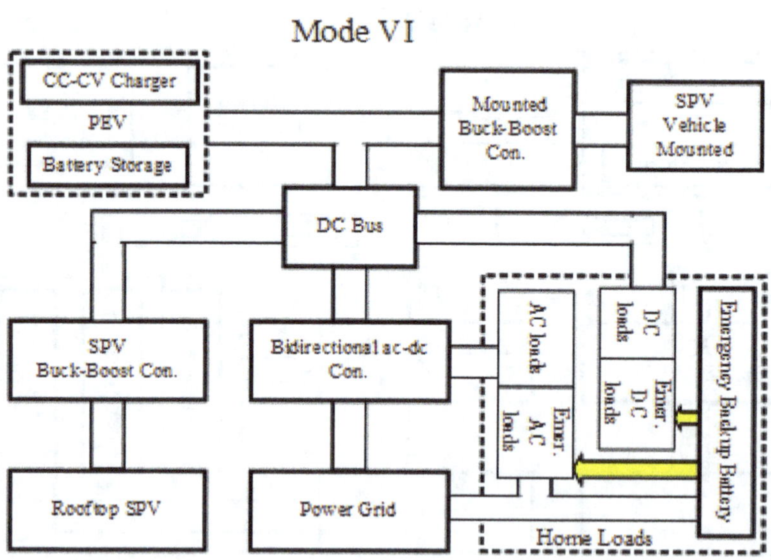

Modeling and Simulation for V2H System

Modeling of Proposed V2H System

In this model, four different energy sources feed energy to the home viz. a solar photovoltaic (SPV) array, electric vehicle (EV), emergency backup battery, and the utility grid. In this model, three types of loads are residential loads, EV load (in charging mode), and emergency backup battery (in charging mode). All sources and loads are connected to the bidirectional AC-DC inverter and the controller to make a heuristics control to optimize power consumption. Residential loads consist of AC loads as well as DC loads and the emergency backup battery. The emergency backup battery is charged by the grid for the reliability of emergency backup, but it feeds the power to the emergency loads of the home in case of non-availability power from all other sources. Figure 10 shows a block diagram of V2H with solar PV and an emergency backup power supply.

Figure 10. Block diagram of V2H with solar PV and emergency backup power source

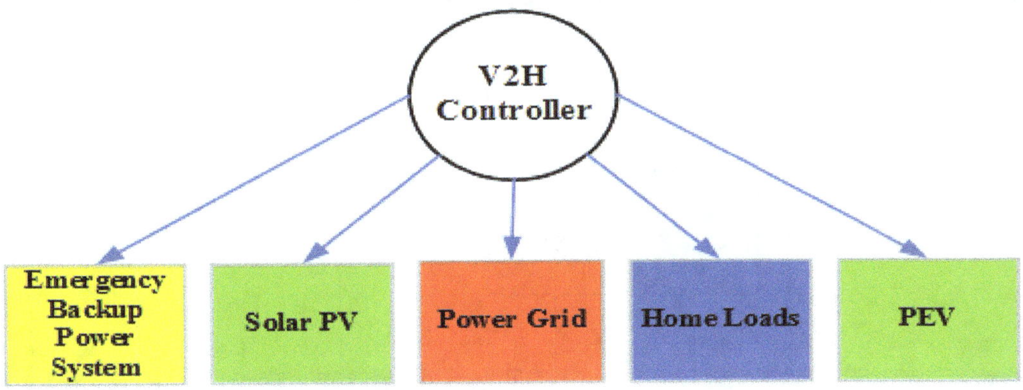

Priority of Energy Sources

To decrease the CO_2 and CO emission and attain cost-benefit, dependency on the utility grid should be reduced. Thus, the first source of supplying power is solar PV source to feed home and charging EV, such that charging the vehicle through the solar PV will be the second source of power, therefore a priority of utilization of considered resources are: 1) Solar PV, 2) EV, 3) Grid, 4) Emergency Backup.

In the first case, while solar power is available, solar PV feeds the generated power to loads. It supplies power for residential AC loads as well as DC loads. The DC loads are connected to the DC link of the system and directly use the output power of solar PV. Therefore, the efficiency of this system will increase due to avoiding DC-AC-DC conversion losses.

In the case of medium or good SoC of EV, the solar PV source also charges the EV at a slow rate. Slow charging plan considered in this case due to size limitation and the high price of solar PV installation which is about Rs (50-70)/W_p and for a high-quality PV panel (monocrystalline) it is Rs 80/W_p. If solar PV power is not available, EV provides electric power for the home until its SOC reaches the preset value. In the absence of both solar PV power and EV (mode II), the grid supplies AC power to residential loads directly and DC power through the bidirectional AC-DC converter to DC bus. The grid also provides power for fast charging the EV whenever the SoC of the EV is very low. In the last case, if all three sources of power are not accessible, then emergency backup battery supplies required power for residential emergency loads such as fans and lights.

Simulation of V2H System in Simulink

For evaluating the proposed model, Simulink®/MATLAB® has a closed-loop. In Figure 11, the whole model is illustrated. In this model, the residential DC loads, EV, solar PV, and emergency backup connected to the DC-link capacitor. The utility grid and residential AC loads are coupled to the other side of the bidirectional AC to DC inverter.

Figure 11. Simulink/MATLAB model of the proposed structure for the V2H application

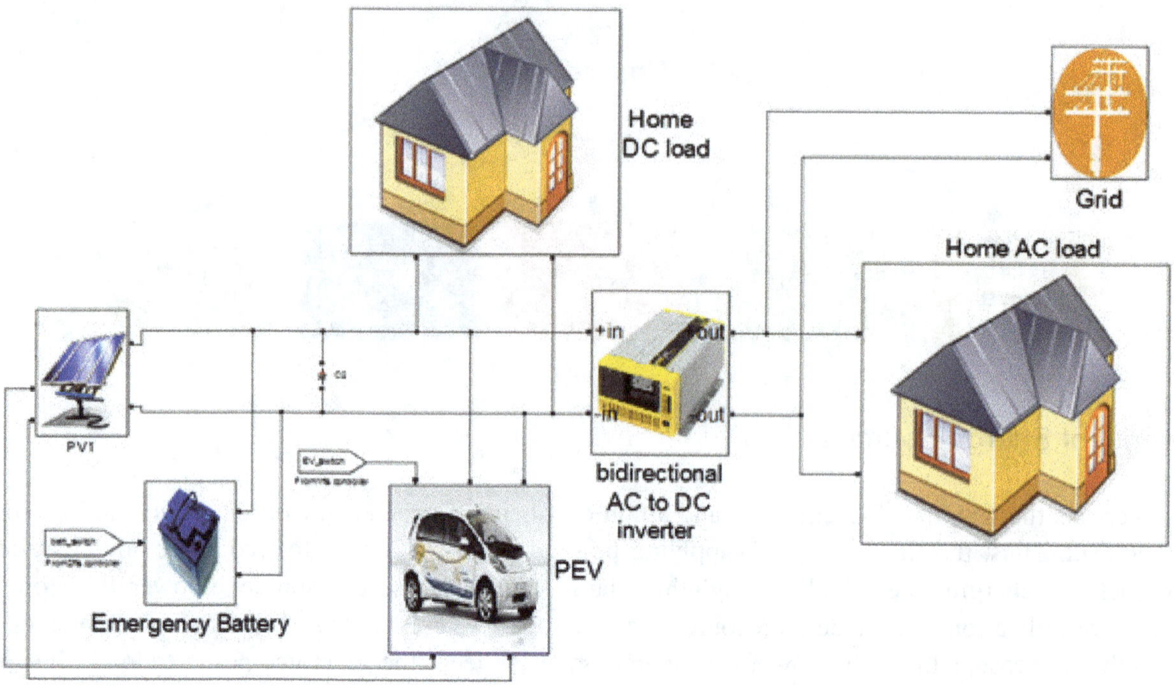

Solar PV System Integration

It has two solar PV arrays where the first solar PV array has a maximum output power of 1200W_p and the second array generates 400 W_p. The first array is connected to DC link through the buck-boost converter and its output is fixed on 24V by using the closed-loop PI controller. The second array is connected to the EV charger through the control switch. C_1 and C_2 are DC link capacitors, and ideal switches are added for the control of these sources. Solar PV module features that have been used for solar PV modeling in SIMULINK are shown in Table 2, the buck-boost converter is used to deliver power to the DC link which provides a 24V output DC voltage as shown in Figure 12 by details.

Table 2. PV Module Specification of Proposed PV system

Maximum Power (W)	195.326
Cell Per Module	50
Voltage at Maximum Power Point V_{mp} (V)	25.4
Current at Maximum Power Point I_{mp} (A)	7.69
Open Circuit Voltage Voc (V)	30.6
Short Circuit Current I_{sc} (A)	8.48
Temperature Coefficient of V_{oc} (%/deg.C)	-0.338
Temperature Coefficient of I_{sc} (%/deg.C)	0.124

Figure 12. The model used for a buck-boost converter (L=25e-6 H and C= 4697e-6 F)

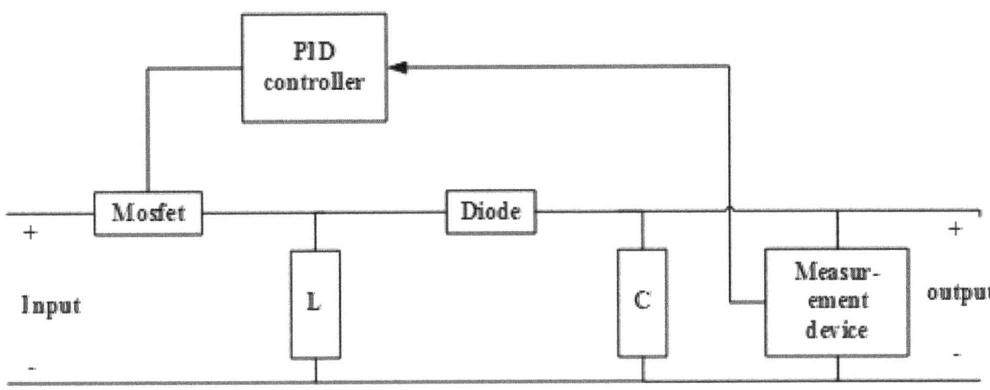

The Emergency Backup Battery Storage System

This block consists of a 200 Ah, 24 V lead-acid battery to deliver power to the emergency loads. An AC/DC/AC inverter that is commercially available in the market for domestic use, charges the battery whenever the grid is available. However, for utilizing the battery power the controller decides by sending the on or off command (0 to make it off and 1 to make it on) to the switches. Figure 13 shows the emergency backup battery block.

As it is shown in Table 3, the emergency backup battery specification which is used in the MATLAB/ SIMULINK model.

Electric Vehicle Battery Storage and Boost Charger Controller

EV battery storage block consists of the lithium-ion battery, battery charger, and control switches for charging and discharging the battery. Figure 14 illustrates the block diagram. This storage device has 4.8 kWh capacity. As it is shown in Table 4 the specification of the EV battery used for simulation. Another part of the EV battery storage block is an onboard battery charger which is shown in Figure 15 in details. The converter is designed to boost voltage to charge the battery by two methods of constant current (CC) and constant voltage (CV).

Figure 13. Emergency backup battery block (200Ah and 24 V)

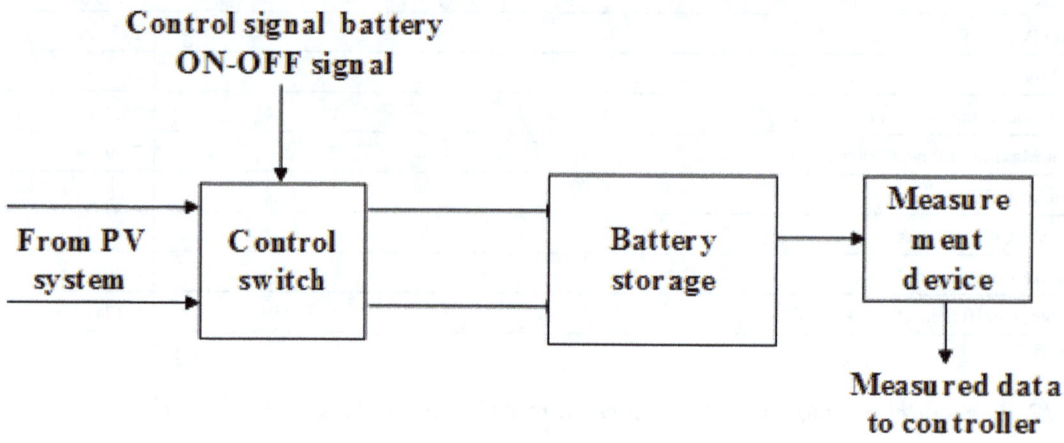

Table 3. Lead-acid Battery Parameters

Battery type	Lead-acid
Rated capacity (Ah)	200
Nominal voltage (V)	24
Maximum capacity (Ah)	208.33
Cut-off voltage (V)	18
Fully charged voltage (V)	26.13
Nominal discharge current (A)	40
Internal resistance (ohms)	0.0012
Capacity (Ah) at nominal voltage	62.05

Figure 14. Electric vehicle battery storage block

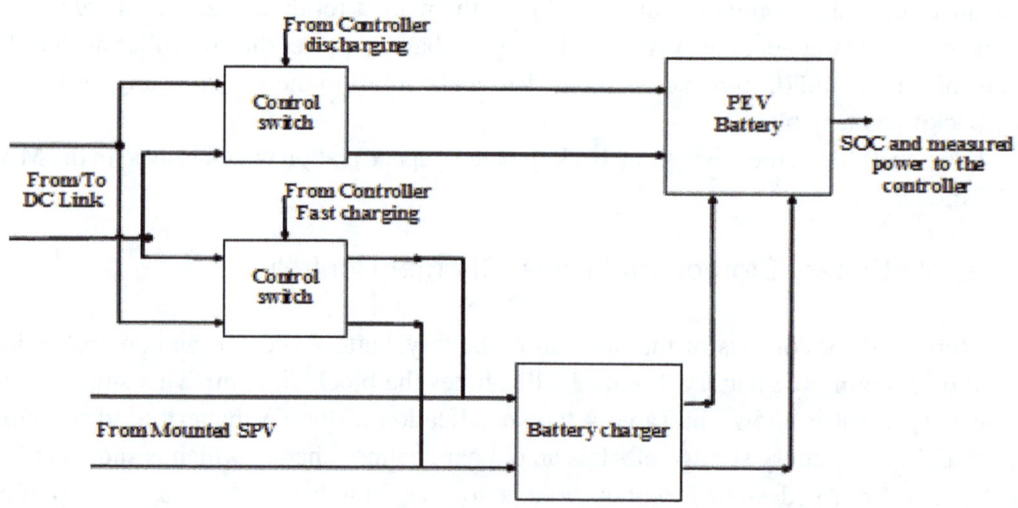

Table 4. EV Battery Specifications

Battery type	Lithium-ion
Rated capacity (Ah)	200
Nominal voltage (V)	24
Maximum capacity (Ah)	200
Cut-off voltage (V)	18
Fully charged voltage (V)	27.93
Nominal discharge current (A)	86.95
Internal resistance (ohms)	0.0012
Capacity (Ah) at nominal voltage	180.86

Another feature of this charger is that it is capable to utilize both PV and grid power to charge the battery by using two modes of charge viz fast charging and slow charging. Fast charging (45A) has been developed to charge the battery by grid power, and slow charging (13A) is designed to charge by solar PV power. Based on the SoC of the battery the controller shifts to CC or CV mode.

Figure 15. Battery charge controller for CV and CC charging modes

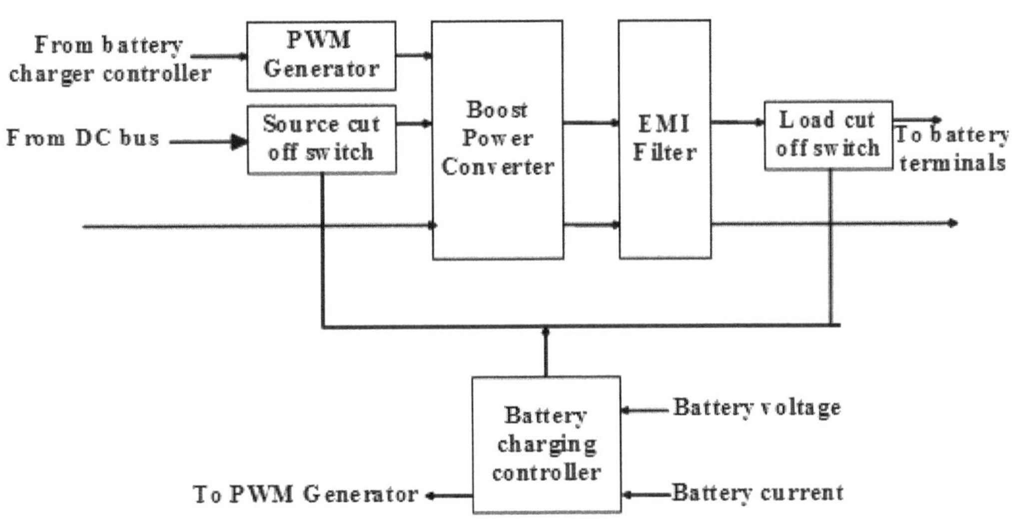

Bidirectional AC to DC Converter

A 24V converter with a 24V/220V transformer is used to step up or step down voltage whenever a change in direction of energy flow is required. In step-down conversion mode, the transformer reduces the AC voltage and zero signal (no signal) is sent to the IGBTs gate. Therefore, it works as a bridge diode and rectifies AC voltage to DC. In step-up conversion, the PWM generator sends four pulses to IGBTs to convert DC input to AC and the transformer boosts the AC voltage. Figure 16 shows the bidirectional ac-dc converter.

Figure 16. Bidirectional AC-DC PWM converter with transformer

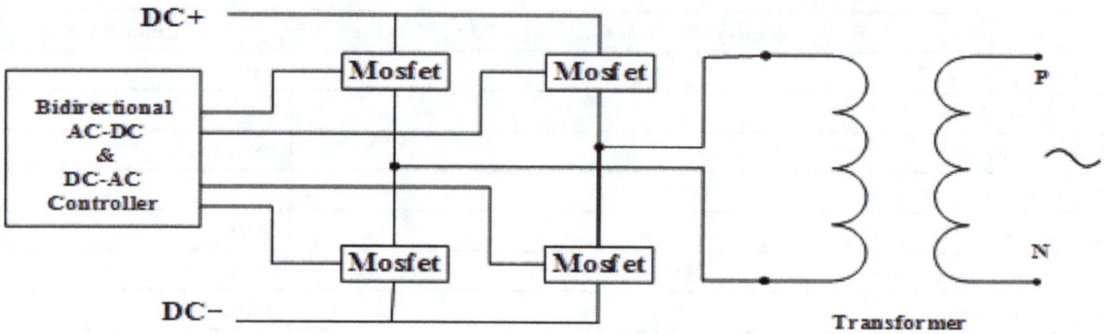

Modeling of Grid Source and Residential AC and DC Loads

The grid block in Figure 17 consists of a single-phase 220V AC source, which is connected to the residential load through the controlling switches. These switches connect or disconnect residential loads to the AC source. Residential AC loads are 700W, and DC loads are 120W where the AC loads are RLC type loads, and DC loads are resistive type loads. Figure 18 shows AC loads that are connected to the AC bus of the home and DC loads that are connected to the DC-link capacitor.

Figure 17. Grid source and its controlling switches

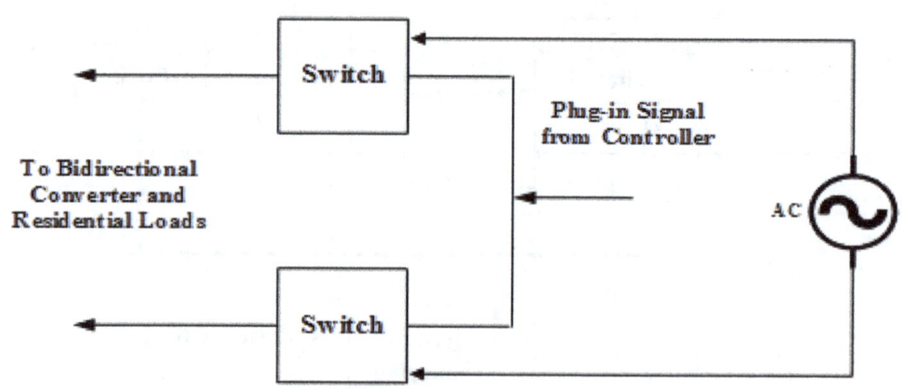

Figure 18. Residential AC loads and residential DC loads

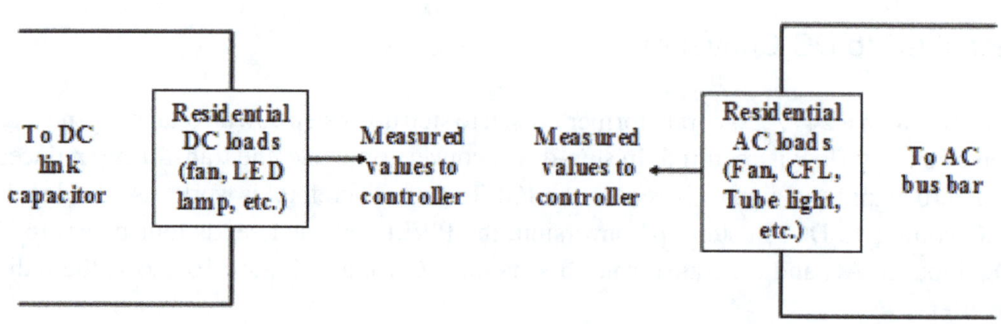

Simulation Results

The output power of solar PV, EV, and emergency backup batteries is zero. The grid provides 830W power which is delivered to residential loads. The load consists of three curves the yellow line represents DC load power and the blue line shows the AC load power and finally, the red line shows the summation of DC and AC load power. Figure 19 shows the output power of each source and consumed power of residential loads for mode I.

Figure 20 shows that the grid provides 2158W power to feed 820W residential load and 1043W for fast charging of the battery. The negative value (-1043W) of battery power consumption shows that it is in charging condition. At this condition PV source is not available, therefore, the corresponding power curve is zero. Moreover, the power by the emergency battery backup is also zero.

Figure 21 shows that the battery is being charged by 330W power from solar PV in slow charging mode and the grid provides power for residential loads which is 830W. Solar PV is generating 350W power to supply the battery charger for charging the battery. Figure 22 shows that solar PV provides power to residential loads and EV. Solar PV provides 1200 W and total residential loads are almost 800 W. The EV takes 350 W from solar PV, the negative value of vehicle battery power shows it is charging.

Figure 23 shows the output power curves of EV and residential loads that EV provides 830W power for residential DC and AC loads. Figure 24 shows that the emergency backup battery supplies the residential loads, while both the other power sources do not supply any power to a residential load.

Figure 19. The output power of each source and consumed power of residential loads for mode I

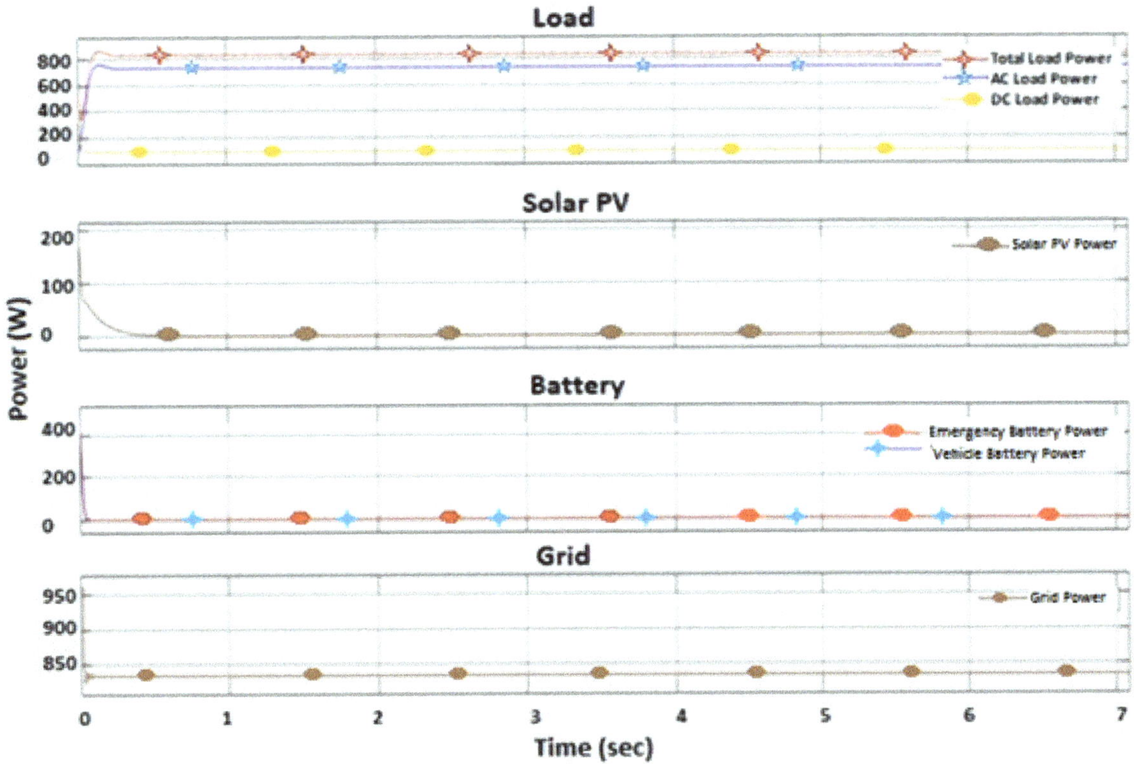

Figure 20. The output power of each source and consumed power of loads for mode II

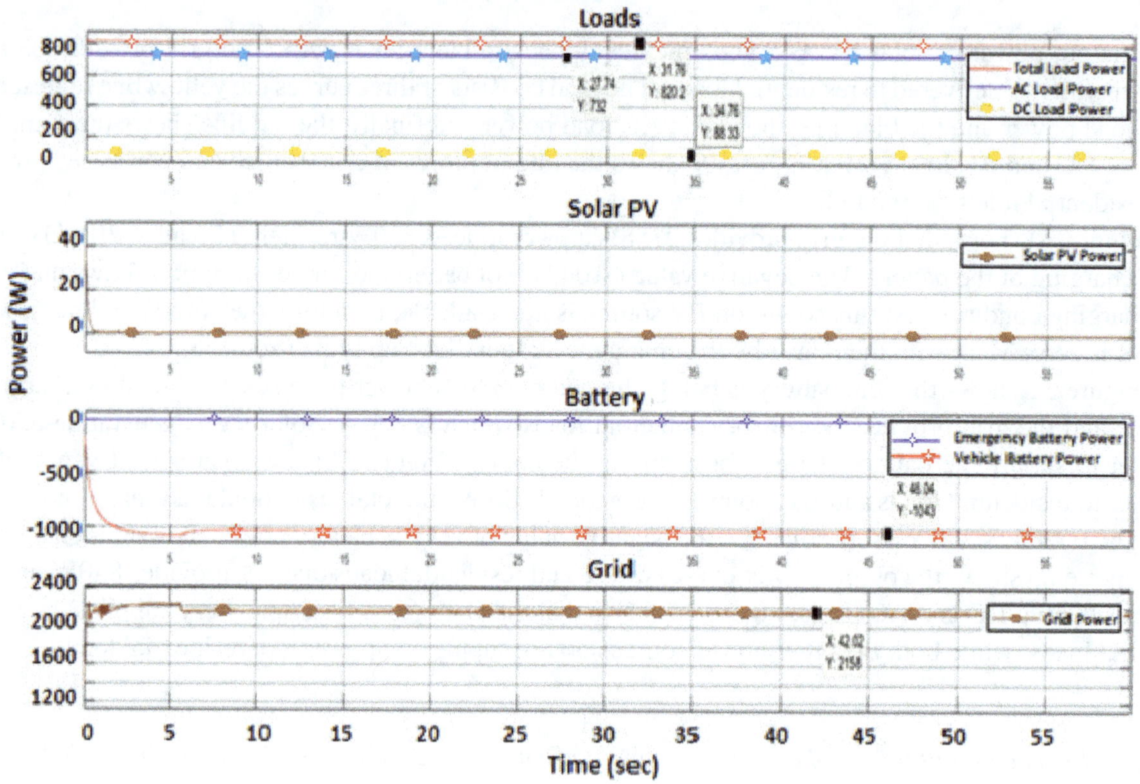

Figure 21. The output power of each source and consumed power of loads for mode III

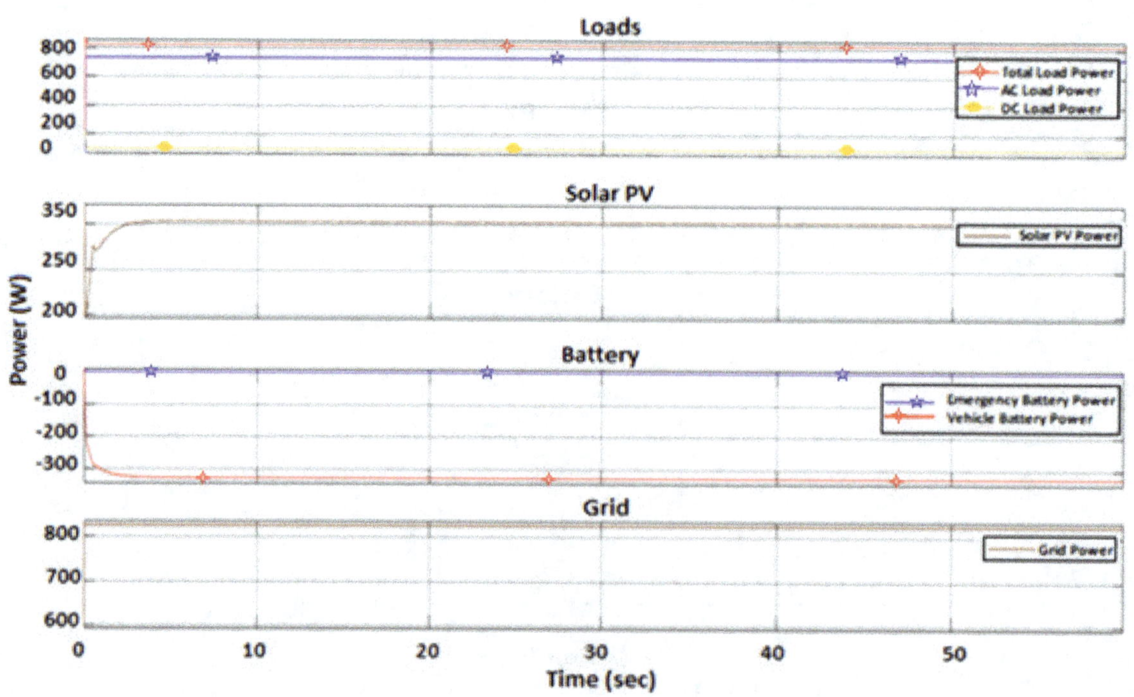

Figure 22. The output power of solar PV consumed power of residential loads and EV for mode IV

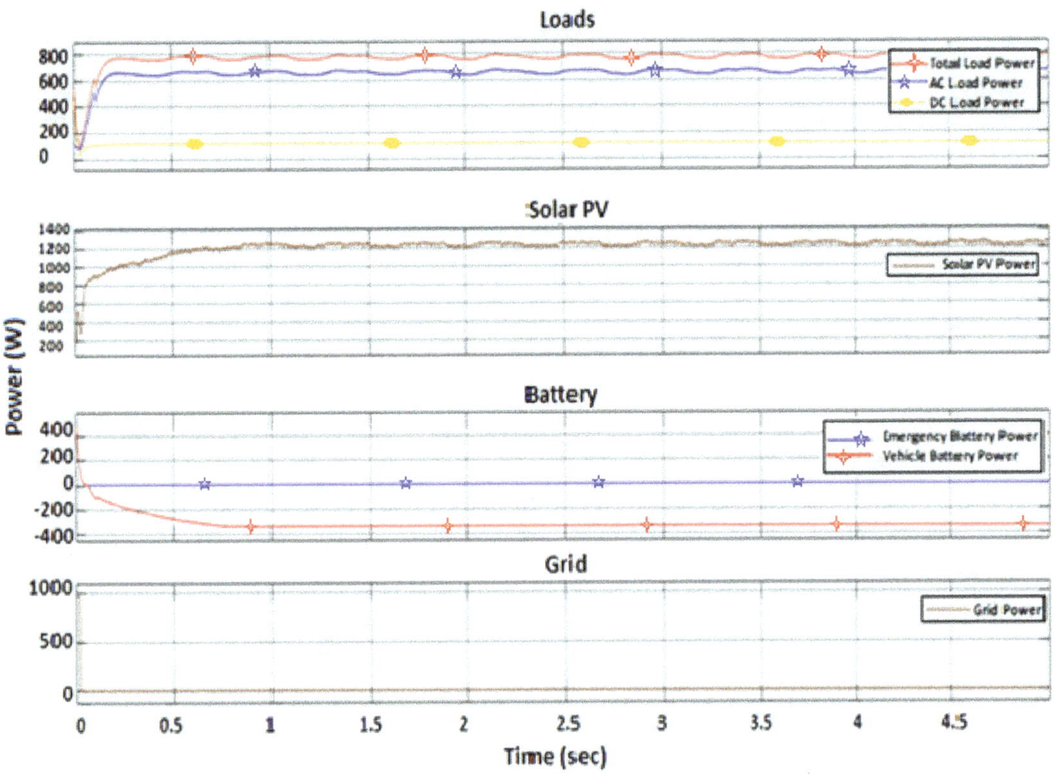

Figure 23. The output power curve of EV and consumed power of residential loads for mode V

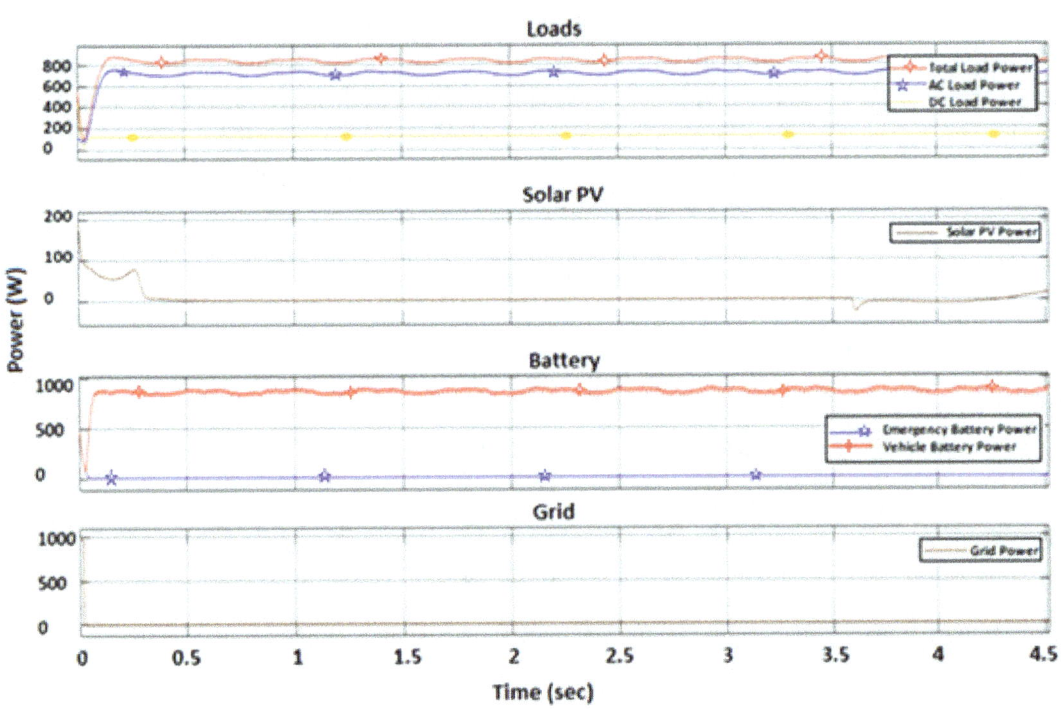

Figure 24. The output of the emergency backup battery and residential emergency loads for mode VI

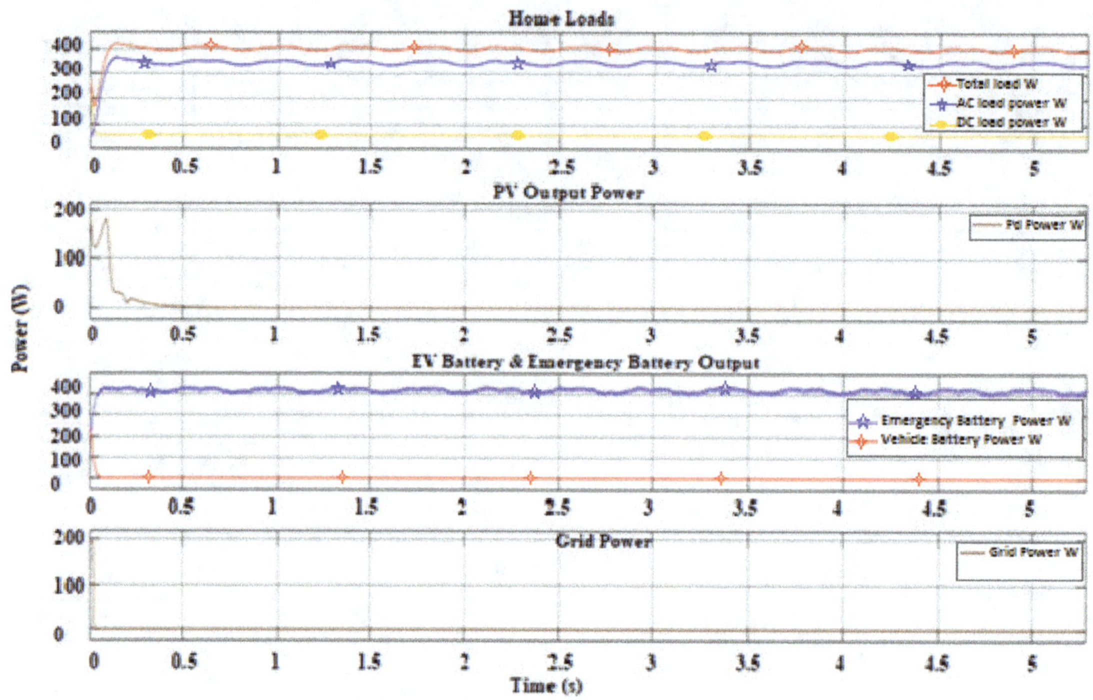

Hardware in the Loop Setup of V2H System

To perform a V2H system, a 1 kW rooftop solar PV panel is selected for a residential house. Instead of higher power rating of solar PV, 1 kW solar PV system is used due to its moderate cost (Rs 50,000 or 700 US $) which the house owner can easily spend this amount of money to reduce the home dependence on the utility grid and to get the uninterruptable power source. Solar radiation and ambient temperature data for Aligarh city are collected for 12 hours for each season. Figure 25 shows the different solar radiation and output power of solar PV (1 kW_p) for four different months respectively.

Figure 25. Solar insolation of different months and output power of 1 kW SPV

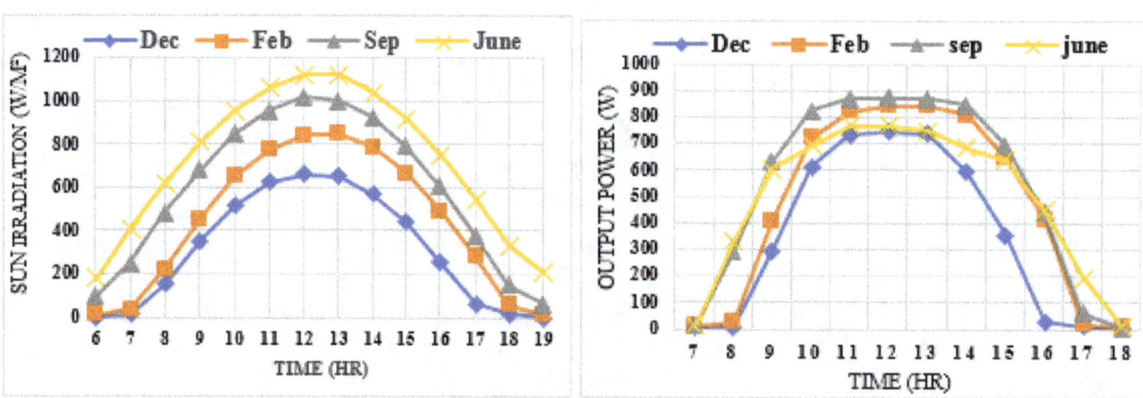

Residential Loads

Residential loads are divided into three (as mentioned previously), high demand electrical loads (e.g., air conditioner, electric oven and heaters, vacuum cleaner) which are connected to the utility grid directly and throughout the power failure (load shedding and blackouts) gets switched off. The fans, lights, room coolers are classified as normal domestic loads, these appliances take power from the V2H system. Emergency light, fans are classified as emergency loads that are linked to an isolated low power battery-based DC-to-AC UPS or inverter (300 to 400 watt). Figure 26, shows the normal loads demand and the solar PV output power for one day. In the daytime, the load's demand is less than the solar PV output power, therefore, the extra power cannot be utilized and it is stored in the EV. Based on the Tesla Powerwall technology, the BSS with 1 kW rooftop solar PV panel can decrease the dependency of households on the utility grid up to 20% (Tesla 2017).

Figure 26. Typical power demands of Aligarh residential load at the end of September and SPV output power

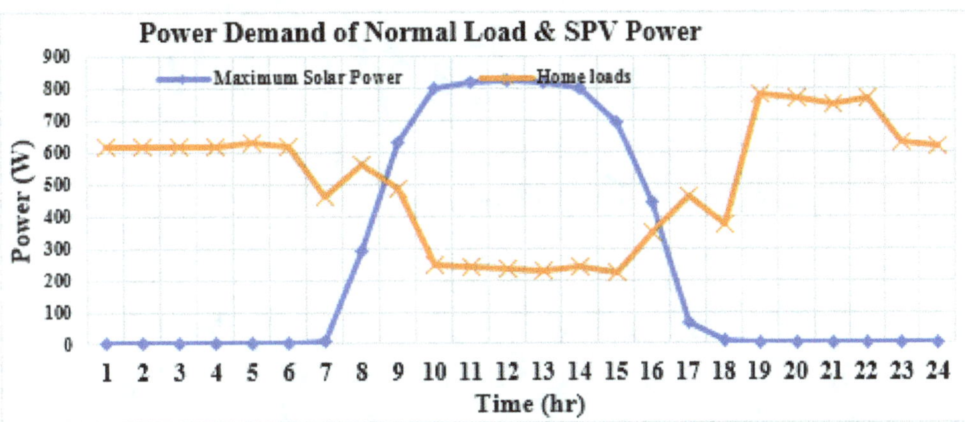

Experimental Setup

The average load of the house for different appliances was estimated and it was about 325W during the daytime. The experimental setup applied for these residential loads (325 W, 0.98 lagging power factor). The loads may grow up to 800 W in the night time because of more extra lighting loads at night time. The rooftop solar PV includes four 250 W_p, constitutes a rooftop solar PV system and four 12 V lead-acid batteries have been used for an energy storage unit (ESU). Both solar PV and ESU are connected to the DC side of the AC-DC-AC converter (commercial inverter). The output of the AC terminal of the inverter is connected to the load. The inverter AC input terminal is connected to the utility supply. The complete experimental setup and distribution board of the vehicle is shown in Figure 27.

Experimental Results

Figure 28 shows the 24 hours collected data for the proposed system. Based on the recorded data, from 1 am to 9 am, the system works in Mode I. After sunrise, the load demand reduces, and solar PV power increases. The operating mode shifts to Mode III (solar PV supplies power to EV). After sometimes, solar PV power increases more than the home demand and EV load. Thus, it supplies power to residential loads and EV (H2V mode). After 4 pm, EV supply power to the home (V2H mode) and it operates in Mode V. At 8 pm, it shifts to Mode I again and finally at 11 pm it operates on Mode VI. Based on the above-collected data from the experimental setup, it is necessary to optimize the V2H system by using control methods.

Figure 27. Experimental setup and distribution board

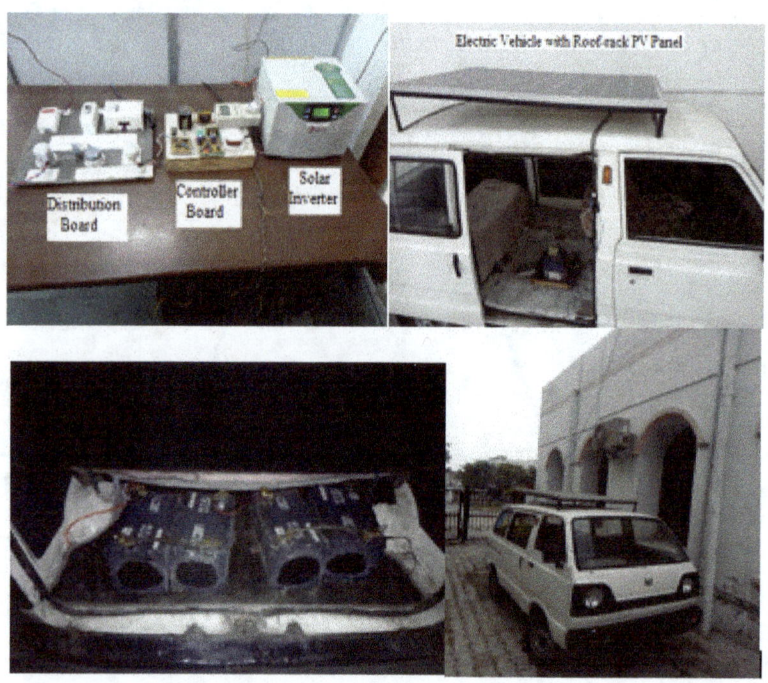

Here, EV acts as a residential load when it is charged by solar PV or grid and it acts as a power source for the home during load shedding or blackout. In addition to the utility grid, the vehicle is also charged from solar PV panels, which enhances the utility and reliability of the power supply. This makes it suitable for poorly maintained power and distribution systems where frequent load shedding is common.

Solar PV charging reduces home billing and load stress on the grid, especially during the day peak-load hours. Thus, it works as a peak power plant and H2G system. Based on the availability of EV at home, three drive patterns of EV are considered i.e. "weekdays", "weekends holiday" and "weekends traveling".

Solar PV charging reduces home billing and load stress on the grid, especially during the day peak-load hours. Thus, it works as a peak power plant and H2G system. Based on the availability of EV at home, three drive patterns of EV are considered i.e. "weekdays", "weekends holiday" and "weekends traveling". It is found that the drive pattern "weekend holiday" is most suitable for the proposed V2H system.

The proposed strategy is successfully implemented for V2H and H2V based back up the power supply. This makes it suitable for poorly maintained, power distribution systems of developing and underdeveloped countries. The controller circuit, thus designed is explained in the next section.

Figure 28. The experimental result showing different modes of operation

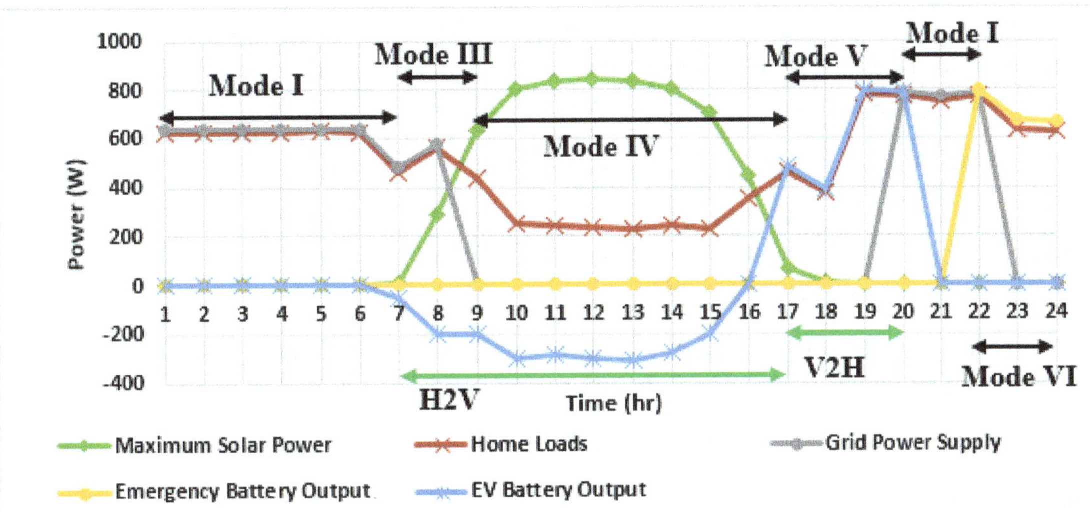

CONCLUSION

The proposed system works as the residential nanogrid, without affecting the operation of EV as an electric vehicle. For this purpose, a V2H system has been utilized. The solar PV-based charging of EV is also employed so, the entire V2H system works as a nanogrid. The EV is considered as a domestic load that is charged by the grid as well as solar PV. However, solar PV panel is preferred to charge the vehicle battery to decrease the dependency of the system on the utility grid. Both fast and slow DC charging is used with CC and CV charging modes for better performance during the daytime to harness and utilize more solar energy.

Six modes of operation for the V2H system have been defined which is based on the requirements of a typical Indian house. Based on these six modes of operation MATLAB modeling and simulation of the designed V2H system have been done. An experimental setup is developed to validate these six modes of operation for the V2H system. Here, EV acts as a residential load when it is charged by solar PV or utility grid and it turns to a source of power for the residential loads during a power outage. The EV besides of power grid is also charged from rooftop solar PV panels which enhances the utility and reliability of the power supply. This makes it suitable for poorly maintained power and distribution systems where frequent load shedding is common. Solar PV charging reduces home billing and load stress on the grid, especially in the daytime on peak-load hours. Thus, it works as a peak power plant and H2G system. The proposed strategy is successfully implemented for V2H and H2V based on the backup power supply. This makes it suitable for poorly maintained, power distribution systems of developing and underdeveloped countries. The system works as a stand-alone solar PV power plant, standby power plant, grid-assisted solar PV power plant, and solar PV peak power plant.

REFERENCES

Agarwal, S., Alam, M. S., Rafat, Y., & Shemami, M. S. (2020). Data Analysis of Grid-Connected Solar Setup and Regression Based Predictive Models. In *2020 5th IEEE International Conference on Recent Advances and Innovations in Engineering, ICRAIE 2020 – Proceeding*. Institute of Electrical and Electronics Engineers Inc.

All Electric Cars by 2030? Nitin Gadkari Inaugurates EV Charging Points at Niti Aayog I India News, The Indian Express. (n.d.). https://indianexpress.com/article/india/all-electric-cars-by-2030-nitin-gadkari-inaugurates-ev-charging-points-at-niti-aayog-5065099/

All-Electric Vehicles. (2017). https://www.fueleconomy.gov/feg/evtech.shtml

CEA. (2017). *Power Sector: Executive Summary for Jun 2017.* Central Electricity Authority, India. http://www.cea.nic.in/reports/monthly/executivesummary/2017/exe_summary-01.pdf%0Ahttp://www.cea.nic.in/reports/monthly/executive_rep/feb14.pdf

DENSO Develops Vehicle-to-Home Power Supply System for Electric Vehicles I News I DENSO Global Website. (n.d.). https://www.denso.com/global/en/news/news-releases/2012/120724-01

Electromobilitysummit.Com. (n.d.). http://electromobilitysummit.com/

ENSTO. (2016). *The Smart Future of Vehicle-to-Grid and Vehicle-to-Home*. ENSTO.

Group, Renewable Energy, Electric Power Industry, and Renewable Energy Unit. (2019). *October 2019 EV Charging Standards and V2X: Which Standard Is Leading Toward V2X Penetration?* Author.

Ha, T., & Zhang, Y. (2017, September 4). Energy Hub Modeling to Minimize Residential Energy Costs Considering Solar Energy and BESS. *Journal of Modern Power Systems and Clean Energy*, *5*(3), 389–399. https://link.springer.com/10.1007/s40565-017-0281-4

High Efficiency Low Emission Coal I HELE I WCA. (n.d.). https://www.worldcoal.org/reducing-co2-emissions/high-efficiency-low-emission-coal

International Emissions Agency. (2015). *CO2 Emissions from Fossil Fuel Combustion: Highlights.* Author.

Jhunjhunwala, A., Lolla, A., & Kaur, P. (2016). Solar-Dc Microgrid for Indian Homes: A Transforming Power Scenario. *IEEE Electrification Magazine*, *4*(2), 10–19.

Mehta, R. (2016). *Optimal Charging Scheduling of Plug-in Electric Vehicles for Maximizing Penetration within a Workplace Car Park.* Academic Press.

Ministry of New and Renewable Energy - Scheme / Documents. (n.d.). http://www.mnre.gov.in/solar-mission/jnnsm/introduction-2/

Morrow, K., Karner, D., & Francfort, J. (2008). U. S. Department of Energy Vehicle Technologies Program - Advanced Vehicle Testing Activity Plug-in Hybrid Electric Vehicle Charging Infrastructure Review. *Power*, (58517), 34.

Nichicon Corporation | News Release | EV Power Station Connection Enabled For Electric Vehicles (EVs) from Mitsubishi Motors Corporation, Thereby Expanding EV Applications. (n.d.). https://www.nichicon.co.jp/english/new/new135.html

Press Release | News･Events | Mitsubishi Motors. (n.d.). https://www.mitsubishi-motors.com/publish/pressrelease_en/corporate/2014/news/detail0954.html

Shemami, M. S., Amrr, S. M., Alam, M. S., & Jamil Asghar, M. S. (2018). Reliable and Economy Modes of Operation for Electric Vehicle-to-Home (V2H) System. *2018 5th IEEE Uttar Pradesh Section International Conference on Electrical, Electronics and Computer Engineering, UPCON 2018*, 1–6. https://www.researchgate.net/publication/328916840%0Ahttps://ieeexplore.ieee.org/abstract/document/8596932

Shemami, S., Mahdi, M. S. A., & Jamil Asghar, M. S. (2018). Fuzzy Control Assisted Vehicle-to-Home (V2H) Energy Management System. *Smart Science, 6*(2), 173–87. https://www.tandfonline.com/doi/full/10.1080/23080477.2017.1417963

Shemami, S., Mahdi, M. S. A., & Jamil Asghar, M. S. (2018). Intelligent Vehicle-to-Home (V2H) and Home-to-Vehicle (H2V) Residential Power Management. *The Intelligent Transportation Society of America 2018 Annual Meeting*, 1–13. https://static1.squarespace.com/static/59c3ed7b197aeabbd2a51a3b/t/5b2a51c90e2e720706986142/1529500105347/TS18_Paper15790.pdf

Tesla. (2017). *Powerwall | The Tesla Home Battery*. https://www.tesla.com/en_GB/powerwall

The Cost of Solar Will Drop Another 25% by 2022. (n.d.). https://futurism.com/the-cost-of-solar-will-drop-another-25-by-2022/

Transportation Overview | Center for Climate and Energy Solutions. (n.d.). https://www.c2es.org/energy/use/transportation

University, B. (2017). *BU-1003: Electric Vehicle (EV) – Battery University*. https://batteryuniversity.com/learn/article/electric_vehicle_ev

Compilation of References

Veldman, E., & Verzijlbergh, R. A. (2015). Distribution grid impacts of smart electric vehicle charging from different perspectives. *IEEE Transactions on Smart Grid*, *6*(1), 333–342. Advance online publication. doi:10.1109/TSG.2014.2355494

Rahman, S., & Shrestha, G. B. (1993). An investigation into the impact of electric vehicle load on the electric utility distribution system. *IEEE Transactions on Power Delivery*, *8*(2), 591–597. Advance online publication. doi:10.1109/61.216865

Rutherford & Yousefzadeh. (2011). *The impact of electric vehicle battery charging on distribution transformers*. . doi:10.1109/APEC.2011.5744627

Zalnidzam, W. I. F. W., Mohamad, H., Salim, N. A., Mokhlis, H., & Yasin, Z. M. (2018). Optimal charging schedule coordination of electric vehicles in smart grid. *Indones. J. Electr. Eng. Comput. Sci.*, *11*(1), 82–89. doi:10.11591/ijeecs.v11.i1.pp82-89

Celli, G., Ghiani, E., Pilo, F., Pisano, G., & Soma, G. G. (2012). *Particle Swarm Optimization for Minimizing the Burden of Electric Vehicles in Active Distribution Networks*. doi:10.1109/PESGM.2012.6345458

Adetunji, K., Hofsajer, I., & Cheng, L. (2020). A coordinated charging model for electric vehicles in a smart grid using whale optimization algorithm. *Proc. 2020 23rd Int. Conf. Inf. Fusion, FUSION 2020*. 10.23919/FUSION45008.2020.9190284

Mavrovouniotis, M., Ellinas, G., & Polycarpou, M. (2019). Electric Vehicle Charging Scheduling Using Ant Colony System. *2019 IEEE Congr. Evol. Comput. CEC 2019 - Proc.*, 2581–2588. 10.1109/CEC.2019.8789989

Alonso, M., Amaris, H., Germain, J. G., & Galan, J. M. (2014). Optimal charging scheduling of electric vehicles in smart grids by heuristic algorithms. *Energies*, *7*(4), 2449–2475. doi:10.3390/en7042449

Alvarez, J. G., González, M. Á., Vela, C. R., & Varela, R. (2018). Electric Vehicle Charging Scheduling by an Enhanced Artificial Bee Colony Algorithm. *Energies*, *11*(10), 1–19. doi:10.3390/en11102752

Mukherjee, J. C., & Gupta, A. (2014). *A Review of Charge Scheduling of Electric Vehicles in Smart Grid*. doi:10.1109/JSYST.2014.2356559

Alishah, R. S., Hosseini, S. H., Babaei, E., & Sabahi, M. (2017). Optimal design of new cascaded switch-ladder multilevel inverter structure. *IEEE Transactions on Industrial Electronics*, *64*(3), 2072–2080. doi:10.1109/TIE.2016.2627019

Brilliantova, V., & Thurner, T. W. (2019). Blockchain and the future of energy. *Technology in Society*, *57*, 38–45. Advance online publication. doi:10.1016/j.techsoc.2018.11.001

How Blockchain Transactions Work (Adding Data to Blockchain). (n.d.). https://youtu.be/I_VBuzSV9vA

K Environmental. (2019). Research Study on Reuse and Recycling of Batteries Employed in Electric Vehicles. *Energy API*.

Naveen, G., Yip, I. H.-I., & Xie, Y. (2014). Modeling and protection of electric vehicle charging station. *Proc. of 2014 6th IEEE Power India International Conference (PIICON)*, 1-6.

Zhang, Y., Dai, X., Han, X., & Jin, X. (2019, June). Analysis of Impact of Electric Vehicle Charging Load on City Grid. *IOP Conference Series. Materials Science and Engineering*, *486*(1), 012139. doi:10.1088/1757-899X/486/1/012139

Xu, Yan, Feng, & Zhao. (2019). Decentralized charging control strategy of the electric vehicle aggregator based on augmented Lagrangian method. *Int. J. Electr. Power Energy Syst.*, *104*, 673–679. . doi:10.1016/j.ijepes.2018.07.024

Vaya, M. G., & Andersson, G. (2012). Centralized and decentralized approaches to smart charging of plug-in Vehicles. *IEEE Power Energy Soc. Gen. Meet.*, 1–8. 10.1109/PESGM.2012.6344902

Pustišek, Kos, & Sedlar. (2018). *Blockchain based autonomous selection of electric vehicle charging station.* . doi:10.1109/IIKI.2016.60

Thakur, S., & Breslin, J. G. (2018). Electric Vehicle Charging Queue Management with Blockchain. Springer International Publishing. doi:10.1007/978-3-030-05081-8_18

Ping, J., Chen, S., Yan, Z., Wang, H., Yao, L., & Qian, M. (2019). EV charging coordination via blockchain-based charging power quota trading. 2019 IEEE PES Innov. Smart Grid Technol. Asia, ISGT 2019, 4362–4367. doi:10.1109/ISGT-Asia.2019.8881070

Liu, C., Chai, K. K., Zhang, X., Lau, E. T., & Chen, Y. (2018). Adaptive Blockchain-Based Electric Vehicle Participation Scheme in Smart Grid Platform. *IEEE Access: Practical Innovations, Open Solutions*, *6*, 25657–25665. doi:10.1109/ACCESS.2018.2835309

Javed & Javaid. (2019). *Scheduling charging of electric vehicles in a secured manner using blockchain technology.* . doi:10.1109/FIT47737.2019.00072

Fu, Z., Dong, P., & Ju, Y. (2020). An intelligent electric vehicle charging system for new energy companies based on consortium blockchain. *Journal of Cleaner Production*, *261*, 121219. Advance online publication. doi:10.1016/j.jclepro.2020.121219

Hunt, T. (2016). *Controlling vehicle features of Nissan LEAFs across the globe via vulnerable APIs.* https://www.troyhunt.com/controlling-vehicle-features-of-nissan/

Acharya, S., Dvorkin, Y., & Karri, R. (2020). Public Plug-in Electric Vehicles + Grid Data: Is a New Cyberattack Vector Viable? *IEEE Transactions on Smart Grid*, *11*(6), 5099–5113. doi:10.1109/TSG.2020.2994177

Alishah, R. S., Nazarpour, D., & Hosseini, S. H. (2015). Reduction of power electronic elements in multilevel converters using a new cascade structure. *IEEE Transactions on Industrial Electronics*, *62*(1), 256–269. doi:10.1109/TIE.2014.2331012

Antonopoulos, A. M., & Wood, G. (n.d.). *Mastering Ethereum.* Available at https://www.docdroid.net/MecfDvO/masteringethereum-pdf#page=17

Monteiro, V., Pinto, J. G., & Afonso, J. L. (2016, March). Operation modes for the electric vehicle in smart grids and smart homes: Present and proposed modes. *IEEE Transactions on Vehicular Technology*, *65*(3), 1007–1020. doi:10.1109/TVT.2015.2481005

Pagliaro, M., & Meneguzzo, F. (2019). Lithium battery reusing and recycling: A circular economy insight. *Heliyon*, *5*(6), e01866–e01866. doi:10.1016/j.heliyon.2019.e01866 PMID:31245638

QZ. (n.d.). https://qz.com/india/1603051/ola-says-electric-vehicles-need-a-whole-new-indian-power-sector/

Underwood, S. (2016). Blockchain beyond bitcoin. *Communications of the ACM*, *59*(11), 15–17. Advance online publication. doi:10.1145/2994581

Antoun, J., Kabir, M. E., Moussa, B., Atallah, R., & Assi, C. (2020). A Detailed Security Assessment of the EV Charging Ecosystem. *IEEE Network*, *34*(3), 200–207. doi:10.1109/MNET.001.1900348

Lee, J.-H., & Kim, H. (2017, July). Security and Privacy Challenges in the Internet of Things [Security and Privacy Matters]. *IEEE Consum. Electron. Mag.*, *6*(3), 134–136. doi:10.1109/MCE.2017.2685019

Li, H., Lu, R., Misic, J., & Mahmoud, M. (2018, May). Security and Privacy of Connected Vehicular Cloud Computing. *IEEE Network*, *32*(3), 4–6. doi:10.1109/MNET.2018.8370870

Wang, Y., Su, Z., Xu, Q., & Zhang, N. (2018). Contract based energy blockchain for secure electric vehicles charging in smart community. *Proc. - IEEE 16th Int. Conf. Dependable, Auton. Secur. Comput. IEEE 16th Int. Conf. Pervasive Intell. Comput. IEEE 4th Int. Conf. Big Data Intell. Comput. IEEE*, *3*, 328–333. 10.1109/DASC/PiCom/DataCom/CyberSciTec.2018.00062

Su, Z., Wang, Y., Xu, Q., Fei, M., Tian, Y. C., & Zhang, N. (2019). A secure charging scheme for electric vehicles with smart communities in energy blockchain. *IEEE Internet of Things Journal*, *6*(3), 4601–4613. doi:10.1109/JIOT.2018.2869297

Chen, X., & Zhang, X. (2019). Secure Electricity Trading and Incentive Contract Model for Electric Vehicle Based on Energy Blockchain. *IEEE Access: Practical Innovations, Open Solutions*, *7*, 178763–178778. doi:10.1109/ACCESS.2019.2958122

Huang, X., Xu, C., Wang, P., & Liu, H. (2018). LNSC: A Security Model for Electric Vehicle and Charging Pile Management Based on Blockchain Ecosystem. *IEEE Access: Practical Innovations, Open Solutions*, *6*, 13565–13574. Advance online publication. doi:10.1109/ACCESS.2018.2812176

Poon & Dryja. (2016). *The Bitcoin Lightning Network: Scalable Off-Chain Instant Payments*. Academic Press.

Wang, Su, & Zhang. (2019). BSIS-Blockchain based Secure Incentive Scheme for Energy Delivery. IEEE Trans. Ind. Informatics. doi:10.1109_TII.2019.2908497

Choubey, Behera, Patel, Mahidhar, & Misra. (2019). *EnergyTradingRank Algorithm for Truthful Auctions among EVs via Blockchain Analytics of Large Scale Transaction Graphs*. . doi:10.1109/COMSNETS.2019.8711249

Babaei, E., Laali, S., & Alilu, S. (2014). Cascaded multilevel inverter with series connection of novel H-bridge basic units. *IEEE Transactions on Industrial Electronics*, *61*(12), 6664–6671. doi:10.1109/TIE.2014.2316264

Crosby, M., & Nachiappan, P. (2016). *Blockchain Technology - Beyond Bitcoin*. Berkley Eng.

Electric Vehicle Regulatory Reference Guide. (2019). Available: https://www.unece.org/

Rajashekara, K. (2013, March). Present status and future trends in electric vehicle propulsion technologies. *IEEE Journal of Emerging and Selected Topics in Power Electronics*, *1*(1), 3–10. doi:10.1109/JESTPE.2013.2259614

Sachan, S., & Kishor, N. (2016). Optimal location for centralized charging of electric vehicle in distribution network. *2016 18th Mediterranean Electrotechnical Conference (MELECON)*, 1-6. 10.1109/MELCON.2016.7495326

YouTube. (n.d.). https://www.youtube.com/watch?v=ZE2HxTmxfrI

The trust machine. (2020). *The Economist*.

University of Waterloo. (2019). *Researchers use blockchain to drive electric-vehicle infrastructure*. https://uwaterloo.ca/math/research/research-stories/researchers-use-blockchain-drive-electric-vehicle

Gorenflo, Golab, & Keshav. (2019). *Mitigating trust issues in electric vehicle charging using a blockchain.* . doi:10.1145/3307772.3328283

Aoun, A. (2019a). Using Blockchain for Smart Metering and Billing in Developing Countries. *The Blockchain Journal.* https://blockchainjournal.blog/2019/12/13/using-blockchain-for-smart-metering-and-billing-in-developing-countries/

Aoun, A. (2019b). Blockchain for Electric Vehicle Charging. *The Blockchain Journal.* https://blockchainjournal.blog/2019/12/05/blockchain-for-electric-vehicle-charging/

Zielinska, A., Skowron, M., & Bien, A. (2019). The concept of the blockchain technology model use to settle the charging process of an electric vehicle. *Appl. Electromagn. Mod. Eng. Med. PTZE, 2019*, 271–274. doi:10.23919/PTZE.2019.8781739

Matsuda, Y., & Tanaka, K. (2020). EV mobility charging service based on blockchain. *Proc. - 2020 IEEE Int. Conf. Environ. Electr. Eng. 2020 IEEE Ind. Commer. Power Syst. Eur. EEEIC / I CPS Eur.* 10.1109/EEEIC/ICPSEurope49358.2020.9160608

ZF. (2017). *Mobile wallet: The electric car which pays for itself.* Blockchain Car eWallet.

Kim, Kang, & Hong. (2017). *Mobile charger billing system using lightweight Blockchain.* . doi:10.1109/APNOMS.2017.8094151

Kohlbrenner, F., Nasirifard, P., Löbel, C., & Jacobsen, H.-A. (2019). A Blockchain-based Payment and Validity Check System for Vehicle Services. *Proceedings of the 20th International Middleware Conference Demos and Posters on ZZZ - Middleware '19*, 17–18. 10.1145/3366627.3368107

Battery Recycling as a Business. (2019). Available: https://batteryuniversity.com

Boora, K., & Kumar, J. (2017). General topology for asymmetrical multilevel inverter with reduced number of switches. IET Power Electronics, 10(15), 2034-2041. doi:10.1049/iet-pel.2016.1011

Casino, F., Dasaklis, T. K., & Patsakis, C. (2019). A systematic literature review of blockchain-based applications: Current status, classification and open issues. *Telematics and Informatics, 36*, 55–81. Advance online publication. doi:10.1016/j.tele.2018.11.006

Erb, D. C., Onar, O. C., & Khaligh, A. (2010). Bi-directional charging topologies for plug-in hybrid electric vehicles. *Proc. 25th Annu. IEEE Appl. Power Electron. Conf.*, 2066–2072. 10.1109/APEC.2010.5433520

Sachan, S., & Kishor, N. (2018). A Probabilistic Approach for DG with Electric Vehicles. *2018 IEEE International Conference on Power Electronics, Drives and Energy Systems (PEDES)*, 1-4. 10.1109/PEDES.2018.8707451

Wang, Ouyang, Yuan, Ni, Han, & Wang. (n.d.). *Blockchain-Enabled Smart Contracts: Architecture, Applications, and Future Trends.* Academic Press.

Chanthong, Ruangsakorn, & Glomglome. (2020). *Blockchain and Smart Contract Payment for Electric Vehicle Charging.* . doi:10.1109/ECTI-CON49241.2020.9158132

Erdin, E., Cebe, M., Akkaya, K., Solak, S., Bulut, E., & Uluagac, S. (2018). Building a Private Bitcoin-Based Payment Network Among Electric Vehicles and Charging Stations. *2018 IEEE International Conference on Internet of Things (iThings) and IEEE Green Computing and Communications (GreenCom) and IEEE Cyber, Physical and Social Computing (CPSCom) and IEEE Smart Data (SmartData)*, 1609–1615. 10.1109/Cybermatics_2018.2018.00269

Shaw, J. (2009). *Exposed: The erosion of privacy in the Internet era*. Harv. Mag. Available http://scholar.google.com/scholar?q=related:IK9kevLZKfkJ:scholar.google.com/&hl=en&num=30&as_sdt=0,5%5Cnfile:///Users/ajacinto/Dropbox/AppDocs/Papers2/Articles/Harvard Magazine/Shaw Shaw 2009.pdf%5Cnpapers2://publication/uuid/60B49E99-EAFA-44EB-833E-94B04

Newman, G. R., & Mcnally, M. M. (2005). Identity theft literature review. *U.S. Dep. Justice*, *210459*, 103. Available: https://www.ncjrs.gov/pdffiles1/nij/grants/210459.pdf

Goodin, D. (2009). *TJX suspect indicted in Heartland, Hannaford breaches - Networks pierced by garden-variety exploit*. http://www.theregister.co.uk/2009/08/17/heartland_payment_suspect/

McCurry, J. (2016). 100 thieves steal $13m in three hours from cash machines across Japan. *The Guardian*. https://www.theguardian.com/world/2016/may/23/japan-cash-machine-100-thieves-steal-13m-dollars-three-hours

Unver, A. H. (2017). Digital Challenges to Democracy: Politics of Automation, Attention, and Engagement. *Journal of International Affairs*.

Shahbaz, A. (2018). *The Rise of Digital Authoritarianism*. Available: https://freedomhouse.org/report/freedom-net/2018/rise-digital-authoritarianism

Westin, A. F. (1966). Science, Privacy, and Freedom: Issues and Proposals for the 1970's. Part II: Balancing the Conflicting Demands of Privacy, Disclosure, and Surveillance. *Columbia Law Review*, *66*(7), 1205. Advance online publication. doi:10.2307/1120983

Kerry, C. F. (2020). *Protecting Privacy in an AI-driven World*. Available: https://www.brookings.edu/research/protecting-privacy-in-an-ai-driven-world/

Ahl, A., Yarime, M., Goto, M., Chopra, S. S., Kumar, N. M., Tanaka, K., & Sagawa, D. (2020, January). Exploring blockchain for the energy transition: Opportunities and challenges based on a case study in Japan. *Renewable & Sustainable Energy Reviews*, *117*, 109488. Advance online publication. doi:10.1016/j.rser.2019.109488

Amarakoon, Smith, & Segal. (2013). *Application of life-cycle assessment to nanoscale technology: Lithium-ion batteries for electric vehicles*. Academic Press.

Banaei, M. R., & Salary, E. (2011). Verification of new family for cascade multilevel inverter switch reduction of components. *Journal of Electrical Engineering & Technology*, *6*(2), 245–254. doi:10.5370/JEET.2011.6.2.245

Gómez, J. R., González, L. G., Vide, P. C., Gameiro, N. S., & Guerron, G. (2017, September). Impact of electric vehicle charging on the distribution system in cuenca, ecuador. In 2017 IEEE PES Innovative Smart Grid Technologies Conference-Latin America (ISGT Latin America) (pp. 1-6). IEEE. doi:10.1109/ISGT-LA.2017.8126756

Mojdehi & Ghosh. (2016). An On-Demand Compensation Function for an EV as a Reactive Power Service Provider. *IEEE Transactions on Vehicular Technology, 65*(6), 4572-4583.

Jouvray, C., Pellischek, G., & Tiguercha, M. (2013). Impact of a smart grid to the electric vehicle ecosystem from a privacy and security perspective. *World Electr. Veh. J.*, *6*(4), 1115–1124. doi:10.3390/wevj6041115

Saxena, N., Grijalva, S., Chukwuka, V., & Vasilakos, A. V. (2017, August). Network Security and Privacy Challenges in Smart Vehicle-to-Grid. *IEEE Wireless Communications*, *24*(4), 88–98. doi:10.1109/MWC.2016.1600039WC

Sweeney, L. (2002, October). k-anonymity: A model for protecting privacy. *International Journal of Uncertainty, Fuzziness and Knowledge-based Systems*, *10*(5), 557–570. doi:10.1142/S0218488502001648

Machanavajjhala, Gehrke, Kifer, & Venkitasubramaniam. (2006). *ℓ-Diversity: Privacy beyond k-anonymity.* . doi:10.1109/ICDE.2006.1

Ninghui, Tiancheng, & Venkatasubramanian. (2007). *t-Closeness: Privacy beyond k-anonymity and ℓ-diversity.* . doi:10.1109/ICDE.2007.367856

Dwork & Roth. (2013). The algorithmic foundations of differential privacy. *Found. Trends Theor. Comput. Sci.* . doi:10.1561/0400000042

Gentry. (2009). *Fully Homomorphic Encryption Using Ideal Lattices.* doi:10.1145/1536414.1536440

Zyskind, G., Nathan, O., & Pentland, A. S. (2015). Decentralizing privacy: Using blockchain to protect personal data. *Proc. - 2015 IEEE Secur. Priv. Work. SPW*, 180–184. doi:10.1109/SPW.2015.27

Ben-Sasson, E. (2014). Zerocash: Decentralized anonymous payments from bitcoin. *Proc. - IEEE Symp. Secur. Priv.*, 459–474. 10.1109/SP.2014.36

Aitzhan, N. Z., & Svetinovic, D. (2018). Security and Privacy in Decentralized Energy Trading Through Multi-Signatures, Blockchain and Anonymous Messaging Streams. *IEEE Transactions on Dependable and Secure Computing*, *15*(5), 840–852. doi:10.1109/TDSC.2016.2616861

Banaei, M. R., Jannati Oskuee, M. R., & Khounjahan, H. (2014, May). Reconfiguration of semi-cascaded multilevel inverter to improve systems performance parameters. *IET Power Electronics*, *7*(5), 1106–1112. doi:10.1049/iet-pel.2013.0277

Ben Jar & Watson. (2016). *Rapid EV chargers: Implementation of a charger.* Electric Power Engineering Centre (EPE-Centre), University of Canterbury.

Creyts, J., & Trbovich, A. (2018). *Can blockchain help us to address the world's energy issues?* World Economic Forum.

Gunes, D., Tekdemir, I., Karaarslan, M., & Alboyaci, B. (2018). *Assessment of the impact of electric vehicle charge station loads on reliability indices.* Academic Press.

Solidity. (n.d.). https://solidity.readthedocs.io/en/v0.7.1/

Tchounwou, P. B., Yedjou, C. G., Patlolla, A. K., & Sutton, D. J. (2012). Heavy metal toxicity and the environment. In Molecular, clinical and environmental toxicology. Springer. doi:10.1007/978-3-7643-8340-4_6

Knirsch, Unterweger, & Engel. (2018). *Privacy-preserving blockchain-based electric vehicle charging with dynamic tariff decisions.* . doi:10.1007/s00450-017-0348-5

Li, L., Liu, J., Cheng, L., Qiu, S., Wang, W., Zhang, X., & Zhang, Z. (2018). CreditCoin: A Privacy-Preserving Blockchain-Based Incentive Announcement Network for Communications of Smart Vehicles. *IEEE Transactions on Intelligent Transportation Systems*, *19*(7), 2204–2220. doi:10.1109/TITS.2017.2777990

Praba Devi, G. S., & Miraclin Joyce Pamila, J. C. (2019). Accident Alert System Application Using a Privacy-Preserving Blockchain-Based Incentive Mechanism. *2019 5th Int. Conf. Adv. Comput. Commun. Syst. ICACCS 2019*, 390–394. 10.1109/ICACCS.2019.8728507

Radi, Lasla, Bakiras, & Mahmoud. (2019). *Privacy-Preserving Electric Vehicle Charging for Peer-to-Peer Energy Trading Ecosystems.* doi:10.1109/ICC.2019.8761788

Kshetri, N. (2017). Blockchain's roles in strengthening cybersecurity and protecting privacy. *Telecommunications Policy*, *41*(10), 1027–1038. Advance online publication. doi:10.1016/j.telpol.2017.09.003

Xu, M., Chen, X., & Kou, G. (2019). A systematic review of blockchain. *Financ. Innov.*, *5*(1), 27. Advance online publication. doi:10.118640854-019-0147-z

Christidis, K., & Devetsikiotis, M. (2016). Blockchains and Smart Contracts for the Internet of Things. *IEEE Access: Practical Innovations, Open Solutions*, *4*, 2292–2303. Advance online publication. doi:10.1109/ACCESS.2016.2566339

Li, X., Jiang, P., Chen, T., Luo, X., & Wen, Q. (2020). A survey on the security of blockchain systems. *Future Generation Computer Systems*, *107*, 841–853. doi:10.1016/j.future.2017.08.020

Tapscott, D., & Tapscott, A. (2016). *Blockchain Revolution*. Portfolio Penguin.

Jensen, O. (2018). *Disruption of the current utility business by utilizing EV charging infrastructure to unlock grid flexibility. Possibilities for Blockchain technology*. KTH Royal Institute of Technology.

Alishah, D. N. R. S., & Hosseini, S. H. (2014). Novel topologies for symmetric, asymmetric, and cascade switched-diode mutilevel converter with minimum number of power electronic components. *IEEE Transactions on Industrial Electronics*, *61*(10), 5300–5310. doi:10.1109/TIE.2013.2297300

Dannen. (n.d.). *Introducing Ethereum and Solidity*. Academic Press.

Karmaker, A. K., Roy, S., & Ahmed, M. R. (2019, February). Analysis of the Impact of Electric Vehicle Charging Station on Power Quality Issues. In *2019 International Conference on Electrical, Computer and Communication Engineering (ECCE)* (pp. 1-6). IEEE. 10.1109/ECACE.2019.8679164

Khan, W., Ahmad, F., & Alam, M. S. (2019, February). Fast EV charging station integration with grid Ensuring optimal and quality power exchange. *Elsevier International Journal of Engineering Science and Technology*, *22*(1), 143–152. doi:10.1016/j.jestch.2018.08.005

Mengelkamp, E., Gärttner, J., Rock, K., Kessler, S., Orsini, L., & Weinhardt, C. (2018). Designing microgrid energy markets: A case study: The Brooklyn Microgrid. *Applied Energy*, *210*, 870–880. doi:10.1016/j.apenergy.2017.06.054

Ordoñez, J., Gago, E., & Girard, A. (2016). Processes and technologies for the recycling and recovery of spent lithium-ion batteries. *Renewable & Sustainable Energy Reviews*, *60*, 195–205. doi:10.1016/j.rser.2015.12.363

ElHusseini, H., Assi, C., Moussa, B., Attallah, R., & Ghrayeb, A. (2020). *Blockchain, AI and Smart Grids: The Three Musketeers to a Decentralized EV Charging Infrastructure. IEEE Internet Things Mag.* doi:10.1109/IOTM.0001.1900081

Cong, L. W., & He, Z. (2019). Blockchain Disruption and Smart Contracts. *Review of Financial Studies*, *32*(5), 1754–1797. doi:10.1093/rfs/hhz007

Ferrag, M. A., Derdour, M., Mukherjee, M., Derhab, A., Maglaras, L., & Janicke, H. (2019). Blockchain technologies for the internet of things: Research issues and challenges. *IEEE Internet of Things Journal*, *6*(2), 2188–2204. Advance online publication. doi:10.1109/JIOT.2018.2882794

Westerström, E. (2020). *Barriers to blockchain adoption in the public electrical vehicle charging market*. Blekinge Institute of Technology.

Zheng, Z., Xie, S., Dai, H. N., Chen, X., & Wang, H. (2018). Blockchain challenges and opportunities: A survey. *International Journal of Web and Grid Services*, *14*(4), 352. Advance online publication. doi:10.1504/IJWGS.2018.095647

PwC. (2018). *PwC's Global Blockchain Survey 2018*. PwC.

Anjum, A., Sporny, M., & Sill, A. (2017). Blockchain Standards for Compliance and Trust. *IEEE Cloud Comput.*, *4*(4), 84–90. doi:10.1109/MCC.2017.3791019

Aristidou, C., & Marcou, E. (2019). Blockchain Standards and Government Applications. *Blockchain Standards and Government Applications*, *7*(3), 287–312. doi:10.13052/jicts2245-800X.736

Institute of Electrical and Electronics Engineers. (2011). *What are Standards? Why are They Important?* Institute of Electrical and Electronics Engineers. http://standardsinsight.com/ieee_company_detail/what-are-standards-why-are-they-important

Mobility Open Blockchain Initiative (MOBI). (2020). *Blockchains for Electric Vehicles & Grid Integration Review Contents.* Available: https://dlt.mobi/wp-content/uploads/2020/10/EVGI-Business-Review.pdf

Alishah, R. S., Nazarpour, D., Hosseini, S. H., & Sabahi, M. (2013). New hybrid structure for multilevel inverter with fewer number of components for high-voltage levels. *IET Power Electronics*, *7*(1), 96–104. doi:10.1049/iet-pel.2013.0156

Hazard Class 9 - Hazard Class Labels Explained. (2019). Available: https://www.general-data.com/about/blog/hazard-class-9-hazard-class-labels-explained

Huang, Y., Bian, Y., Li, R., Zhao, J., & Shi, P. (2019). Smart contract security: A software lifecycle perspective. *IEEE Access: Practical Innovations, Open Solutions*, *7*, 150184–150202. doi:10.1109/ACCESS.2019.2946988

Manríquez, F., Sauma, E., Aguado, J., de la Torre, S., & Contreras, J. (2020). The impact of electric vehicle charging schemes in power system expansion planning. *Applied Energy*, *262*, 114527. doi:10.1016/j.apenergy.2020.114527

Sun, Chen, Zhu, & Li. (2018). *Research on the Application of Blockchain Technology in Energy Internet.* . doi:10.1109/EI2.2018.8582599

Van den Bossche, Turcksin, Omar, & Van Mierlo. (2016). Developments and Challenges for EV Charging Infrastructure Standardization. *World Electric Vehicle Journal*, *8*, 1–7.

König, L., Korobeinikova, Y., Tjoa, S., & Kieseberg, P. (2020). *Comparing Blockchain Standards and Recommendations.* Futur. Internet. doi:10.3390/fi1212022

Dika & Nowostawski. (n.d.). *Security Vulnerabilities in Ethereum Smart Contracts.* Academic Press.

Gray, M. K., & Morsi, W. G. (2015). Power quality assessment in distribution systems embedded with plug-in hybrid and battery electric vehicles. *IEEE Transactions on Power Systems*, *30*(2), 663–671. doi:10.1109/TPWRS.2014.2332058

Hosseinian, Shahinzadeh, Gharehpetian, Azani, & Shaneh. (2020). Blockchain outlook for deployment of IoT in distribution networks and smart homes. *Int. J. Electr. Comput. Eng.* . doi:10.11591/ijece.v10i3.pp2787-2796

McManus, M. C. (2012). Environmental consequences of the use of batteries in low carbon systems: The impact of battery production. *Applied Energy*, *93*, 288–295. doi:10.1016/j.apenergy.2011.12.062

Pinto, D. R., Arioli, V. T., Hax, G. R., Taniguchi, F. K., Torquato, R., Teixeira, W. W., & Eduardo, P. L. (2018, May). Field investigation of the power quality impact of electric vehicles in secondary residential systems. In *2018 18th International Conference on Harmonics and Quality of Power (ICHQP)* (pp. 1-6). IEEE. 10.1109/ICHQP.2018.8378904

Wang, L., Wu, Q. H., & Tang, W. (2017, December). Novel Cascaded Switched-Diode Multilevel Inverter for Renewable Energy Integration. *IEEE Transactions on Energy Conversion*, *32*(4), 1574–1582. doi:10.1109/TEC.2017.2710352

Abdullah, R., Rahim, N. A., Sheikh Raihan, S. R., & Ahmad, A. Z. (2014, October). Five-level diode-clamped inverter with three-level boost converter. *IEEE Transactions on Industrial Electronics*, *61*(10), 5155–5163. doi:10.1109/TIE.2013.2297315

Araoz, M. (2017). *Introducing zeppelinos: The operating system for smart contract applications.* https://blog.zeppelin.solutions/introducingzeppelinos-the-operating-system-for-smart-contract-applications-82b042514aa8

Francis, S. (n.d.). *Top 20 Electric Vehicle Charging Station Companies*. Robotics and automation news report. https://roboticsandautomationnews.com/

Kim, M. (2019). *A secure charging system for electric vehicles based on blockchain*. Sensors. doi:10.339019133028

Meshram, P., Mishra, A., & Sahu, R. (2019). Environmental impact of spent lithium ION batteries and green recycling perspectives by organic acids–A review. *Chemosphere*, 125291. PMID:31896181

Susuki, Y., Mizuta, N., Ishigame, A., Ota, Y., Kawashima, A., Inagaki, S., & Suzuki, T. (2017, October). A continuum approach to assessing the impact of spatio-temporal EV charging to distribution grids. In *2017 IEEE 20th International Conference on Intelligent Transportation Systems (ITSC)* (pp. 1-6). IEEE. 10.1109/ITSC.2017.8317625

Arfeen, Khairuddin, Munir, Azam, Faisal, & Arif. (n.d.). En route of electric vehicles with the vehicle to grid technique in distribution networks–Status and technological review. *Energy Storage*.

El Kharbachi, A., Zavorotynska, O., Latroche, M., Cuevas, F., Yartys, V., & Fichtner, M. (2020). Exploits, Advances and Challenges Benefiting Beyond Li-Ion Battery Technologies. *Journal of Alloys and Compounds*, *817*, 153261. doi:10.1016/j.jallcom.2019.153261

Guarnieri, M. (2012). Looking back to electric cars. *3rd Reg. IEEE Hist. Electro - Technol. Conf. Orig. Electrotechnol. HISTELCON 2012 - Conf. Proc.* 10.1109/HISTELCON.2012.6487583

Leou, R. C., Su, C. L., & Lu, C. N. (2013). Stochastic analyses of electric vehicle charging impacts on distribution network. *IEEE Transactions on Power Systems*, *29*(3), 1055–1063. doi:10.1109/TPWRS.2013.2291556

Nabae, I. T., & Akagi, H. (1981). A New Neutral-Point-Clamped PWM Inverter. *IEEE Transactions on Industry Applications*, *IA-17*(5), 518–523. doi:10.1109/TIA.1981.4503992

Nakamoto, S. (2008). *Bitcoin: A peer-to-peer electronic cash system*. www.bitcoin.org/bitcoin.pdf

Hao, H., Mu, Z., Jiang, S., Liu, Z., & Zhao, F. (2017). GHG Emissions from the production of lithium-ion batteries for electric vehicles in China. *Sustainability*, *9*(4), 504. doi:10.3390u9040504

IEEE Recommended Practice for Powering and Grounding Electronic Equipment-Red-line, in IEEE Std. 1100-2005 (Revision of IEEE Std. 1100-1999)- Redline ed., pp. 1-703, 2006.

Khazraei, M., Sepahvand, H., Corzine, K. A., & Ferdowsi, M. (2012, February). Active capacitor voltage balancing in single-phase flying-capacitor multilevel power converters. *IEEE Transactions on Industrial Electronics*, *59*(2), 769–778. doi:10.1109/TIE.2011.2157290

Pothinun, T., & Premrudeepreechacharn, S. (2018, September). Power quality impact of charging station on mv distribution networks: A case study in pea electrical power system. In *2018 53rd International Universities Power Engineering Conference (UPEC)* (pp. 1-6). IEEE. 10.1109/UPEC.2018.8541921

Zaheer, H. (2018). *Vehicle Charging: Is blockchain a driver of EV adoption? | Power Technology Research*. Power Technol. Res. Available https://powertechresearch.com/p2p-vehicle-charging-is-blockchain-a-driver-of-ev-adoption/

eMotorWerks. (2017). eMotorWerks and Share&Charge Deliver North America's First Peer to Peer Electric Vehicle Charging Network with Blockchain Payments. *PRNewswire*. https://emotorwerks.com/about/enewsso/press-releases/329-emotorwerks-and-share-charge-deliver-north-americas-first-peer-to-peer-electric-vehicle-charging-network-with-blockchain-payments

Fuchs, E., & Masoum, M. A. (2015). *Power Quality in Power Systems and Electrical Machines* (2nd ed.). Elsevier.

Satarworn, S., & Hooncharеon, N. (2017, June). Impact of EV home charger on distribution transformer overloading in an urban area. In *2017 14th International Conference on Electrical Engineering/Electronics, Computer, Telecommunications and Information Technology (ECTI-CON)* (pp. 469-472). IEEE. 10.1109/ECTICon.2017.8096276

Shukla, A., Ghosh, A., & Joshi, A. (2012, November). Control of dc capacitor voltages in diode-clamped multilevel inverter using bidirectional buck-boost choppers. *IET Power Electronics*, *5*(9), 1723–1732. doi:10.1049/iet-pel.2012.0237

Stephens, D., Shawcross, P., Stout, G., Sullivan, E., Saunders, J., & Risser, S. (2017). *Lithium-ion battery safety issues for electric and plug-in hybrid vehicles.* National Highway Traffic Safety Administration, Report No. *DOT HS*, *812*, 418.

de MattosAffonso, CKezunovic, M. (2018). Technical and economic impact of PV-BESS charging station on transformer life: A case study. *IEEE Transactions on Smart Grid*, *10*(4), 4683–4692.

Duffey, C. K., & Stratford, R. D. (1989). Update of Harmonic Standard IEEE-519: IEEE Recommended Practices and Requirements for Harmonic Control in Electric Power Systems. *IEEE Transactions on Industry Applications*, *25*(6), 1025–1034. doi:10.1109/28.44238

Marco, S. W., & Chau Chanand, K. T. (2007, December). A New Switched-Capacitor Boost-Multilevel Inverter Using Partial Charging. *IEEE Transactions on Circuits and Wystems. II, Express Briefs*, *54*(12), 1145–1149. doi:10.1109/TCSII.2007.905352

Shepard, J. (2017). Blockchain Payment Technology for EV Charging. *eepower.* https://eepower.com/news/blockchain-payment-technology-for-ev-charging/#

Yang, J., Gu, F., & Guo, J. (2020). Environmental feasibility of secondary use of electric vehicle lithium-ion batteries in communication base stations. *Resources, Conservation and Recycling*, *156*, 104713. doi:10.1016/j.resconrec.2020.104713

Axelrod, B., Berkovich, Y., & Ioinovici, A. (2005, December). A cascade boost switched-capacitor-converter two-level inverter with an optimized multilevel output waveform. *IEEE Transactions on Circuits and Systems. I, Fundamental Theory and Applications*, *52*(12), 2763–2770. doi:10.1109/TCSI.2005.852205

Buterin, V. (2014). *A next-generation smart contract and decentralized application platform.* Etherum.

Gaines, L., & Nelson, P. (2009). Lithium-ion batteries: possible materials issues. *13th international battery materials recycling seminar and exhibit,* 16.

Wamburu, J., Lee, S., Shenoy, P., & Irwin, D. (2018, June). Analyzing distribution transformers at city scale and the impact of evs and storage. In *Proceedings of the Ninth International Conference on Future Energy Systems* (pp. 157-167). 10.1145/3208903.3208925

Watson & Watson. (2017). Impact of Electric Vehicle chargers on harmonic levels in New Zealand. *Proc. of 2017 7th IEEE Innovative Smart Grid Technologies (ISGT Asia 2017) annual premier smart grid technology meeting.*

Hinago, Y., & Koizumi, H. (2012, February). A Switched-Capacitor Inverter Using Series/Parallel Conversion With Inductive Load. *IEEE Transactions on Industrial Electronics*, *59*(2), 878–887. doi:10.1109/TIE.2011.2158768

Iyer, V. M., Gulur, S., Gohil, G., & Bhattacharya, S. (2018). Extreme fast charging station architecture for electric vehicles with partial power processing. *2018 IEEE Applied Power Electronics Conference and Exposition (APEC)*, 659-665. 10.1109/APEC.2018.8341082

Rohr, S., Müller, S., Baumann, M., Kerler, M., Ebert, F., Kaden, D., & Lienkamp, M. (2017). Quantifying uncertainties in reusing lithium-ion batteries from electric vehicles. *Procedia Manufacturing*, *8*, 603–610. doi:10.1016/j.promfg.2017.02.077

Utakrue, M., & Hongesombut, K. (2018, June). Impact analysis of electric vehicle quick charging to power transformer life time in distribution system. In 2018 IEEE Transportation Electrification Conference and Expo, Asia-Pacific (ITEC Asia-Pacific) (pp. 1-5). IEEE. doi:10.1109/ITEC-AP.2018.8433281

Watanabe, H., Fujimura, S., Nakadaira, A., Miyazaki, Y., Akutsu, A., & Kishigami, J. J. (2016). Blockchain contract: A complete consensus using blockchain. *2015 IEEE 4th Glob. Conf. Consum. Electron. GCCE 2015*, 577–578. 10.1109/GCCE.2015.7398721

Dong, M., Nassif, A. B., & Li, B. (2018). A data-driven residential transformer overloading risk assessment method. *IEEE Transactions on Power Delivery, 34*(1), 387–396. doi:10.1109/TPWRD.2018.2882215

Florea, B. C., & Taralunga, D. D. (2020). *Blockchain IoT for smart electric vehicles battery management.* Sustain. doi:10.3390u12103984

Sharma, Sood, Alam, & Shariff. (2020). Comparison of Common DC and AC Bus Architectures for EV Fast Charging Stations and Impact on Power Quality. *Elsevier eTransportation, 5.*

Sonoc, A., Jeswiet, J., & Soo, V. K. (2015). Opportunities to improve recycling of automotive lithium ion batteries. *Procedia CIRP, 29*, 752–757. doi:10.1016/j.procir.2015.02.039

Ye, Y., Cheng, K. W. E., Liu, J., & Ding, K. (2014, December). A Step-Up Switched-Capacitor Multilevel Inverter With Self-Voltage Balancing. *IEEE Transactions on Industrial Electronics, 61*(12), 6672–6680. doi:10.1109/TIE.2014.2314052

Bangarraju, J., Rajagopal, V., & Jayalaxmi, A. (2014). Unit template synchronous reference frame theory based Control algorithm for DSTATCOM. *J.Inst. Eng. India Ser. B, 95*(2), 135–141. doi:10.100740031-014-0087-y

Gai, K., Wu, Y., Zhu, L., Choo, K.-K. R., & Xiao, B. (2020). Blockchain-Enabled Trustworthy Group Communications in UAV Networks. *IEEE Transactions on Intelligent Transportation Systems*, 1–13. doi:10.1109/TITS.2020.3015862

Hosseini, S. H., Alishah, R. S., & Gharehkoushan, A. Z. (2015). Enhancement of extracted maximum power from partially shaded multi-string PV panels using a new cascaded high step-up DC-DC-AC converter. *2015 9th International Conference on Electrical and Electronics Engineering (ELECO)*, 644-648. 10.1109/ELECO.2015.7394619

Kundur, P., Balu, N. J., & Lauby, M. G. (1994). *Power system stability and control* (Vol. 7). McGraw-Hill.

Weisman, J. (2019). *Tesla LIB and PV Recycling Programs.* Available: www2.calrecycle.ca.gov

Chowdhury, A., & Koval, D. (2015). *Power Distribution System Reliability: Practical Methods and Applications.* John Wiley & Sons.

Ngo, B.-B., Nguyen, M.-K., Kim, J.-H., & Zare, F. (2018). Single-phase multilevel inverter based on switched-capacitor structure. *IET Power Electronics, 11*(11), 1858–1865. doi:10.1049/iet-pel.2017.0857

Qian, L. P., Wu, Y., Xu, X., Ji, B., Shi, Z., & Jia, W. (2020). Distributed Charging-Record Management for Electric Vehicle Networks via Blockchain. *IEEE Internet of Things Journal*, Advance online publication. doi:10.1109/JIOT.2020.3027482

Richa, K., Babbitt, C. W., Nenadic, N. G., & Gaustad, G. (2017). Environmental trade-offs across cascading lithium-ion battery life cycles. *The International Journal of Life Cycle Assessment, 22*(1), 66–81. doi:10.100711367-015-0942-3

Ahmadi, L., Young, S. B., Fowler, M., Fraser, R. A., & Achachlouei, M. A. (2017). A cascaded life cycle: Reuse of electric vehicle lithium-ion battery packs in energy storage systems. *The International Journal of Life Cycle Assessment, 22*(1), 111–124. doi:10.100711367-015-0959-7

Billinton, R., & Alen, R. (1994). *Reliability Evaluation of Power Systems* (2nd ed.). Plenum.

Taghvaie, J. A., & Rezanejad, M. (2018). A Self-Balanced Step-Up Multilevel Inverter Based on Switched-Capacitor Structure. *IEEE Transactions on Power Electronics*, *33*(1), 199–209. doi:10.1109/TPEL.2017.2669377

Zhou, G., Lv, L., Li, G., Huang, Y., Zhen, H., & Xiang, Z. (2019). Design of Energy Block Chain System Supporting Electric Vehicle Charge and Discharge Trading. *2019 IEEE 3rd International Electrical and Energy Conference (CIEEC)*, 526–533. 10.1109/CIEEC47146.2019.CIEEC-2019218

Anonymous. (2019). Partnership for Traceable Cobalt for Batteries: Engineering, Geology, Mineralogy, Metallurgy, Chemistry, etc. *Engineering and Mining Journal*.

Barzegarkhoo, R., Moradzadeh, M., Zamiri, E., Madadi Kojabadi, H., & Blaabjerg, F. (2018). A New Boost Switched-Capacitor Multilevel Converter with Reduced Circuit Devices. *IEEE Transactions on Power Electronics*, *33*(8), 6738–6754. doi:10.1109/TPEL.2017.2751419

Bhadra, J., & Chattopadhyay, T. K. (2015). Analysis of distribution network by reliability indices. *Proceedings of the International Conference on Energy, Power, and Environment: Towards Sustainable Growth (ICEPE)*, 1–5. 10.1109/EPETSG.2015.7510094

Hendrickson, T. P., Kavvada, O., Shah, N., Sathre, R., & Scown, C. D. (2015). Life-cycle implications and supply chain logistics of electric vehicle battery recycling in California. *Environmental Research Letters*, *10*(1), 014011. doi:10.1088/1748-9326/10/1/014011

Behrens, C. E., & Glover, C. (2012). U.S. energy: Overview and key statistics. In *Fossil Fuels*. Competition and Elasticities.

Deb, S., Tammi, K., Kalita, K., & Mahanta, P. (2018a). Impact of electric vehicle charging station load on distribution network. *Energies*, *11*(1), 178. doi:10.3390/en11010178

Huang, Li, Lai, Zhao, Wu, Li, Tong, & Lai. (2021). *A Novel Power Market Mechanism Based on Blockchain for Electric Vehicle Charging Stations*. doi:10.3390/electronics10030307

Irle, R. (2019). *Global BEV & PHEV Sales for 2019*. Available: www. EV-volumes.com

Miao, Y., Hynan, P., von Jouanne, A., & Yokochi, A. (2019). Current Li-Ion Battery Technologies in Electric Vehicles and Opportunities for Advancements. *Energies*, *12*(6), 1074. doi:10.3390/en12061074

Shalchi Alishah, R., Sabahi, M., Nazarpour, D., & Hosseini, S. H. (2014). Novel multilevel inverter topologies for medium and high-voltage applications with lower values of blocked voltage by switches. *IET Power Electronics*, *7*(12), 3062–3071. doi:10.1049/iet-pel.2013.0670

Dewulf, J., Van der Vorst, G., Denturck, K., Van Langenhove, H., Ghyoot, W., Tytgat, J., & Vandeputte, K. (2010). Recycling rechargeable lithium ion batteries: Critical analysis of natural resource savings. *Resources, Conservation and Recycling*, *54*(4), 229–234. doi:10.1016/j.resconrec.2009.08.004

GrandView Research. (2019). *Global Blockchain Technology Market Size Report, 2019-2025*. GVR-1-68038-329-4.

Liu, J., Zhu, X., & Zeng, J. (n.d.). A Seven-level Inverter with Self-balancing and Low Voltage Stress. IEEE Journal of Emerging and Selected Topics in Power Electronics. doi:10.1109/JESTPE.2018.2879890

Rupa, J. A. M., & Ganesh, S. (2014). Power flow analysis for radial distribution system using backward/forward sweep method. *Int. J. Electr. Comput. Electron. Commun. Eng.*, *10*, 1540–1544.

Arrillaga, J., & Watson, N. R. (2004). *Power system harmonics*. John Wiley & Sons.

Das, A., Li, D., Williams, D., & Greenwood, D. (2018). Joining technologies for automotive battery systems manufacturing. *World Electric Vehicle Journal*, *9*(2), 22. doi:10.3390/wevj9020022

Kuo, T. T., Kim, H. E., & Ohno-Machado, L. (2017). Blockchain distributed ledger technologies for biomedical and health care applications. *Journal of the American Medical Informatics Association*, *24*(6), 1211–1220. Advance online publication. doi:10.1093/jamia/ocx068 PMID:29016974

Nakagawa, Y., & Koizumi, H. (2017). A hybrid nine-level inverter with series/parallel conversion. *Proc. IEEE Int. Symp. Circuits Syst.*, 1-4. 10.1109/ISCAS.2017.8050652

Chen, G., Xu, B., Lu, M., & Chen, N.-S. (2018). *Exploring blockchain technology and its potential applications for education*. Smart Learn. Environ. doi:10.118640561-017-0050-x

Gaines, L. (2014). The future of automotive lithium-ion battery recycling: Charting a sustainable course. *Sustainable Materials and Technologies*, *1*, 2–7. doi:10.1016/j.susmat.2014.10.001

Hadley, S. W. (2006). *Impact of plug-in hybrid vehicles on the electric grid.* ORNL Report.

Nakagawa, Y., & Koizumi, H. (2019, July). A Boost-Type Nine-Level Switched Capacitor Inverter. *IEEE Transactions on Power Electronics*, *34*(7), 6522–6532. doi:10.1109/TPEL.2018.2876158

Andoni, M., Robu, V., Flynn, D., Abram, S., Geach, D., Jenkins, D., McCallum, P., & Peacock, A. (2019). Blockchain technology in the energy sector: A systematic review of challenges and opportunities. *Renewable & Sustainable Energy Reviews*, *100*, 143–174. Advance online publication. doi:10.1016/j.rser.2018.10.014

Ghosh, D., Deb, S., & Mohanta, D. K. (2017). Reliability evaluation and enhancement of microgrid incorporating the effect of distributed generation. In *Handbook of Distributed Generation* (pp. 685–730). Springer. doi:10.1007/978-3-319-51343-0_20

Lai & Peng. (1996). Multilevel converters-a new breed of power converters. *IAS '95. Conf. Rec. 1995 IEEE Ind. Appl. Conf. Thirtieth IAS Annu. Meet.*, *3*(3), 2348–2356.

Li, X., Mo, Y., Qing, W., Shao, S., Tang, C. Y., & Li, J. (2019). Membrane-based technologies for lithium recovery from water lithium resources: A review. *Journal of Membrane Science*, *591*, 117317. doi:10.1016/j.memsci.2019.117317

Ciez, R. E., & Whitacre, J. F. (2019). Examining different recycling processes for lithium-ion batteries. Nature Sustainability, 2, 148-156. doi:10.103841893-019-0222-5

Forum of Regulators. (n.d.). http://www.forumofregulators.gov.in/Data/study/EV.pdf

Khoukha, C. H., & Madjid, B. E. (2007). Multilevel selective harmonic elimination PWM technique in the nine level voltage inverter. *2007 International Conference on Computer Engineering & Systems*, 387-392. 10.1109/ICCES.2007.4447075

Ølnes, S., Ubacht, J., & Janssen, M. (2017). Blockchain in government: Benefits and implications of distributed ledger technology for information sharing. *Government Information Quarterly*, *34*(3), 355–364. Advance online publication. doi:10.1016/j.giq.2017.09.007

Meshram, P. M., & Borghate, V. B. (2015, January). A Simplified Nearest Level Control (NLC) Voltage Balancing Method for Modular Multilevel Converter (MMC). *IEEE Transactions on Power Electronics*, *30*(1), 450–462. doi:10.1109/TPEL.2014.2317705

Saberi, S., Kouhizadeh, M., Sarkis, J., & Shen, L. (2019). Blockchain technology and its relationships to sustainable supply chain management. *International Journal of Production Research*, *57*(7), 2117–2135. Advance online publication. doi:10.1080/00207543.2018.1533261

Sharma, A., Kapoor, M. A., & Chakrabarti, S. (2019). *Impact of Plug-in Electric Vehicles on Power Distribution System of Major Cities of India: A Case Study*. Academic Press.

Shi, Y., Chen, G., Liu, F., Yue, X., & Chen, Z. (2018). Resolving the Compositional and Structural Defects of Degraded LiNi x Co y Mn z O2 Particles to Directly Regenerate High-Performance Lithium-Ion Battery Cathodes. *ACS Energy Letters*, *3*(7), 1683–1692. doi:10.1021/acsenergylett.8b00833

Conoscenti, Vetro, & De Martin. (2016). *Blockchain for the Internet of Things: A systematic literature review.*. doi:10.1109/AICCSA.2016.7945805

Deb, S., Tammi, K., Kalita, K., & Mahanta, P. (2018b). Review of recent trends in charging infrastructure planning for electric vehicles. *Wiley Interdisciplinary Reviews. Energy and Environment*, *7*(6), e306. doi:10.1002/wene.306

Gaines, L., Sullivan, J., Burnham, A., & Belharouak, I. (2011). Life-cycle analysis of production and recycling of lithium ion batteries. *Transportation Research Record: Journal of the Transportation Research Board*, *2252*(1), 57–65. doi:10.3141/2252-08

Kolomyjski, W., Malinowski, M., & Kazmierkowski, M. P. (2006). *Adaptive space vector modulator for three-level inverter-fed induction motor. 9th IEEE Int. Work. Adv. Motion Control.*

Narayanan, Bonneau, Felten, & Miller. (2016). Bitcoin and cryptocurrency technologies. Princeton University Press.

Recycling Process. (2020). Available: https://csm.umicore.com/en/recycling/battery-recycling/our-recycling-process

Wang, F. (2002). Sine-triangle versus space-vector modulation for three-level PWM voltage77 source inverters. *IEEE Transactions on Industry Applications*, *38*(2), 500–506. doi:10.1109/28.993172

Du, M., Ma, X., Zhang, Z., Wang, X., & Chen, Q. (2017). A review on consensus algorithm of blockchain. *2017 IEEE Int. Conf. Syst. Man, Cybern. SMC 2017*, 2567–2572. 10.1109/SMC.2017.8123011

Popper, Lanfrankie, & Caro. (2019). *The (Potential) Value of Labeling in the Lithium Ion Battery Supply Chain.* Available: https://blogs.anderson.ucla.edu/

Trzynadlowski, A. M. (2016). Introduction to Modern Power Electronics. John Wiley & Sons.

Angulo, M., Lezana, P., Kouro, S., & Mar, S. (2007). *Level-shifted PWM for Cascaded Multilevel Inverters with Even Power Distribution.* doi:10.1109/PESC.2007.4342382

Swain, B. (2017). Recovery and recycling of lithium: A review. *Separation and Purification Technology*, *172*, 388–403. doi:10.1016/j.seppur.2016.08.031

Zheng, Z., Xie, S., Dai, H., Chen, X., & Wang, H. (2017). An Overview of Blockchain Technology: Architecture, Consensus, and Future Trends. *Proceedings - 2017 IEEE 6th International Congress on Big Data, BigData Congress 2017*, 557–564. 10.1109/BigDataCongress.2017.85

BNEF. (n.d.). https://about.bnef.com/electric-vehicle-outlook/

González, L. G., Siavichay, E., & Espinoza, J. L. (2019). Impact of EV fast charging stations on the power distribution network of a Latin American intermediate city. *Renewable & Sustainable Energy Reviews*, *107*, 309–318. doi:10.1016/j.rser.2019.03.017

Larminie, J., & Lowry, J. (2012). *Electric Vehicle Technology Explained* (2nd ed.). doi:10.1002/9781118361146

Mahmood. (2021). *Blockchain Technology: Initiatives and Use Cases in the Industry.* doi:10.4018/978-1-7998-6650-3.ch009

Roy, T., Sadhu, P. K., & Dasgupta, A. (2019, November). Cross-Switched Multilevel Inverter Using Novel Switched Capacitor Converters. *IEEE Transactions on Industrial Electronics*, *66*(11), 8521–8532. doi:10.1109/TIE.2018.2889632

Slowik, P., Pavlenko, N., & Lutsey, N. (2016). *Assessment of next-generation electric vehicle technologies.* White paper, International Council on Clean Transportation.

Heelan, J., Gratz, E., Zheng, Z., Wang, Q., Chen, M., Apelian, D., & Wang, Y. (2016). Current and prospective Li-ion battery recycling and recovery processes. *JOM, 68*(10), 2632–2638. doi:10.100711837-016-1994-y

Karaca, H. (2015). GA Based Selective Harmonic Elimination for Multilevel Inverter with Reduced Number of Switches (Vol. I). Academic Press.

Nakamoto. (2008). *Bitcoin: A Peer-to-Peer Electronic Cash System.* Available: https://bitcoin.org/bitcoin.pdf

King & Nadal. (2012). *PPCoin: Peer-to-Peer Crypto-Currency with Proof-of-Stake.* Academic Press.

Lithium-Ion. (2020). Available: https://www.retrievtech.com/lithiumion

Sreedhar. (n.d.). *Genetic Algorithm Based Selective Harmonic Elimination Technique For Multilevel Inverters With Unequal Voltage Sources.* Academic Press.

BitShares. (2014). https://bitshares.org/

Hyperledger project. (2015). Academic Press.

Buterin, V. (2015). On Public and Private Blockchains. Ethereum Blog Crypto Renaiss. Salon, 1–24.

Pilkington, M. (2020). Blockchain technology: principles and applications. In *Research Handbook on Digital Transformations* (pp. 225–253). Edward Elgar Publishing.

Guidehouse Insight. (2020). *Market Data: Blockchain-Enabled EV Charging and Integration.* Author.

Beresford, C. (2020). GM Unveils Battery with Capacity Twice as Big as Tesla's. *Car and Driver.* https://www.caranddriver.com/news/a31226611/gm-ultium-electric-vehicle-battery-revealed/#:~:text=Rivian%2C

Autovista Group. (2020). *EV charging anxiety is the new range anxiety.* Autovista Group. https://autovistagroup.com/news-and-insights/ev-charging-anxiety-new-range-anxiety?fbclid=IwAR252W27xd4WfvupAdsGt0K2cjbIuewsMJ5BXuF0wptsvE9oM9JmIWsUgyE

Olson, E. B. (2020). *Federal Investment in EV Charging Infrastructure for Economic Recovery.* Climate, and Public Health.

Bhatti, A. R., & Salam, Z. (2018). A rule-based energy management scheme for uninterrupted electric vehicles charging at constant price using photovoltaic-grid system. *Renewable Energy, 125*, 384–400. doi:10.1016/j.renene.2018.02.126

Bunjaku, F., Gjorgieva-Trajkovska, O., & Miteva-Kacarski, E. (2017). Cryptocurrency - Advantages and Disadvantages. *Journal of Economics, 2*(1).

Deb, S., Kalita, K., & Mahanta, P. (2017, December). Review of impact of electric vehicle charging station on the power grid. In *2017 International Conference on Technological Advancements in Power and Energy (TAP Energy)* (pp. 1-6). IEEE. 10.1109/TAPENERGY.2017.8397215

Globe Newswire. (n.d.). https://www.globenewswire.com/

Kobashi, T. (2020). Smart city and ICT infrastructure with vehicle to X applications toward urban decarbonization. Urban Systems Design. doi:10.1016/B978-0-12-816055-8.00009-9

Mak, O. C., & Ioinovici, A. (1998, April). Switched-capacitor inverter with high power density and enhanced regulation capability. *IEEE Transactions on Circuits and Systems. I, Fundamental Theory and Applications, 45*(4), 336–347. doi:10.1109/81.669056

Iea. (2004). Electric vehicles in Europe. *Energy*, *3*, 33. Available: http://wrs.region-stuttgart.de/sixcms/media.php/923/ElectricMobilityPilotRegion_english.pdf%5Cn

Hanley, S. (2020). China Invests. In *EV Charging Infrastructure To Offset Coronavirus Economic Slump*. Clean Technica.

Aljabri, Aldhaheri, Mohammed, Alkaabi, Abdella, & Shuaib. (2019). *Facilitating Electric Vehicle Charging Across the UAE Using Blockchain.* . doi:10.1109/ICECTA48151.2019.8959705

Jeong, S., Dao, N.-N., Lee, Y., Lee, C., & Cho, S. (2018). Blockchain Based Billing System for Electric Vehicle and Charging Station. *2018 Tenth International Conference on Ubiquitous and Future Networks (ICUFN)*, 308–310. 10.1109/ICUFN.2018.8436987

Kobashi, T., Yoshida, T., Yamagata, Y., Naito, K., Pfenninger, S., Say, K., Takeda, Y., Ahl, A., Yarime, M., & Hara, K. (2020). On the potential of 'Photovoltaics + Electric vehicles' for deep decarbonization of Kyoto's power systems: Techno-economic-social considerations. *Applied Energy*, *275*, 115419. Advance online publication. doi:10.1016/j.apenergy.2020.115419

Akin, Y., Dikkollu, C., Kaplan, B. B., Yayan, U., & Yolacan, E. N. (2019). Ethereum Blockchain Network-based Electrical Vehicle Charging Platform with Multi-Criteria. *Decision Support Systems*, 1–5. Advance online publication. doi:10.1109/UBMYK48245.2019.8965557

Share & Charge. (2020). Cross-provider smart charging based on OCPI Unlocking the innovation potential in EV charging. *Share & Charge*. https://shareandcharge.com/

IRENA. (2020). *Electricity trading arrangements*. Available: https://www.youtube.com/watch?v=UDv04q3U6e0

Zhang, C., Wu, J., Long, C., & Cheng, M. (2017). Review of Existing Peer-to-Peer Energy Trading Projects. *Energy Procedia*, *105*, 2563–2568. doi:10.1016/j.egypro.2017.03.737

Centrica. (2018). *Centrica and LO3 Energy to deploy blockchain technology as part of Local Energy Market trial in Cornwall*. www.centrica.com/news/centrica-and-lo3-%0Aenergy-deploy-blockchain-technology-part-localenergy-%0Amarket-trial-cornwall

Babaei, E., & Hosseini, S. H. (2009). New cascaded multilevel inverter topology with minimum number of switches. *Elsevier J. Energy Convers. Manage*, *50*(11), 2761–2767. doi:10.1016/j.enconman.2009.06.032

Expansion, challenges, and opportunities in the EV market and EVSE industry blog on e-mobility. (n.d.). https://circontrol.com/expansion-challenges-ev-market-evse-industry/

International Energy Agency (IEA). (2020). Global EV Outlook 2020. Advance online publication. doi:10.1787/d394399e-en

Lü, X., Wu, Y., Lian, J., Zhang, Y., Chen, C., Wang, P., & Meng, L. (2020). Energy management of hybrid electric vehicles: A review of energy optimization of fuel cell hybrid power system based on genetic algorithm. *Energy Conversion and Management*, *205*, 112474. doi:10.1016/j.enconman.2020.112474

Sachan, S., & Adnan, N. (2018). Stochastic charging of electric vehicles in smart power distribution grids. *Sustainable Cities and Society, 40,* 91-100.

Xu, L., Chen, L., Gao, Z., Carranco, L., Fan, X., Shah, N., Diallo, N., & Shi, W. (2019). *Supporting blockchain-based cryptocurrency mobile payment with smart devices*. IEEE.

Energie, E. (2019). *A saving, generating and exchanging energy in Eemnes*. https://eemnesenergie.nl/

Electrify. (2020). *The Big Singapore's Energy Story*. https://electrify.asia/

Electron. (2020). *EmpoEnergy markets*. https://electron.net/

Gridplus. (2017). *Grid+ Raises $29 Million as Blockchain Fever Grows*. https://gridplus.io/

LichtBlick. (2016). *German green utility goes international with 'sharing economy' concept for distributed energy*. https://www.lichtblick.de/privatkunden/schwarm-energie

Lumenaza. (2020). *The Software for the energy revolution*. www.lumenaza.de/en

Brandherm, Baus, & Frey. (2012). *Peer energy cloud-Civil marketplace for trading renewable energies*. . doi:10.1109/IE.2012.46

Kingdom. (2015). *Piclo peer-to-peer energy trading*. Academic Press.

PowerLedger. (2019). *Power Ledger completes blockchain-enabled P2P trial with KEPCO*. https://www.powerledger.io/

Open Utility. (2016). *A glimpse into the future of Britain's energy economy*. White Pap. *Open Utility Ltd.*

Ahmad, A., Alam, M. S., & Chabaan, R. (2017). A Comprehensive Review of Wireless Charging Technologies for Electric Vehicles. *IEEE Trans. Transp. Electrif.*, *4*(1), 38–63. doi:10.1109/TTE.2017.2771619

Better World Solutions. (2019). Available: https://www.betterworldsolutions.eu/repurpose-ev-batteries/

Deb, S., Kalita, K., & Mahanta, P. (2019). Distribution Network planning considering the impact of Electric Vehicle charging station load. In *Smart Power Distribution Systems* (pp. 529–553). Academic Press. doi:10.1016/B978-0-12-812154-2.00022-5

Kangarlu, M. F., Babaei, E., & Laali, S. (2012). Symmetric multilevel inverter with reduced components based on non-insulated dc voltage sources. *IET Power Electronics*, *5*(5), 571–581. doi:10.1049/iet-pel.2011.0263

Lee. (2019). *Rise of Anonymous Cryptocurrencies: Brief Introduction*. IEEE.

Moradewicz, A. J. (2019). On / Off – board chargers for electric vehicles. *Przegląd Elektrotechniczny, 95*(2).

Sun Contract. (2017). *An energy trading platform that utilises blockchain technology to create a new disruptive model for buying and selling electricity*. Available: https://suncontract.org/res/whitepaper.pdf

UNFCCC. (2019). Me Solshare: Peer-To-Peer Smart Village Grids | Bangladesh. *United Nations Framework Convention on Climate Change*. http://unfccc.int/secretariat/momentum_for_change/items/9940.php

Sonnen. (2019). *What Is the Sonnen Community?* https://sonnengroup.com/sonnencommunity

Orsini, L., Kessler, S., Wei, J., & Field, H. (2018). How the brooklyn microgrid and transactive grid are paving the way to next-gen energy markets. The Energy Internet: An Open Energy Platform to Transform Legacy Power Systems into Open Innovation and Global Economic Engines.

Vandebron. (2020). *Duurzame Energie Van Nederlandse Bodem*. https://vandebron.nl

Energycentral. (2016). *Lights Out for Yeloha: Why We Shut Down the Solar Sharing Network*. Available: https://energycentral.com/c/um/lights-out-yeloha-why-we-shut-down-solar-sharing-network

Denholm, P., Margolis, R., & Milford, J. (2009). Production cost modeling for high levels of photovoltaics penetration. *Renew. Energy Grid Interact. Bus. Photovoltaics*, 87–118.

Pal. (2020). *Solving Duck Curve Problem Due to Solar Integration Using Blockchain Technology*. . doi:10.1007/978-981-32-9119-5_17

Mao, T., Zhang, X., & Zhou, B. (2018). Modeling and solving method for supporting 'vehicle-to-anything' EV charging mode. *Applied Sciences (Basel, Switzerland)*, *8*(7), 1048. Advance online publication. doi:10.3390/app8071048

Alvaro-Hermana, R., Fraile-Ardanuy, J., Zufiria, P. J., Knapen, L., & Janssens, D. (2016). Peer to Peer Energy Trading with Electric Vehicles. *IEEE Intell. Transp. Syst. Mag.*, *8*(3), 33–44. doi:10.1109/MITS.2016.2573178

Loi, L., Duc-Hiep, C., & Hrishi, O. (2016). Making Smart Contracts Smarter. *ACM SIGSAC Conference on Computer and Communications Security (CCS '16).*

Ma, G., Jiang, L., Chen, Y., Dai, C., & Ju, R. (2017). Study on the impact of electric vehicle charging load on nodal voltage deviation. *Archives of Electrical Engineering*, *66*(3), 495–505. doi:10.1515/aee-2017-0037

Patil, D., McDonough, M. K., Miller, J. M., Fahimi, B., & Balsara, P. T. (2017). *Wireless Power Transfer for Vehicular Applications: Overview and Challenges. IEEE Trans. Transp. Electrif.* doi:10.1109/TTE.2017.2780627

Shaukat, N. (2018, January). A survey on electric vehicle transportation within smart grid system. *Renew. Sustain. Energy Rev.*, *81*, 1329–1349. Advance online publication. doi:10.1016/j.rser.2017

Zeeshan Ahmad Arfeen, M. P. A. (2020). *Energy Storage Usages: Engineering Reactions, Economic-Technological Values for Electric Vehicles – A Technological Outlook.* International Transactions on Electrical Energy Systems.

Xue, Mao, Pan, Qin, Li, & Shi. (2020). *Design of blockchain-based trading mechanism under sharing mode of electric vehicle under smart grid.* doi:10.1109/ACPEE48638.2020.9136233

Sun, G., Dai, M., Zhang, F., Yu, H., Du, X., & Guizani, M. (2020). Blockchain-Enhanced High-Confidence Energy Sharing in Internet of Electric Vehicles. *IEEE Internet of Things Journal*, *7*(9), 7868–7882. Advance online publication. doi:10.1109/JIOT.2020.2992994

Mannaro, K., Pinna, A., & Marchesi, M. (2017). Crypto-Trading: Blockchain-oriented energy market. *2017 AEIT Int. Annu. Conf. Infrastructures Energy ICT Oppor. Foster. Innov. AEIT 2017*, 1–5. 10.23919/AEIT.2017.8240547

Thakur, Hayes, & Breslin. (2018). *A unified model of peer to peer energy trade and electric vehicle charging using blockchains.* . doi:10.1049/cp.2018.1909

Liu, Chai, Lau, & Chen. (2018). *Blockchain based energy trading model for electric vehicle charging schemes.* . doi:10.1007/978-3-319-94965-9_7

Zhou, Tan, & Xu. (2018). *Blockchain and Edge Computing Based Vehicle-to-Grid Energy Trading in Energy Internet.* . doi:10.1109/EI2.2018.8582652

Umoren, Jaffary, Shakir, Katzis, & Ahmadi. (2020). Blockchain-based energy trading in electric-vehicle-enabled microgrids. *IEEE Consum. Electron. Mag.* doi:10.1109/MCE.2020.2988904

Kang, J., Yu, R., Huang, X., Maharjan, S., Zhang, Y., & Hossain, E. (2017, December). Enabling Localized Peer-to-Peer Electricity Trading among Plug-in Hybrid Electric Vehicles Using Consortium Blockchains. *IEEE Transactions on Industrial Informatics*, *13*(6), 3154–3164. doi:10.1109/TII.2017.2709784

Silva, F. C., Ahmed, M. A., Martínez, J. M., & Kim, Y. C. (2019). Design and implementation of a blockchain-based energy trading platform for electric vehicles in smart campus parking lots. *Energies*. Advance online publication. doi:10.3390/en12244814

Nagatsuka, Sano, & Yamaguchi. (2018). *Decentralized Transaction System of Surplus PV Output Using Blockchain.* Academic Press.

Dubey, A., & Santoso, S. (2015). Electric vehicle charging on residential distribution systems: Impacts and mitigations. *IEEE Access: Practical Innovations, Open Solutions*, *3*, 1871–1893. doi:10.1109/ACCESS.2015.2476996

Falk, Nedjalkov, Angelmahr, & Schade. (n.d.). *Applying Lithium-Ion Second Life Batteries for Off-Grid Solar Powered System—A Socio-Economic Case Study for Rural Development*. Academic Press.

Lee & Clark. (2018). *Charging the Future: Challenges and Opportunities for Electric Vehicle Adoption*. Academic Press.

Tran, V. T., Sutanto, D., & Muttaqi, K. M. (2017). The state of the art of battery charging infrastructure for electrical vehicles: Topologies, power control strategies, and future trend. *Proc. of AUPEC*, 1-6. 10.1109/AUPEC.2017.8282421

Wöhrer & Zdun. (2018). *Smart Contracts: Security patterns in the Ethereum ecosystem and Solidity*. IEEE.

Chen, X., Zhang, T., Ye, W., Wang, Z., & Iu, H. H.-C. (2020). Blockchain-based Electric Vehicle Incentive System for Renewable Energy Consumption. *IEEE Transactions on Circuits and Wystems. II, Express Briefs*. Advance online publication. doi:10.1109/TCSII.2020.2996161

Islam, Shahjalal, Hasan, & Jang. (2020). *Blockchain-based Energy Transaction Model for Electric Vehicles in V2G Network*. . doi:10.1109/ICAIIC48513.2020.9065221

Alskaif, T., Holthuizen, B., Schram, W., Lampropoulos, I., & Van Sark, W. (2020, December). A blockchain-based configuration for balancing the electricity grid with distributed assets. *World Electr. Veh. J.*, *11*(4), 1–17. doi:10.3390/wevj11040062

Douceur, J. R. (2002). The sybil attack. Lecture Notes in Computer Science, 2429, 251–260. doi:10.1007/3-540-45748-8_24

Asfia, U., Kamuni, V., Sutavani, S., Sheikh, A., Wagh, S., & Singh, N. M. (2019). A blockchain construct for energy trading against sybil attacks. *27th Mediterranean Conference on Control and Automation, MED 2019 - Proceedings*, 422–427. 10.1109/MED.2019.8798489

Waseda & Tanaka. (2020). *Bidding Agent for Electric Vehicles in Peer-to-Peer Electricity Trading Market considering uncertainty*. . doi:10.1109/EEEIC/ICPSEurope49358.2020.9160846

Xia, S., Lin, F., Chen, Z., Tang, C., Ma, Y., & Yu, X. (2020). A Bayesian Game Based Vehicle-to-Vehicle Electricity Trading Scheme for Blockchain-Enabled Internet of Vehicles. *IEEE Transactions on Vehicular Technology*, *69*(7), 6856–6868. Advance online publication. doi:10.1109/TVT.2020.2990443

Tai, X., Sun, H., & Guo, Q. (2016, December). Electricity transactions and congestion management based on blockchain in energy internet. *Dianwang Jishu/Power Syst. Technol.*, *40*(12), 3630–3638. doi:10.13335/j.1000-3673.pst.2016.12.002

Anastasiadis, A. G., Kondylis, G. P., Polyzakis, A., & Vokas, G. (2019). Effects of Increased Electric Vehicles into a Distribution Network. *Energy Procedia*, *157*, 586–593. doi:10.1016/j.egypro.2018.11.223

Canals Casals, L., & Amante García, B. (2016). Assessing electric vehicles battery second life remanufacture and management. *Journal of Green Engineering*, *6*, 77–98. doi:10.13052/jge1904-4720.614

Das, H. S., Rahman, M. M., Li, S., & Tan, C. W. (2020). Electric vehicles standards, charging infrastructure, and impact on grid integration: A technological review. *Renewable & Sustainable Energy Reviews*, *120*, 109618. Advance online publication. doi:10.1016/j.rser.2019.109618

Savjee. (n.d.). https://www.savjee.be/videos/simply-explained/how-does-a-blockchain-work/

Tan, L., Wu, B., Yaramasu, V., Rivera, S., & Guo, X. (2016). Effective Voltage Balance Control for Bipolar-DC-Bus-Fed EV Charging Station with Three-Level DC–DC Fast Charger. *IEEE Transactions on Industrial Electronics*, *63*(7), 4031–4041. doi:10.1109/TIE.2016.2539248

Waltrich, G., & Barbi, I. (2010). Three-phase cascaded multilevel inverter using power cells with two inverter legs in series. *IEEE Transactions on Industrial Electronics*, *57*(8), 2605–2612. doi:10.1109/TIE.2010.2043040

Abdel-Khalik, A. S., Massoud, A., & Ahmed, S. (2018). Interior permanent magnet motor-based isolated on-board integrated battery charger for electric vehicles. *IET Electric Power Applications*, *12*(1), 124–134. doi:10.1049/iet-epa.2017.0319

Achour, H., Carton, J. G., & Olabi, A. G. (2011). Estimating vehicle emissions from road transport, case study: Dublin City. *Applied Energy*, *88*(5), 1957–1964. doi:10.1016/j.apenergy.2010.12.032

Achour, H., Marashly, A., & Olabi, A.-G. (2012). Assessing energy consumption of the transport sector in Aleppo, Syria. *Sustainable Manufacturing and Renewable Energy*, *1*(3–4), 87–102.

Achour, H., & Olabi, A. G. (2016). Driving cycle developments and their impacts on energy consumption of transportation. *Journal of Cleaner Production*, *112*, 1778–1788. doi:10.1016/j.jclepro.2015.08.007

Agarwal, S., Alam, M. S., Rafat, Y., & Shemami, M. S. (2020). Data Analysis of Grid-Connected Solar Setup and Regression Based Predictive Models. In *2020 5th IEEE International Conference on Recent Advances and Innovations in Engineering, ICRAIE 2020 – Proceeding*. Institute of Electrical and Electronics Engineers Inc.

Ahmad, F., Alam, M. S., & Shahidehpour, M. (2018). Optimal placement of electric, hybrid and plug-in hybrid electric vehicles (xEVs) in Indian power market. 2017 Saudi Arabia Smart Grid Conference, 1–7.

Ahmad, F., Alam, M.S., & Shariff, S.M. (2019a). A Cost-efficient Energy Management System for Battery Swapping Station. *IEEE Syst J*, 1–10. doi:10.1109/JSYST.2018.2890569

Ahmad, F., & Alam, M. S. (2017). Feasibility study, design and implementation of smart polygeneration microgrid at AMU. *Sustainable Cities and Society*, *35*, 309–322. doi:10.1016/j.scs.2017.08.007

Ahmad, F., & Alam, M. S. (2018a). Economic and Ecological Aspects for Microgrids Deployment in India. *Sustainable Cities and Society*, *37*, 407–419. doi:10.1016/j.scs.2017.11.027

Ahmad, F., & Alam, M. S. (2018b). Optimal Sizing and Analysis of Solar PV, Wind, and Energy Storage Hybrid System for Campus Microgrid. *Smart Sci*, *6*(2), 150–157. doi:10.1080/23080477.2017.1417005

Ahmad, F., & Alam, M. S. (2019). Assessment of power exchange based electricity market in India. *Energy Strateg Rev*, *23*, 163–177. doi:10.1016/j.esr.2018.12.012

Ahmad, F., Alam, M. S., & Asaad, M. (2017). Developments in xEVs charging infrastructure and energy management system for smart microgrids including xEVs. *Sustainable Cities and Society*, *35*, 552–564. doi:10.1016/j.scs.2017.09.008

Ahmad, F., Alam, M., Alsaidan, I., & Shariff, S. M. (2020). *Battery Swapping Station for Electric Vehicles: Opportunities and Challenges*. IET Smart Grid. doi:10.1049/iet-stg.2019.0059

Ahmad, F., Alam, M., Shariff, S., & Krishnamurthy, M. (2019). *A Cost-efficient Approach to EV Charging Station Integrated Community Microgrid. IEEE Transactions on Transportation Electrification*. doi:10.1109/TTE.2019.2893766

Ajzen, I. (1991). The theory of planned behavior. *Organizational Behavior and Human Decision Processes*, *50*(2), 179–211. doi:10.1016/0749-5978(91)90020-T

Alaswad, A., Baroutaji, A., Achour, H., Carton, J., Al Makky, A., & Olabi, A. G. (2016a). Developments in fuel cell technologies in the transport sector. *International Journal of Hydrogen Energy*, *41*(37), 16499–16508. doi:10.1016/j.ijhydene.2016.03.164

Alaswad, A., Dassisti, M., Prescott, T., & Olabi, A. G. (2015). Technologies and developments of third generation biofuel production. *Renewable & Sustainable Energy Reviews*, *51*, 1446–1460. doi:10.1016/j.rser.2015.07.058

Al-Awami, A. T., Amleh, N. A., & Muqbel, A. M. (2017). Optimal demand response bidding and pricing mechanism with fuzzy optimization: Application for a virtual power plant. *IEEE Transactions on Industry Applications*, *53*(5), 5051–5061. doi:10.1109/TIA.2017.2723338

Al-Hamad & Qamber. (2019). GCC electrical long-term peak load forecasting modeling using ANFIS and MLR methods. *International Arab Journal of Basic and Applied Sciences, 26*, 269–282. doi:10.1080/25765299.2019.1565464

Al-Hamad, M., & Qamber, I. (2018). The GCC Countries Power Trading Modeling and Challenges. In *Saudi Arabia Smart Grid Conference on Smart Grids and Green Energy (SASG 2018)*, Jeddah, Saudi Arabia.

Ali, Mascarella, Joos, & Moschopoulos. (2015). Circulating current minimization for dual three phase motor integrated battery charger. *2015 IEEE Energy Conversion Congress and Exposition (ECCE)*, 3651–3658. doi:10.1109/ECCE.2015.7310176

Ali, S. Q., Mascarella, D., & Joos, G. (2014). Integrated battery charger for delta connected machines in plug-in hybrid electric vehicles. *2014 IEEE 27th Canadian Conference on Electrical and Computer Engineering (CCECE)*, 1–6. 10.1109/CCECE.2014.6901072

Aliasghari, P., Mohammadi-Ivatloo, B., Alipour, M., Abapour, M., & Zare, K. (2018a). Optimal scheduling of plug-in electric vehicles and renewable micro-grid in energy and reserve markets considering demand response program. *Journal of Cleaner Production*, *186*, 293–303. doi:10.1016/j.jclepro.2018.03.058

Ali, S. Q. (2018). *Integrated Battery Chargers for Electric Vehicle Traction Drives with Synchronous Machines*. McGill University Libraries.

Ali, S. Q., Mascarella, D., Joos, G., & Tan, L. (2017). Torque elimination for integrated battery charger based on two permanent magnet synchronous motor drives for electric vehicles. *IET Electric Power Applications*, *11*(9), 1627–1635. doi:10.1049/iet-epa.2017.0302

Ali, S. Q., Mascarella, D., Joos, G., & Tan, L. (2018). Torque Cancelation of Integrated Battery Charger Based on Six-Phase Permanent Magnet Synchronous Motor Drives for Electric Vehicles. *IEEE Transactions on Transportation Electrification*, *4*(2), 344–354. doi:10.1109/TTE.2017.2788190

All Electric Cars by 2030? Nitin Gadkari Inaugurates EV Charging Points at Niti Aayog | India News, The Indian Express. (n.d.). https://indianexpress.com/article/india/all-electric-cars-by-2030-nitin-gadkari-inaugurates-ev-charging-points-at-niti-aayog-5065099/

All-Electric Vehicles. (2017). https://www.fueleconomy.gov/feg/evtech.shtml

Al-Shariff, S., Alam, M., Ahmad, Z., & Ahmad, F. (2019). Smart Transportation System: Mobility solution for Smart Cities. *Second Smart Cities Symposium (SCS 2019, University of Bahrain)*.

Amini, M. H., Moghaddam, M. P., & Karabasoglu, O. (2017). Simultaneous allocation of electric vehicles' parking lots and distributed renewable resources in smart power distribution networks. *Sustainable Cities and Society*, *28*, 332–342. Advance online publication. doi:10.1016/j.scs.2016.10.006

Amjad, M., Ahmad, A., Husain, M., & Umer, T. (2018). A review of EVs charging : From the perspective of energy optimization, optimization approaches, and charging techniques. *Transportation Research*, *62*(Part D), 386–417. doi:10.1016/j.trd.2018.03.006

Amrr, S. M., Alam, M. S., Asghar, M. S. J., & Ahmad, F. (2018). Low cost residential microgrid system based home to grid (H2G) back up power management. *Sustainable Cities and Society*, *36*, 204–214. doi:10.1016/j.scs.2017.10.016

Andersen, P. B., Marinelli, M., Olesen, O. J., Andersen, C. A., Poilasne, G., Christensen, B., & Alm, O. (2014, October). The Nikola project Intelligent electric vehicle integration. In IEEE PES Innovative Smart Grid Technologies, Europe (pp. 1-6). IEEE. doi:10.1109/ISGTEurope.2014.7028765

Andújar, J. M., & Segura, F. (2009). Fuel cells: History and updating. A walk along two centuries. *Renewable & Sustainable Energy Reviews*, *13*(9), 2309–2322. doi:10.1016/j.rser.2009.03.015

Angulakshmi, N., Kumar, R. S., Kulandainathan, M. A., & Stephan, A. M. (2014). Composite polymer electrolytes encompassing metal organic frame works: A new strategy for all-solid-state lithium batteries. *The Journal of Physical Chemistry C*, *118*(42), 24240–24247. doi:10.1021/jp506464v

Antunes, R. A., Oliveira, M. C. L., Ett, G., & Ett, V. (2010). Corrosion of metal bipolar plates for PEM fuel cells: A review. *International Journal of Hydrogen Energy*, *35*(8), 3632–3647. doi:10.1016/j.ijhydene.2010.01.059

Anwar, S., Zhang, W., Wang, F., & Costinett, D. J. (2016). Integrated DC-DC converter design for Electric Vehicle powertrains. *2016 IEEE Applied Power Electronics Conference and Exposition (APEC)*, 424–431. 10.1109/APEC.2016.7467907

Arafat, M. Y., Khan, A. M., Saleem, I., Khan, N. A., & Khan, M. M. (2020). Intellectual and cognitive aspects of women entrepreneurs in India. *International Journal of Knowledge Management Studies*, *11*(3), 278–297. doi:10.1504/IJKMS.2020.109092

Arafat, M. Y., & Saleem, I. (2017). Examining start-up intention of Indians through cognitive approach: A study using GEM data. *Journal of Global Entrepreneurship Research*, *7*(1), 1–11. doi:10.118640497-017-0073-3

Arafat, M. Y., Saleem, I., & Dwivedi, A. K. (2020). Understanding entrepreneurial intention among Indian youth aspiring for self-employment. *International Journal of Knowledge and Learning*, *13*(3), 185–200. doi:10.1504/IJKL.2020.109882

Arafat, M. Y., Saleem, I., Dwivedi, A. K., & Khan, A. (2020). Determinants of agricultural entrepreneurship: A GEM data based study. *The International Entrepreneurship and Management Journal*, *16*(1), 345–370. doi:10.100711365-018-0536-1

Arenius, P., & Minniti, M. (2005). Perceptual variables and nascent entrepreneurship. *Small Business Economics*, *24*(3), 233–247. doi:10.100711187-005-1984-x

Armitage, C. J., & Conner, M. (2001). Efficacy of the theory of planned behaviour: A meta-analytic review. *British Journal of Social Psychology*, *40*(4), 471–499. doi:10.1348/014466601164939 PMID:11795063

Arslan, O., & Karasan, O. E. (2013). Cost and emission impacts of virtual power plant formation in plug-in hybrid electric vehicle penetrated networks. *Energy*, *60*, 116–124. doi:10.1016/j.energy.2013.08.039

Asaad, M., Ahmad, F., Alam, M. S., & Rafat, Y. (2017a). IoT Enabled Monitoring of an Optimized Electric Vehicle's Battery System. *Mobile Networks and Applications*, *23*(4), 994–1005. doi:10.100711036-017-0957-z

Asaad, M., Ahmad, F., & Saad Alam, M. (2017b). *IoT Enabled Electric Vehicle's Battery Monitoring System. EAI.* SGIOT. doi:10.4108/eai.7-8-2017.152984

Asensio, M., De Quevedo, P. M., Muñoz-Delgado, G., & Contreras, J. (2018a). Joint distribution network and renewable energy expansion planning considering demand response and energy storage-part I: Stochastic programming model. *IEEE Transactions on Smart Grid*, *9*(2), 655–666. doi:10.1109/TSG.2016.2560339

Asensio, M., De Quevedo, P. M., Muñoz-Delgado, G., & Contreras, J. (2018b). Joint distribution network and renewable energy expansion planning considering demand response and energy storage-part II: Numerical results. *IEEE Transactions on Smart Grid*, *9*(2), 667–675. doi:10.1109/TSG.2016.2560341

Ashique, R., Salam, Z., Bin Abdul Aziz, M., & Bhatti, A. R. (2017). Integrated photovoltaic-grid dc fast charging system for electric vehicle: A review of the architecture and control. *Renewable & Sustainable Energy Reviews*, *69*, 1243–1257. doi:10.1016/j.rser.2016.11.245

Atallah, R. F., Assi, C. M., Fawaz, W., Tushar, M. H. K., & Khabbaz, M. J. (2018). Optimal Supercharge Scheduling of Electric Vehicles: Centralized Versus Decentralized Methods. *IEEE Transactions on Vehicular Technology*, *67*(9), 7896–7909. doi:10.1109/TVT.2018.2842128

Aujla, G. S., Kumar, N., Singh, M., & Zomaya, A. Y. (2019). Energy trading with dynamic pricing for electric vehicles in a smart city environment. *Journal of Parallel and Distributed Computing*, *127*, 169–183. doi:10.1016/j.jpdc.2018.06.010

Azar, A. G., & Jacobsen, R. H. (2016). Agent-based charging scheduling of electric vehicles. *2016 IEEE Online Conference on Green Communications (OnlineGreenComm)*, 64–69. 10.1109/OnlineGreenCom.2016.7805408

Balachandra, P. (2011). Modern energy access to all in rural India: An integrated implementation strategy. *Energy Policy*, *39*(12), 7803–7814. doi:10.1016/j.enpol.2011.09.026

Ball, M., & Weeda, M. (2015). The hydrogen economy – Vision or reality? *International Journal of Hydrogen Energy*, *40*(25), 7903–7919. doi:10.1016/j.ijhydene.2015.04.032

Bandura, A. (1977). Self-efficacy: Toward a unifying theory of behavioral change. *Psychological Review*, *84*(2), 191–215. doi:10.1037/0033-295X.84.2.191 PMID:847061

Baroutaji, A., Carton, J. G., Stokes, J., & Olabi, A. G. (n.d.). *Design and Development of Proton Exchange Membrane Fuel Cell using Open Pore Cellular Foam as Flow Plate Material*. Academic Press.

Binding, C., Gantenbein, D., Jansen, B., Sundström, O., Andersen, P. B., Marra, F., & Træholt, C. (2010, July). Electric vehicle fleet integration in the Danish EDISON project-a virtual power plant on the island of Bornholm. In *IEEE PES general meeting* (pp. 1–8). IEEE. doi:10.1109/PES.2010.5589605

Bıyıkoğlu, A. (2005). *RETRACTED: Review of proton exchange membrane fuel cell models*. Elsevier.

Bodo, N., Levi, E., Subotic, I., Espina, J., Empringham, L., & Johnson, C. M. (2017). Efficiency Evaluation of Fully Integrated On-Board EV Battery Chargers With Nine-Phase Machines. *IEEE Transactions on Energy Conversion*, *32*(1), 257–266. doi:10.1109/TEC.2016.2606657

Bosma, N., Acs, Z. J., Autio, E., Coduras, A., & Levie, J. (2008). *Global entrepreneurship monitor executive report*. Babson Park.

Briane, B., & Loudot, S. (2011). Rapid reversible charging device for an electric vehicle. *Patent* No. US20110254494, A1.

Bruye, X., Re, A., De Sousa, L., Bouchez, B., Sandulescu, P., Kestelyn, X., & Semail, E. (2010). *A multiphase traction/fast-battery-charger drive for electric or plug-in hybrid vehicles: Solutions for control in traction mode*. doi:10.1109/VPPC.2010.5729220

Cacciola, G., Antonucci, V., & Freni, S. (2001). Technology up date and new strategies on fuel cells. *Journal of Power Sources*, *100*(1–2), 67–79. doi:10.1016/S0378-7753(01)00884-9

Cacciotti, G., & Hayton, J. C. (2015). Fear and entrepreneurship: A review and research agenda. *International Journal of Management Reviews*, *17*(2), 165–190. doi:10.1111/ijmr.12052

Calearo, L., Thingvad, A., Ipsen, H. H., & Marinelli, M. (2019). *Economic Value and User Remuneration for EV Based Distribution Grid Services. In 2019 IEEE PES Innovative Smart Grid Technologies Europe*. IEEE.

Carton, J. G., & Olabi, A.-G. (2017). Three-dimensional proton exchange membrane fuel cell model: Comparison of double channel and open pore cellular foam flow plates. *Energy*, *136*, 185–195. doi:10.1016/j.energy.2016.02.010

CEA. (2017). *Power Sector: Executive Summary for Jun 2017.* Central Electricity Authority, India. http://www.cea.nic.in/reports/monthly/executivesummary/2017/exe_summary-01.pdf%0Ahttp://www.cea.nic.in/reports/monthly/executive_rep/feb14.pdf

Cele, N. P., Sinha-Ray, S., Munda, J., & Jimoh, A. A. (2010). The state of the art proton exchange membrane fuel cells for clean energy. *2010 IEEE International Energy Conference*, 865–869. 10.1109/ENERGYCON.2010.5771804

Center for Sustainable Energy. (2020). *Electric Vehicle Charging 101(Learn more about different charging options for electric vehicles (EVs)).* https://calevip.org/electric-vehicle-charging-101#:~:text=One%20distinction%20between%20these%20three,between%20200%20and%20600%20volts

Change, I. C. (2014). *Mitigation of climate change.* Contribution of Working Group III to the Fifth Assessment Report of the Intergovernmental Panel on Climate Change, 1454.

Charging Technology, E. V. (2020). *Vehicle-to-Grid: Everything you need to know, V2G technology.* Liikennevirta Oy (Ltd.). https://www.virta.global/vehicle-to-grid-v2g

Chinmaya, K. A., & Singh, G. K. (2018a). A Multifunctional Integrated Onboard Battery Charger for Plug-in Electric Vehicles (PEVs). *2018 IEEE 18th International Power Electronics and Motion Control Conference (PEMC)*, 28–33. 10.1109/EPEPEMC.2018.8521990

Chinmaya, K. A., & Singh, G. K. (2018b). A Single-stage Integrated Charger for Electric Vehicles (EVs) and Plug—In Electric Vehicles (PEVs) Incorporating Induction Motor Drive. *IECON 2018 - 44th Annual Conference of the IEEE Industrial Electronics Society*, 954–959. 10.1109/IECON.2018.8591669

Chinmaya, K. A., & Singh, G. K. (2019). Integrated onboard single-stage battery charger for PEVs incorporating asymmetrical six-phase induction machine. *IET Electrical Systems in Transportation*, *9*(1), 8–15. doi:10.1049/iet-est.2018.5015

Cho, E. A., Jeon, U.-S., Hong, S.-A., Oh, I.-H., & Kang, S.-G. (2005). Performance of a 1kW-class PEMFC stack using TiN-coated 316 stainless steel bipolar plates. *Journal of Power Sources*, *142*(1–2), 177–183. doi:10.1016/j.jpowsour.2004.10.032

Choi, H.-J., Jung, S.-M., Seo, J.-M., Chang, D. W., Dai, L., & Baek, J.-B. (2012). Graphene for energy conversion and storage in fuel cells and supercapacitors. *Nano Energy*, *1*(4), 534–551. doi:10.1016/j.nanoen.2012.05.001

Cocconi, A. G. (1994). *Combined Motor Drive and Battery Recharge System.* US Patent No. US 5341075.

Costamagna, P., & Srinivasan, S. (2001). Quantum jumps in the PEMFC science and technology from the 1960s to the year 2000: Part II. Engineering, technology development and application aspects. *Journal of Power Sources*, *102*(1–2), 253–269. doi:10.1016/S0378-7753(01)00808-4

De Sousa, L., & Bouchez, B. (2011). *Combined Electric Device for Powering and Charging.* Patent No. 0221363 A1.

De Sousa, L., Silvestre, B., & Bouchez, B. (2010). *A combined multiphase electric drive and fast battery charger for Electric Vehicles.* doi:10.1109/VPPC.2010.5729057

Deb, S., Kalita, K., & Mahanta, P. (2017b, December). Impact of electric vehicle charging stations on reliability of distribution network. In *2017 International Conference on Technological Advancements in Power and Energy (TAP Energy)* (pp. 1-6). IEEE. 10.1109/TAPENERGY.2017.8397272

Deivanayagam, R. (2017). Vehicle-to-grid Technology: Concept, Status, and Challenges. *The Journal of Undergraduate Research at the University of Illinois at Chicago*, *10*(1). Advance online publication. doi:10.5210/jur.v10i1.8013

Delhi, S. I. N. (2015). Fame-India Scheme—Putting E-Mobility on Road. *Auto Tech Review*, *5*(4), 22–27.

DENSO Develops Vehicle-to-Home Power Supply System for Electric Vehicles | News | DENSO Global Website. (n.d.). https://www.denso.com/global/en/news/news-releases/2012/120724-01

Development of Auxiliary Power Unit for Recreational yachts. (n.d.). Retrieved September 7, 2020, from https://www.fch.europa.eu/project/development-auxiliary-power-unit-recreational-yachts

Deza, A., Huang, K., & Mete, M. R. (2020). Charging station optimization for balanced electric car sharing. *Discrete Applied Mathematics*. Advance online publication. doi:10.1016/j.dam.2020.01.042

Dixon & Bell. (2020). Electric vehicles: Battery capacity, charger power, access to charging and the impacts on distribution networks. *eTransportation, 4*, 1-18. doi:10.1016/j.etran.2020.100059

Douglas, J. (2019). *Tesla's Musk says solar panels on cars make little sense, but that's not stopping Toyota, Hyundai.* CNBC. https://www.cnbc.com/2019/09/28/teslas-musk-rejects-solar-on-cars-its-not-stopping-toyota-hyundai.html

Duan, Feng, Wei, Yongxing, Dong, & Xuhui. (2014). Analysis of charging torque in a three-phase integrated charger of electrical vehicle. *2014 IEEE Conference and Expo Transportation Electrification Asia-Pacific (ITEC Asia-Pacific)*, 1–5. 10.1109/ITEC-AP.2014.6940908

Dusmez, S., & Khaligh, A. (2013). A Compact and Integrated Multifunctional Power Electronic Interface for Plug-in Electric Vehicles. *IEEE Transactions on Power Electronics*, *28*(12), 5690–5701. doi:10.1109/TPEL.2012.2233763

Dusmez, S., & Khaligh, A. (2014). A Charge-Nonlinear-Carrier-Controlled Reduced-Part Single-Stage Integrated Power Electronics Interface for Automotive Applications. *IEEE Transactions on Vehicular Technology*, *63*(3), 1091–1103. doi:10.1109/TVT.2013.2284592

EDF. (2020). *Electric cars and the environment.* Électricité de France S.A. https://www.edfenergy.com/electric-cars/environment

Ehsani, M., Gao, Y., Longo, S., & Ebrahimi, K. (2018). *Modern electric, hybrid electric, and fuel cell vehicles.* CRC Press.

Electric Vehicle Database. (2020). *2020 EV Database - v4.2.* https://ev-database.org/#sort:path~type~order=.rank~number~desc|range-slider-range:prev~next=0~1200|range-slider-acceleration:prev~next=2~23|range-slider-topspeed:prev~next=110~450|range-slider-battery:prev~next=10~200|range-slider-eff:prev~next=100~300|range-slider-fastcharge:prev~next=0~1500|paging:currentPage=0|paging:number=9

Electrical Technology. (2020). *How to Calculate the Battery Charging Time & Battery Charging Current – Example.* https://www.electricaltechnology.org/2013/03/easy-charging-time-formula-for.html

Electromobilitysummit.Com. (n.d.). http://electromobilitysummit.com/

Ellingsen, L. A.-W., Majeau-Bettez, G., Singh, B., Srivastava, A. K., Valøen, L. O., & Strømman, A. H. (2014). Life cycle assessment of a lithium-ion battery vehicle pack. *Journal of Industrial Ecology*, *18*(1), 113–124. doi:10.1111/jiec.12072

El-Taweel, N., Khani, H., & Farag, H. (2019). *Analytical Size Estimation Methodologies for Electrified Transportation Fueling Infrastructures Using Public–Domain Market Data.* IEEE Transactions on Transportation Electrification. doi:10.1109/TTE.2019.2927802

Elzen, B., Geels, F. W., Hofman, P. S., & Green, K. (2004). Socio-technical scenarios as a tool for transition policy: An example from the traffic and transport domain. *System Innovation and the Transition to Sustainability: Theory. Evidence & Policy*, 251–281.

Emilio, M. D. P. (2020, March 12). EETimes—EC to Bet on Hydrogen Fuel Cell Vehicles. *EETimes*. https://www.eetimes.com/ec-to-bet-on-hydrogen-fuel-cell-vehicles/

ENSTO. (2016). *The Smart Future of Vehicle-to-Grid and Vehicle-to-Home*. ENSTO.

Erdinc, O., Vural, B., & Uzunoglu, M. (2009). A dynamic lithium-ion battery model considering the effects of temperature and capacity fading. *2009 International Conference on Clean Electrical Power*, 383–386. 10.1109/ICCEP.2009.5212025

Falvo, M., Sbordone, D., Bayram, I., & Devetsikiotis, M. (2014). EV charging stations and modes: International standards. Academic Press.

Fayolle, A., & Gailly, B. (2015). The impact of entrepreneurship education on entrepreneurial attitudes and intention: Hysteresis and persistence. *Journal of Small Business Management*, *53*(1), 75–93. doi:10.1111/jsbm.12065

FCT - Fuel Cell Applications—Fuel and Infrastructure. (n.d.). Retrieved September 7, 2020, from http://www.fuelcelltoday.com/applications/fuel-and-infrastructure

Ferdowsi, M. (2007). Plug-in Hybrid Vehicles - A Vision for the Future. *Proceedings of the IEEE Vehicle Power and Propulsion Conference*, 457-462. 10.1109/VPPC.2007.4544169

Fernández, J., Liñán, F., & Santos, F. J. (2009). Cognitive aspects of potential entrepreneurs in Southern and Northern Europe: An analysis using GEM-data. *Revista de Economía Mundial*, (23), 151–178.

Ferreira, J. J., Raposo, M. L., Rodrigues, R. G., Dinis, A., & do Paço, A. (2012). A model of entrepreneurial intention: An application of the psychological and behavioral approaches. *Journal of Small Business and Enterprise Development.*

Ferrero, E., Alessandrini, S., & Balanzino, A. (2016). Impact of the electric vehicles on the air pollution from a highway. *Applied Energy*, *169*(x), 450–459. doi:10.1016/j.apenergy.2016.01.098

Ferro, G., Laureri, F., Minciardi, R., & Robba, M. (2019a). A predictive discrete event approach for the optimal charging of electric vehicles in microgrids. *Control Engineering Practice*, *86*(September), 11–23. doi:10.1016/j.conengprac.2019.02.004

Ferro, G., Laureri, F., Minciardi, R., & Robba, M. (2019b). Optimal charging of electric vehicles in microgrids through discrete event optimization. *2019 18th European Control Conference (ECC)*, 916–921. 10.23919/ECC.2019.8795999

Ferro, G., Laureri, F., Minciardi, R., & Robba, M. (2018). An optimization model for electrical vehicles scheduling in a smart grid. *Sustainable Energy. Grids and Networks*, *14*, 62–70. doi:10.1016/j.segan.2018.04.002

Fontenot, H., & Dong, B. (2019). Modeling and control of building-integrated microgrids for optimal energy management – A review. *Applied Energy*, *254*(July), 113689. doi:10.1016/j.apenergy.2019.113689

Fu, Y., Hou, M., Xu, H., Hou, Z., Ming, P., Shao, Z., & Yi, B. (2008). Ag–polytetrafluoroethylene composite coating on stainless steel as bipolar plate of proton exchange membrane fuel cell. *Journal of Power Sources*, *182*(2), 580–584. doi:10.1016/j.jpowsour.2008.04.051

Gabriel, C. A., & Kirkwood, J. (2016). Business models for model businesses: Lessons from renewable energy entrepreneurs in developing countries. *Energy Policy*, *95*, 336–349. doi:10.1016/j.enpol.2016.05.006

Gamez, M. R., Perez, A. V., Sera, A. S., & Ronquillo, Z. M. (2017). Renewable energy sources and local development. *International Journal of Social Science and Humanity*, *1*(2), 10–19.

Garche, J., & Jossen, A. (2000). Battery management systems (bms) for increasing battery lifetime. *TELESCON 2000. Third International Telecommunications Energy Special Conference*, 81–88.

Gasteiger, H. A., Markovic, N., Ross, P. N. Jr, & Cairns, E. J. (1993). Methanol electrooxidation on well-characterized platinum-ruthenium bulk alloys. *Journal of Physical Chemistry*, *97*(46), 12020–12029. doi:10.1021/j100148a030

Godarzi, A., Niaki, S. N., Ahmadkhanlou, F., & Iravani, R. (2011, January). Local and global optimization of exportable vehicle power based smart microgrid. In *ISGT 2011* (pp. 1–7). IEEE. doi:10.1109/ISGT.2011.5759144

Gong, X., Xiong, R., & Mi, C. C. (2014). Study of the characteristics of battery packs in electric vehicles with parallel-connected lithium-ion battery cells. *IEEE Transactions on Industry Applications*, *51*(2), 1872–1879. doi:10.1109/TIA.2014.2345951

Grigsby, L.L. (2017). *Power system stability and control*. Academic Press.

Group, Renewable Energy, Electric Power Industry, and Renewable Energy Unit. (2019). *October 2019 EV Charging Standards and V2X: Which Standard Is Leading Toward V2X Penetration?* Author.

Guena, T., & Leblanc, P. (2006). How depth of discharge affects the cycle life of lithium-metal-polymer batteries. *INTELEC 06-Twenty-Eighth International Telecommunications Energy Conference*, 1–8. 10.1109/INTLEC.2006.251641

Gui-Jia, S., & Lixin, T. (2011). *Current source inverter based traction drive for EV battery charging applications*. doi:10.1109/VPPC.2011.6043143

Gui-Jia, S., & Lixin, T. (2013). *An integrated onboard charger and accessory power converter for plug-in electric vehicles*. doi:10.1109/ECCE.2013.6646895

Gulati, V. (2012). *National electric mobility mission plan 2020*. Department of Heavy Industry, Ministry of Heavy Industries & Public Enterprises, Government of India.

Gupta, M., Rafat, Y., & Alam, M. S. (2017). *Well to wheel cum tailpipe emission analysis: ICE vs xEV* (No. 2017-26-0107). SAE Technical Paper.

Haghbin, S. (2013). *Integrated Motor Drives and Battery Chargers for Electric or Plug-in Hybrid Electric Vehicles*. Chalmers University of Technology. https://publications.lib.chalmers.se/records/fulltext/173471/173471.pdf

Haghbin, S., Khan, K., Zhao, S., Alakula, M., Lundmark, S., & Carlson, O. (2013). An Integrated 20-kW Motor Drive and Isolated Battery Charger for Plug-in Vehicles. *IEEE Transactions on Power Electronics*, *28*(8), 4013–4029. doi:10.1109/TPEL.2012.2230274

Hanauer, D. (2018). Technical Note: Mode 2 Charging—Testing and Certification for International Market Access. *World Electric Vehicle Journal*, *9*(2), 26. doi:10.3390/wevj9020026

Hao, H., Wang, H., Song, L., Li, X., & Ouyang, M. (2010). Energy consumption and GHG emissions of GTL fuel by LCA: Results from eight demonstration transit buses in Beijing. *Applied Energy*, *87*(10), 3212–3217. doi:10.1016/j.apenergy.2010.03.029

Hassan, A. S., Marmaras, C. E., Xydas, E. S., Cipcigan, L. M., & Jenkins, N. (2013). *Integration of wind power using V2G as a flexible storage*. Academic Press.

Ha, T., & Zhang, Y. (2017, September 4). Energy Hub Modeling to Minimize Residential Energy Costs Considering Solar Energy and BESS. *Journal of Modern Power Systems and Clean Energy*, *5*(3), 389–399. https://link.springer.com/10.1007/s40565-017-0281-4

Hegazy, O., Barrero, R., Van Mierlo, J., Lataire, P., Omar, N., & Coosemans, T. (2013). An advanced power electronics interface for electric vehicles applications. *IEEE Transactions on Power Electronics*, *28*(12), 5508–5521. doi:10.1109/TPEL.2013.2256469

Hentall, P. L., Lakeman, J. B., Mepsted, G. O., Adcock, P. L., & Moore, J. M. (1999). New materials for polymer electrolyte membrane fuel cell current collectors. *Journal of Power Sources*, *80*(1–2), 235–241. doi:10.1016/S0378-7753(98)00264-X

Hermann, A., Chaudhuri, T., & Spagnol, P. (2005). Bipolar plates for PEM fuel cells: A review. *International Journal of Hydrogen Energy*, *30*(12), 1297–1302. doi:10.1016/j.ijhydene.2005.04.016

He, S., Xu, Z., Chen, M., Yang, H., & Li, W. (2020). General Derivation Law with Torque-Free Achieving of Integral On-Board Charger on Compact Powertrains. *IEEE Transactions on Industrial Electronics*, 1–1. doi:10.1109/TIE.2020.3005063

High Efficiency Low Emission Coal I HELE I WCA. (n.d.). https://www.worldcoal.org/reducing-co2-emissions/high-efficiency-low-emission-coal

Hirscher, M., Panella, B., & Schmitz, B. (2010). Metal-organic frameworks for hydrogen storage. *Microporous and Mesoporous Materials*, *129*(3), 335–339. doi:10.1016/j.micromeso.2009.06.005 PMID:21226135

Hofman, P. S., Elzen, B. E., & Geels, F. W. (2004). Sociotechnical scenarios as a new policy tool to explore system innovations: Co-evolution of technology and society in the Netherland's electricity domain. *Innovation*, *6*(2), 344–360. doi:10.5172/impp.2004.6.2.344

Hosseini, S. S., Badri, A., & Parvania, M. (2012, September). The plug-in electric vehicles for power system applications: The vehicle to grid (V2G) concept. In *2012 IEEE International Energy Conference and Exhibition (ENERGYCON)* (pp. 1101-1106). IEEE. 10.1109/EnergyCon.2012.6347734

Høyer, K. G. (2008). The history of alternative fuels in transportation: The case of electric and hybrid cars. *Utilities Policy*, *16*(2), 63–71. doi:10.1016/j.jup.2007.11.001

Huang, X., Ke, S., Lv, H., & Liu, Y. (2018). A dynamic capacity fading model with thermal evolution considering variable electrode thickness for lithium-ion batteries. *Ionics*, *24*(11), 3439–3450. doi:10.100711581-018-2476-8

Hung-Chun, C., & Chang-Ming, L. (2009). Development of a Compact Switched-Reluctance Motor Drive for EV Propulsion With Voltage-Boosting and PFC Charging Capabilities. *IEEE Transactions on Vehicular Technology*, *58*(7), 3198–3215. doi:10.1109/TVT.2009.2017546

Hung-Chun, C., & Chang-Ming, L. (2011). An Integrated Driving/Charging Switched Reluctance Motor Drive Using Three-Phase Power Module. *IEEE Transactions on Industrial Electronics*, *58*(5), 1763–1775. doi:10.1109/TIE.2010.2051938

Hung, Y., Tawfik, H., & Mahajan, D. (2009). Durability and characterization studies of polymer electrolyte membrane fuel cell's coated aluminum bipolar plates and membrane electrode assembly. *Journal of Power Sources*, *186*(1), 123–127. doi:10.1016/j.jpowsour.2008.09.079

Hydrogen Fuel Cell That's As Durable As A Conventional Engine. (n.d.). Retrieved September 7, 2020, from http://www.acalenergy.co.uk/news/release/acal-energy-system-breaks-the-10000-hour-endurance-barrier/en/

IEC 62196-1:2014, Plugs, socket-outlets, vehicle connectors and vehicle inlets - Conductive charging of electric vehicles - Part 1: General requirements. Retrieve Web Page: https://webstore.iec.ch/preview/info_iec62196-1%7Bed3.0%7Db.pdf

India's Current Account Deficit to Touch $30 Billion in 2018. (2017). *The Economic Times*.

International Emissions Agency. (2015). *CO2 Emissions from Fossil Fuel Combustion: Highlights*. Author.

Ito, A., Kawashima, A., Suzuki, T., Inagaki, S., Yamaguchi, T., & Zhou, Z. (2018). Model Predictive Charging Control of In-Vehicle Batteries for Home Energy Management Based on Vehicle State Prediction. *IEEE Transactions on Control Systems Technology*, *26*(1), 51–64. doi:10.1109/TCST.2017.2664727

James, B. D., Kalinoski, J. A., & Baum, K. N. (2010). *Mass Production Cost Estimation For Direct H2 PEM Fuel Cell Systesm for Automotive Applications. 2010 Update*. Directed Technologies. doi:10.2172/1218888

Jhunjhunwala, A., Lolla, A., & Kaur, P. (2016). Solar-Dc Microgrid for Indian Homes: A Transforming Power Scenario. *IEEE Electrification Magazine*, *4*(2), 10–19.

Jinhao, G., Hui, W., Tengxin, W., & Yubin, W. (2017). An integrated topology for on-board charger and driven of electric vehicle. *2017 IEEE International Conference on Industrial Technology (ICIT)*, 504–508. 10.1109/ICIT.2017.7913283

Joseph, S., Mcclure, J., Chianelli, R., Pich, P., & Sebastian, P. (2005). Conducting polymer-coated stainless steel bipolar plates for proton exchange membrane fuel cells (PEMFC). *International Journal of Hydrogen Energy*, *30*(12), 1339–1344. doi:10.1016/j.ijhydene.2005.04.011

Ju, L., Li, H., Zhao, J., Chen, K., Tan, Q., & Tan, Z. (2016). Multi-objective stochastic scheduling optimization model for connecting a virtual power plant to wind-photovoltaic-electric vehicles considering uncertainties and demand response. *Energy Conversion and Management*, *128*, 160–177. doi:10.1016/j.enconman.2016.09.072

Kahlen, M. T., Ketter, W., & van Dalen, J. (2018). Electric vehicle virtual power plant dilemma: Grid balancing versus customer mobility. *Production and Operations Management*, *27*(11), 2054–2070. doi:10.1111/poms.12876

Kahlen, M., & Ketter, W. (2015, February). Aggregating electric cars to sustainable virtual power plants: The value of flexibility in future electricity markets. *Twenty-Ninth AAAI Conference on Artificial Intelligence*.

Kalkhambkar, V., Kumar, R., & Bhakar, R. (2016). Energy loss minimization through peak shaving using energy storage. *Perspectives on Science*, *8*, 162–165. doi:10.1016/j.pisc.2016.04.022

Kempton, W., Marra, F., Andersen, P. B., & Garcia-Valle, R. (2013). Business models and control and management architectures for EV electrical grid integration. In *Electric Vehicle Integration into Modern Power Networks* (pp. 87–105). Springer. doi:10.1007/978-1-4614-0134-6_4

Kempton, W., & Tomić, J. (2005). Vehicle-to-grid power fundamentals: Calculating capacity and net revenue. *Journal of Power Sources*, *144*(1), 268–279. doi:10.1016/j.jpowsour.2004.12.025

Kempton, W., & Tomić, J. (2005). Vehicle-to-grid power implementation: From stabilizing the grid to supporting large-scale renewable energy. *Journal of Power Sources*, *144*(1), 280–294. doi:10.1016/j.jpowsour.2004.12.022

Kester, J., Noel, L., Lin, X., de Rubens, G. Z., & Sovacool, B. K. (2019). The coproduction of electric mobility: Selectivity, conformity and fragmentation in the sociotechnical acceptance of vehicle-to-grid (V2G) standards. *Journal of Cleaner Production*, *207*, 400–410. doi:10.1016/j.jclepro.2018.10.018

Khalife. (2020). *Brighter Future: UAE is investing in electric cars, green oasis & solar parks*. Retrieve Web Page: https://stepfeed.com/brighter-future-uae-is-investing-in-electric-cars-green-oasis-and-solar-parks-6928

Khaligh, A., & Li, Z. (2010). Battery, ultracapacitor, fuel cell, and hybrid energy storage systems for electric, hybrid electric, fuel cell, and plug-in hybrid electric vehicles: State of the art. *IEEE Transactions on Vehicular Technology*, *59*(6), 2806–2814. doi:10.1109/TVT.2010.2047877

Khan, A. M., Arafat, M. Y., Raushan, M. A., Khan, M. S., Dwivedi, A. K., Khan, M. M., & Saleem, I. (2019). Role of intellectual capital in augmenting the start-up intentions of Indians—An analysis using GEM data. *Journal of Global Entrepreneurship Research*, *9*(1), 1–14.

Khan, A. M., Arafat, M. Y., Raushan, M. A., & Saleem, I. (2020). Examining the relevance of intellectual capital in improving the entrepreneurial propensity among Indians. *International Journal of Knowledge Management*, *16*(1), 123–140.

Khan, S., Ahmad, A., & Ahmad, F. (2017). *A Comprehensive Review on Solar Powered Electric Vehicle Charging System*. Smart Science. doi:10.1080/23080477.2017.1419054

Kinoshita, S. (1996). *Electric System for Electrical Vehicle—Has Power Converter which Rectifies AC Power to DC Power so that DC Power is Regenerative to the Secondary Battery*. Patent No. US 5504414.

Kolvereid, L., & Isaksen, E. (2006). New business start-up and subsequent entry into self-employment. *Journal of Business Venturing*, *21*(6), 866–885.

Krueger, N. F., Reilly, M. D., & Carsrud, A. L. (2000). Entrepreneurial intentions: A competing models approach. *Journal of Business Venturing*, *15*(5/6), 411–432.

Kurdve, M., Zackrisson, M., Johansson, M., Ebin, B., & Harlin, U. (2019). Considerations when Modelling EV Battery Circularity Systems. *Batteries*, *5*(2), 1–20. doi:10.3390/batteries5020040

Lacressonniere, F., & Cassoret, B. (2005). *Converter Used as a Battery Charger and a Motor Speed Controller in an Industrial Truck*. doi:10.1109/EPE.2005.219286

Lacroix, S., Laboure, E., & Hilairet, M. (2010). *An Integrated Fast Battery Charger for Electric Vehicle*. doi:10.1109/VPPC.2010.5729063

Latifi, M., Rastegarnia, A., Khalili, A., & Sanei, S. (2019). Agent-Based Decentralized Optimal Charging Strategy for Plug-in Electric Vehicles. *IEEE Transactions on Industrial Electronics*, *66*(5), 3668–3680. doi:10.1109/TIE.2018.2853609

Lazzeroni, P., Olivero, S., Repetto, M., Stirano, F., & Vallet, M. (2019). Optimal battery management for vehicle-to-home and vehicle-to-grid operations in a residential case study. *Energy*, *175*, 704–721. doi:10.1016/j.energy.2019.03.113

Lee, J. Y., Yu, D. M., Kim, T.-H., Yoon, S. J., & Hong, Y. T. (2015). Multi-block copolymers based on poly(p-phenylene)s with excellent durability and fuel cell performance. *Journal of Membrane Science*, *492*, 209–219. doi:10.1016/j.memsci.2015.04.013

Lee, Y.-J., Khaligh, A., & Emadi, A. (2009). Advanced integrated bidirectional AC/DC and DC/DC converter for plug-in hybrid electric vehicles. *IEEE Transactions on Vehicular Technology*, *58*(8), 3970–3980. doi:10.1109/TVT.2009.2028070

Leo, T. J., Durango, J. A., & Navarro, E. (2010). Exergy analysis of PEM fuel cells for marine applications. *Energy*, *35*(2), 1164–1171. doi:10.1016/j.energy.2009.06.010

Letendre, S. E., Kempton, W., & Stavros, R. (2002). The V2G concept: A new model for power? *Public Utilities Fortnightly*, *140*(4), 16–16.

Lezama, F., Soares, J., Canizes, B., & Vale, Z. (2020). Flexibility management model of home appliances to support DSO requests in smart grids. *Sustainable Cities and Society*, *55*, 102048. doi:10.1016/j.scs.2020.102048

Li, D., Chiu, W.-Y. Y., Sun, H., & Poor, H. V. (2018). Multiobjective optimization for demand side management program in smart grid. *IEEE Transactions on Industrial Informatics*, *14*(4), 1482–1490. doi:10.1109/TII.2017.2776104

Liñán, F., & Chen, Y. W. (2009). Development and cross–cultural application of a specific instrument to measure entrepreneurial intentions. *Entrepreneurship Theory and Practice*, *33*(3), 593–617.

Liñán, F., Urbano, D., & Guerrero, M. (2011). Regional variations in entrepreneurial cognitions: Start-up intentions of university students in Spain. *Entrepreneurship and Regional Development*, *23*(3-4), 187–215.

Litster, S., & McLean, G. (2004). PEM fuel cell electrodes. *Journal of Power Sources*, *130*(1–2), 61–76. doi:10.1016/j.jpowsour.2003.12.055

Liu, H., Wang, B., Wang, N., Wu, Q., Yang, Y., Wei, H., & Li, C. (2018). Enabling strategies of electric vehicles for under frequency load shedding. *Applied Energy*, *228*, 843–851. doi:10.1016/j.apenergy.2018.06.122

Liu, J., Gao, D., & Cao, J. (2012). *Study on the effects of temperature on lifepo 4 battery life. In 2012 IEEE Vehicle Power and Propulsion Conference*. IEEE.

Liu, T. H., Chen, Y., Yi, P. H., & Chen, J. L. (2015). Integrated battery charger with power factor correction for electric-propulsion systems. *IET Electric Power Applications*, *9*(3), 229–238. doi:10.1049/iet-epa.2014.0168

Liu, Z., Wu, Q., Huang, S., Wang, L., Shahidehpour, M., & Xue, Y. (2018). Optimal Day-Ahead Charging Scheduling of Electric Vehicles Through an Aggregative Game Model. *IEEE Transactions on Smart Grid*, *9*(5), 5173–5184. doi:10.1109/TSG.2017.2682340

Liu, Z., Wu, Q., Ma, K., Shahidehpour, M., Xue, Y., & Huang, S. (2019). Two-Stage Optimal Scheduling of Electric Vehicle Charging Based on Transactive Control. *IEEE Transactions on Smart Grid*, *10*(3), 2948–2958. doi:10.1109/TSG.2018.2815593

Liu, Z., Zheng, W., Qi, F., Wang, L., Zou, B., Wen, F., & Xue, Y. (2018). Optimal dispatch of a virtual power plant considering demand response and carbon trading. *Energies*, *11*(6), 1488. doi:10.3390/en11061488

Li, W., Feng, G., Lai, C., Li, Z., Tian, J., & Kar, N. C. (2019). Demagnetization Analysis of Interior Permanent Magnet Machines Under Integrated Charging Operation. *IEEE Transactions on Industry Applications*, *55*(5), 5204–5213. doi:10.1109/TIA.2019.2922209

Long, J. W., Stroud, R. M., Swider-Lyons, K. E., & Rolison, D. R. (2000). How To Make Electrocatalysts More Active for Direct Methanol OxidationAvoid PtRu Bimetallic Alloys! *The Journal of Physical Chemistry B*, *104*(42), 9772–9776. doi:10.1021/jp001954e

Loock, M. (2012). Going beyond best technology and lowest price: On renewable energy investors' preference for service-driven business models. *Energy Policy*, *40*, 21–27.

Loudot, S., Briane, B., Ploix, O., & Villeneuve, A. (2014). *Fast Charging Device For An Electric Vehicle*. US Patent No. US 8847555 B2.

Mahoor, M., Hosseini, Z. S., & Khodaei, A. D. K. (2017). Electric Vehicle Battery Swapping Station. *CIGRE Grid of the Future Symposium*, 1–5.

Malinowski, J., & Kaderly, K. (2004). Peak shaving - A method to reduce utility costs. *IEEE Region 5 Conference: Annual Technical and Leadership Workshop*.

Marcinkoski, J., James, B. D., Kalinoski, J. A., Podolski, W., Benjamin, T., & Kopasz, J. (2011). Manufacturing process assumptions used in fuel cell system cost analyses. *Journal of Power Sources*, *196*(12), 5282–5292. doi:10.1016/j.jpowsour.2011.02.035

Marnay, C. (2013). *Los Angeles air force base vehicle to grid pilot project*. Academic Press.

Matulka, R. (2014). The history of the electric car. US Department of Energy.

Mawdsley, J. R., Carter, J. D., Wang, X., Niyogi, S., Fan, C. Q., Koc, R., & Osterhout, G. (2013). Composite-coated aluminum bipolar plates for PEM fuel cells. *Journal of Power Sources*, *231*, 106–112. doi:10.1016/j.jpowsour.2012.12.074

Meher, S. R., Banerjee, S., Vankayalapati, B., & Singh, R. K. (2020). A Reconfigurable On-board Power Converter for Electric Vehicle with Reduced Switch Count. *IEEE Transactions on Vehicular Technology*, *69*(4), 1–1. doi:10.1109/TVT.2020.2973316

Mehmood, K. K., Khan, S. U., Lee, S.-J., Haider, Z. M., Rafique, M. K., & Kim, C.-H. (2017). Optimal sizing and allocation of battery energy storage systems with wind and solar power dgs in a distribution network for voltage regulation considering the lifespan of batteries. *IET Renewable Power Generation*, *11*(10), 1305–1315. doi:10.1049/iet-rpg.2016.0938

Mehrjerdi, H., & Hemmati, R. (2019). Electric vehicle charging station with multilevel charging infrastructure and hybrid solar-battery-diesel generation incorporating comfort of drivers. *Journal of Energy Storage*, *26*, 1–9. doi:10.1016/j.est.2019.100924

Mehta, R. (2016). *Optimal Charging Scheduling of Plug-in Electric Vehicles for Maximizing Penetration within a Workplace Car Park*. Academic Press.

Mehta, V., & Cooper, J. S. (2003). Review and analysis of PEM fuel cell design and manufacturing. *Journal of Power Sources*, *114*(1), 32–53. doi:10.1016/S0378-7753(02)00542-6

Milanovic, M., Roskaric, A., & Auda, M. (1999). Battery Charger Based on Double-Buck and Boost Converter. doi:10.1109/ISIE.1999.798706

Ministry of New and Renewable Energy - Scheme / Documents. (n.d.). http://www.mnre.gov.in/solar-mission/jnnsm/introduction-2/

Mohamed, N., Aymen, F., Ben Hamed, M., & Lassaad, S. (2019). *Analysis of battery-EV state of charge for a dynamic wireless charging system*. Energy Storage. doi:10.1002/est2.117

Morais, H., Sousa, T., Vale, Z., & Faria, P. (2014). Evaluation of the electric vehicle impact in the power demand curve in a smart grid environment. *Energy Conversion and Management*, *82*, 268–282. doi:10.1016/j.enconman.2014.03.032

Morrow, K., Karner, D., & Francfort, J. (2008). U. S. Department of Energy Vehicle Technologies Program - Advanced Vehicle Testing Activity Plug-in Hybrid Electric Vehicle Charging Infrastructure Review. *Power*, (58517), 34.

Muellerlanger, F., Tzimas, E., Kaltschmitt, M., & Peteves, S. (2007). Techno-economic assessment of hydrogen production processes for the hydrogen economy for the short and medium term. *International Journal of Hydrogen Energy*, *32*(16), 3797–3810. doi:10.1016/j.ijhydene.2007.05.027

Mukundan, S., Dhulipati, H., Lai, C., Mukherjee, K., Tjong, J., & Kar, N. C. (2018). Design and Optimization of Traction IPMSM With Asymmetrical Damper Bars for Integrated Charging Capability Using Evolutionary Algorithm. *IEEE Transactions on Energy Conversion*, *33*(4), 2060–2069. doi:10.1109/TEC.2018.2877104

Musio, M., & Damiano, A. (2012, October). A virtual power plant management model based on electric vehicle charging infrastructure distribution. In 2012 3rd IEEE PES Innovative Smart Grid Technologies Europe (ISGT Europe) (pp. 1-7). IEEE. doi:10.1109/ISGTEurope.2012.6465684

Musio, M., Lombardi, P., & Damiano, A. (2010, September). Vehicles to grid (V2G) concept applied to a virtual power plant structure. In *The XIX International Conference on Electrical Machines-ICEM 2010* (pp. 1-6). IEEE. 10.1109/ICELMACH.2010.5608261

Mustafa, M. A., Zhang, N., Kalogridis, G., & Fan, Z. (2013, February). Smart electric vehicle charging: Security analysis. In 2013 IEEE PES Innovative Smart Grid Technologies Conference (ISGT) (pp. 1-6). IEEE.

Nagabhushana, K. S., Weidenthaler, C., Hocevar, S., Strmcnik, D., Gaberšček, M., Antozzi, A. L., & Martelli, G. N. (2006). *Preparation, characterization and properties of Pt-CuCo-reduced and Pt-on-Cu skin type bimetallic carbon-supported (Vulcan XC72) electrocatalysts.* /paper/Preparation%2C-characterization-and-properties-of-and-Nagabhushana-Weidenthaler/65e5dd1b1d89bf0f657ff1e8007e0b74d5640168

Napoli, G., Polimeni, A., Micari, S., Andaloro, L., & Antonucci, V. (2020). Optimal allocation of electric vehicle charging stations in a highway network: Part 1. Methodology and test application. *Journal of Energy Storage*, *27*, 1–9. doi:10.1016/j.est.2019.101102

Napoli, G., Polimeni, A., Micari, S., Dispenza, G., & Antonucci, V. (2019). Optimal allocation of electric vehicle charging stations in a highway network: Part 2. The Italian case study. *Journal of Energy Storage*, *26*, 1–23. doi:10.1016/j.est.2019.101015

Nichicon Corporation | News Release | EV Power Station Connection Enabled For Electric Vehicles (EVs) from Mitsubishi Motors Corporation, Thereby Expanding EV Applications. (n.d.). https://www.nichicon.co.jp/english/new/new135.html

Nikki Gordon-Bloomfield. (2010). *What Is EVSE And Why Does Your Electric Car Charger Need It?* https://www.greencarreports.com/news/1050948_what-is-evse-and-why-does-your-electric-car-charger-need-it

Noel, L., de Rubens, G. Z., Kester, J., & Sovacool, B. K. (2019). Navigating expert skepticism and consumer distrust: Rethinking the barriers to vehicle-to-grid (V2G) in the Nordic region. *Transport Policy*, *76*, 67–77. doi:10.1016/j.tranpol.2019.02.002

Noel, L., & McCormack, R. (2014). A cost benefit analysis of a V2G-capable electric school bus compared to a traditional diesel school bus. *Applied Energy*, *126*, 246–255. doi:10.1016/j.apenergy.2014.04.009

Odeh, A. O., Osifo, P., & Noemagus, H. (2013). Chitosan: A low cost material for the production of membrane for use in PEMFC-A review. *Energy Sources. Part A, Recovery, Utilization, and Environmental Effects*, *35*(2), 152–163. doi:10.1080/15567036.2010.523760

Ogden, J., & Yang, C. (2005). *Implementing a Hydrogen Energy Infrastructure: Storage Options and System Design.* https://escholarship.org/uc/item/4xb940vg

Olabi, A. G. (2010). The 3rd international conference on sustainable energy and environmental protection SEEP 2009-Guest Editor's Introduction. *Energy*, *35*(12), 4508–4509. doi:10.1016/j.energy.2010.09.053

Olabi, A. G. (2012). *Developments in sustainable energy and environmental protection.* Elsevier. doi:10.1016/j.energy.2011.12.037

Olabi, A. G. (2013). *State of the art on renewable and sustainable energy*. Elsevier. doi:10.1016/j.energy.2013.10.013

Olabi, A. G. (2014). 100% sustainable energy. *Energy*, *77*, 1–5. doi:10.1016/j.energy.2014.10.083

Ou, X., Yan, X., Zhang, X., & Liu, Z. (2012). Life-cycle analysis on energy consumption and GHG emission intensities of alternative vehicle fuels in China. *Applied Energy*, *90*(1), 218–224. doi:10.1016/j.apenergy.2011.03.032

Owejan, J. P., Gagliardo, J. J., Sergi, J. M., & Trabold, T. A. (2008). Two-Phase Flow Considerations in PEMFC Design and Operation. *ASME 2008 6th International Conference on Nanochannels, Microchannels, and Minichannels*, 1733–1740. 10.1115/ICNMM2008-62037

Pandžić, H., Morales, J. M., Conejo, A. J., & Kuzle, I. (2013). Offering model for a virtual power plant based on stochastic programming. *Applied Energy*, *105*, 282–292. doi:10.1016/j.apenergy.2012.12.077

Parsons, G. R., Hidrue, M. K., Kempton, W., & Gardner, M. P. (2014). Willingness to pay for vehicle-to-grid (V2G) electric vehicles and their contract terms. *Energy Economics*, *42*, 313–324. doi:10.1016/j.eneco.2013.12.018

Pasetti, M., Rinaldi, S., & Manerba, D. (2018). A virtual power plant architecture for the demand-side management of smart prosumers. *Applied Sciences (Basel, Switzerland)*, *8*(3), 432. doi:10.3390/app8030432

Pellegrino, G., Armando, E., & Guglielmi, P. (2010). An Integral Battery Charger With Power Factor Correction for Electric Scooter. *IEEE Transactions on Power Electronics*, *25*(3), 751–759. doi:10.1109/TPEL.2009.2033187

Performance update on ITM Power's fuel cell membrane. (2013). *Fuel Cells Bulletin*, *11*(1), 11. doi:10.1016/S1464-2859(13)70026-9

Pollet, B. G., Staffell, I., & Shang, J. L. (2012). Current status of hybrid, battery and fuel cell electric vehicles: From electrochemistry to market prospects. *Electrochimica Acta*, *84*, 235–249. doi:10.1016/j.electacta.2012.03.172

Pottmaier, D., & Baricco, M. (2015). Materials for hydrogen storage and the Na-Mg-B-H system. *AIMS Energy*, *3*(1), 75–100. doi:10.3934/energy.2015.1.75

Pozio, A., Silva, R. F., De Francesco, M., & Giorgi, L. (2003). Nafion degradation in PEFCs from end plate iron contamination. *Electrochimica Acta*, *48*(11), 1543–1549. doi:10.1016/S0013-4686(03)00026-4

Press Release | News · Events | Mitsubishi Motors. (n.d.). https://www.mitsubishi-motors.com/publish/pressrelease_en/corporate/2014/news/detail0954.html

Qamber, I. (2020). Power System Control and Reliability: Electric Power Design and Enhancement. AAP & CRC, Taylor & Francis.

Qian, K., Zhou, C., & Yuan, Y. (2015). Impacts of high penetration level of fully electric vehicles charging loads on the thermal ageing of power transformers. *International Journal of Electrical Power & Energy Systems*, *65*, 102–112. doi:10.1016/j.ijepes.2014.09.040

Qian, K., Zhou, C., Yuan, Y., & Allan, M. (2010). Temperature effect on electric vehicle battery cycle life in vehicle-to-grid applications. In *CICED 2010 Proceedings* (pp. 1–6). IEEE.

Raab, A. F., Ferdowsi, M., Karfopoulos, E., Unda, I. G., Skarvelis-Kazakos, S., Papadopoulos, P., . . . Strunz, K. (2011, September). Virtual power plant control concepts with electric vehicles. In *2011 16th International Conference on Intelligent System Applications to Power Systems* (pp. 1-6). IEEE. 10.1109/ISAP.2011.6082214

Rady, H. M. (1992). Renewable energy in rural areas of developing countries: Some recommendations for a sustainable strategy. *Energy Policy*, *20*(6), 581–588.

Rahman, I., Vasant, P. M., Singh, B. S. M., Abdullah-Al-Wadud, M., & Adnan, N. (2016). Review of recent trends in optimization techniques for plug-in hybrid, and electric vehicle charging infrastructures. *Renewable & Sustainable Energy Reviews*, *58*, 1039–1047. doi:10.1016/j.rser.2015.12.353

Ramos-Rodríguez, A. R., Medina-Garrido, J. A., & Ruiz-Navarro, J. (2012). Determinants of hotels and restaurants entrepreneurship: A study using GEM data. *International Journal of Hospitality Management*, *31*(2), 579–587.

Reed, P. (2011). *Electric-Vehicle Battery Basics*. https://www.myev.com/research/ev-101/electric-vehicle-battery-basics

Ren, Z., Zhang, D., & Wang, Z. (2012). Stacks with TiN/titanium as the bipolar plate for PEMFCs. *Energy*, *48*(1), 577–581. doi:10.1016/j.energy.2012.10.020

Richter, M. (2012). Utilities' business models for renewable energy: A review. *Renewable & Sustainable Energy Reviews*, *16*(5), 2483–2493.

Richter, M. (2013). Business model innovation for sustainable energy: German utilities and renewable energy. *Energy Policy, 62*, 1226–1237.

Rippel, W. E. (1990). *Integrated Traction Inverter and Battery Charger Apparatus*. Patent No. US 4920475.

Rippel, W. E., & Cocconi, A. G. (1992). *Integrated Motor Drive and Recharge System*. Patent, US Patent No. US 5099186.

Roche & Sabrià. (2015). An Accessible Pre-design Calculation Tool to Support the Definition of EV Components. *World Electric Vehicle Journal, 7*, 1-13.

Romare, M. & Dahll¨of, L. (2017). The life cycle energy consumption and greenhouse gas emissions from lithium-ion batteries. *Stockholm. Zugriff am, 23*.

Saboohi, Y., & Farzaneh, H. (2009). Model for developing an eco-driving strategy of a passenger vehicle based on the least fuel consumption. *Applied Energy, 86*(10), 1925–1932. doi:10.1016/j.apenergy.2008.12.017

Sachan, S., & Adnan, N. (2018). Stochastic charging of electric vehicles in smart power distribution grids. *Sustainable Cities and Society, 40*, 91–100. Advance online publication. doi:10.1016/j.scs.2018.03.031

Sachan, S., Deb, S., & Singh, S. N. (2020). Different charging infrastructures along with smart charging strategies for electric vehicles. *Sustainable Cities and Society, 60*, 102238. doi:10.1016/j.scs.2020.102238

Sarikprueck, P., Lee, W.-J., Kulvanitchaiyanunt, A., Chen, V. C. P., & Rosenberger, J. M. (2018). Bounds for Optimal Control of a Regional Plug-in Electric Vehicle Charging Station System. *IEEE Transactions on Industry Applications, 54*(2), 977–986. doi:10.1109/TIA.2017.2766230

Sarker, M. R., Pandžić, H., & Ortega-Vazquez, M. A. (2015). Optimal operation and services scheduling for an electric vehicle battery swapping station. *IEEE Transactions on Power Systems, 30*(2), 901–910. doi:10.1109/TPWRS.2014.2331560

Sbordone, D., Bertini, I., Di Pietra, B., Falvo, M. C., Genovese, A., & Martirano, L. (2015). EV fast charging stations and energy storage technologies: A real implementation in the smart micro grid paradigm. *Electric Power Systems Research, 120*, 96–108. doi:10.1016/j.epsr.2014.07.033

Seung-Ki, S., & Sang-Joon, L. (1995). An Integral Battery Charger for Four-Wheel Drive Electric Vehicle. *IEEE Transactions on Industry Applications, 31*(5), 1096–1099. doi:10.1109/28.464524

Shapero, A., & Sokol, L. (1982). *The social dimensions of entrepreneurship*. University of Illinois at Urbana-Champaign's Academy for Entrepreneurial Leadership Historical Research Reference in Entrepreneurship.

Sharaf, O. Z., & Orhan, M. F. (2014). An overview of fuel cell technology: Fundamentals and applications. *Renewable & Sustainable Energy Reviews, 32*, 810–853. doi:10.1016/j.rser.2014.01.012

Shareef, H., Islam, M. M., & Mohamed, A. (2016). A review of the stage-of-the-art charging technologies, placement methodologies, and impacts of electric vehicles. *Renewable & Sustainable Energy Reviews, 64*, 403–420. doi:10.1016/j.rser.2016.06.033

Shariff, S., Iqbal, D., Alam, M., & Ahmad, F. (2019). A State-of-the-Art Review of Electric Vehicle to Grid (V2G) technology. *IOP Conf. Series: Materials Science and Engineering, 561*. 10.1088/1757-899X/561/1/012103

Shariff, S.M., Alam, M.S., & Ahmad, F. (2019). System Design and Realization of a Solar-Powered Electric Vehicle Charging Station. *IEEE Syst J*, 1–12. doi:10.1109/JSYST.2019.2931880

Sharma, Aware, & Bhowate. (2020). Integrated Battery Charger for EV by using Three-Phase Induction Motor Stator Windings as Filter. *IEEE Transactions on Transportation Electrification*, 1–1. doi:10.1109/TTE.2020.2972765

Shemami, M. S., Amrr, S. M., Alam, M. S., & Jamil Asghar, M. S. (2018). Reliable and Economy Modes of Operation for Electric Vehicle-to-Home (V2H) System. *2018 5th IEEE Uttar Pradesh Section International Conference on Electrical, Electronics and Computer Engineering, UPCON 2018*, 1–6. https://www.researchgate.net/publication/328916840%0Ahttps://ieeexplore.ieee.org/abstract/document/8596932

Shemami, S., Mahdi, M. S. A., & Jamil Asghar, M. S. (2018). Fuzzy Control Assisted Vehicle-to-Home (V2H) Energy Management System. *Smart Science, 6*(2), 173–87. https://www.tandfonline.com/doi/full/10.1080/23080477.2017.1417963

Shemami, S., Mahdi, M. S. A., & Jamil Asghar, M. S. (2018). Intelligent Vehicle-to-Home (V2H) and Home-to-Vehicle (H2V) Residential Power Management. *The Intelligent Transportation Society of America 2018 Annual Meeting*, 1–13. https://static1.squarespace.com/static/59c3ed7b197aeabbd2a51a3b/t/5b2a51c90e2e720706986142/1529500105347/TS18_Paper15790.pdf

Shemami, M., Alam, M., Asghar, M., & Shariff, S. (2019). Adaptive Neuro-Fuzzy Inference System (ANFIS) for Optimization of Solar Based Electric Vehicle-to-Home (V2H) Fuzzy Inference System (FIS) Controller. *IEEE Conference Paper*. 10.1109/ITEC.2019.8790579

Shi, C., & Khaligh, A. (2018). A Two-Stage Three-Phase Integrated Charger for Electric Vehicles With Dual Cascaded Control Strategy. *IEEE Journal of Emerging and Selected Topics in Power Electronics*, *6*(2), 898–909. doi:10.1109/JESTPE.2018.2797913

Shi, C., Tang, Y., & Khaligh, A. (2017). A Single-Phase Integrated Onboard Battery Charger Using Propulsion System for Plug-in Electric Vehicles. *IEEE Transactions on Vehicular Technology*, *66*(12), 10899–10910. doi:10.1109/TVT.2017.2729345

Shi, C., Tang, Y., & Khaligh, A. (2018). A Three-Phase Integrated Onboard Charger for Plug-In Electric Vehicles. *IEEE Transactions on Power Electronics*, *33*(6), 4716–4725. doi:10.1109/TPEL.2017.2727398

Shinzaki, S., Sadano, H., Maruyama, Y., & Kempton, W. (2015). *Deployment of vehicle-to-grid technology and related issues* (No. 2015-01-0306). SAE Technical Paper.

Shi, R., Semsar, S., & Lehn, P. W. (2017). Constant Current Fast Charging of Electric Vehicles via a DC Grid Using a Dual-Inverter Drive. *IEEE Transactions on Industrial Electronics*, *64*(9), 6940–6949. doi:10.1109/TIE.2017.2686362

Show, Y. (2007). Electrically conductive amorphous carbon coating on metal bipolar plates for PEFC. *Surface and Coatings Technology*, *202*(4–7), 1252–1255. doi:10.1016/j.surfcoat.2007.07.065

Singh, K., Amir, M., & Member, S. (2020). *An Integral Tilt Derivative Control Strategy for Frequency Control in Multimicrogrid System*. Academic Press.

Singh, M., Kumar, P., & Kar, I. (2010, September). Analysis of vehicle to grid concept in Indian scenario. In *Proceedings of 14th International Power Electronics and Motion Control Conference EPE-PEMC 2010* (pp. T6-149). IEEE. 10.1109/EPEPEMC.2010.5606641

Singh, S., Jain, S., Ps, V., Tiwari, A. K., Nouni, M. R., Pandey, J. K., & Goel, S. (2015). Hydrogen: A sustainable fuel for future of the transport sector. *Renewable & Sustainable Energy Reviews*, *51*, 623–633. doi:10.1016/j.rser.2015.06.040

Slicker, J. M. (1985). *Pulse Width Modulation Inverter With Battery Charger*. Patent No. US 4491768. https://www.google.com/patents/US4491768

Society of Manufacturers of Electric Vehicles. (n.d.). *EV Industry*. Author.

Solar, J. O. J. U. (2018). *EV Facts and Figures, Batteries and Smart Grids*. https://www.jojusolar.co.uk/ev-facts-and-figures/

Solero, L. (2001). Nonconventional On-board Charger for Electric Vehicle Propulsion Batteries. *IEEE Transactions on Vehicular Technology*, *50*(1), 144–149. doi:10.1109/25.917904

Somasekhar, V. T., Srinivas, S., & Kumar, K. K. (2008). Effect of Zero-Vector Placement in a Dual-Inverter Fed Open-End Winding Induction-Motor Drive With a Decoupled Space-Vector PWM Strategy. *IEEE Transactions on Industrial Electronics*, *55*(6), 2497–2505. doi:10.1109/TIE.2008.918644

Srinivasan, S., Palmie, G., Srinivasan, S., Palmieri, G., Del Vecchio, C., & Glielmo, L. (2016). Demand Side Management for heating controls in Microgrids. *IFAC-PapersOnLine*, *49*(1), 611–616. doi:10.1016/j.ifacol.2016.03.123

Stolten, D., & Grube, T. (Eds.). (2010). *18th World Hydrogen Energy Conference 2010 - WHEC 2010: Proceedings. 4: Parallel sessions book: storage systems, policy perspectives, initiatives and cooperations*. Forschungszentrum, IEF-3.

Subotic, I., Bodo, N., & Levi, E. (2016). Single-Phase On-Board Integrated Battery Chargers for EVs Based on Multiphase Machines. *IEEE Transactions on Power Electronics*, *31*(9), 6511–6523. doi:10.1109/TPEL.2015.2504400

Subotic, I., Bodo, N., & Levi, E. (2017). Integration of Six-Phase EV Drivetrains into Battery Charging Process with Direct Grid Connection. *IEEE Transactions on Energy Conversion*, *32*(99), 1–1. doi:10.1109/TEC.2017.2679280

Subotic, I., Bodo, N., Levi, E., Dumnic, B., Milicevic, D., & Katic, V. (2016). Overview of fast on-board integrated battery chargers for electric vehicles based on multiphase machines and power electronics. *IET Electric Power Applications*, *10*(3), 217–229. doi:10.1049/iet-epa.2015.0292

Subotic, I., Bodo, N., Levi, E., & Jones, M. (2015). On-board Integrated Battery Charger for EVs Using an Asymmetrical Nine-Phase Machine. *IEEE Transactions on Industrial Electronics*, *62*(99), 1–1. doi:10.1109/TIE.2014.2345341

Subotic, I., Bodo, N., Levi, E., Jones, M., & Levi, V. (2016). Isolated Chargers for EVs Incorporating Six-Phase Machines. *IEEE Transactions on Industrial Electronics*, *63*(1), 653–664. doi:10.1109/TIE.2015.2412516

Sun, Y., Delucchi, M., & Ogden, J. (2011). The impact of widespread deployment of fuel cell vehicles on platinum demand and price. *International Journal of Hydrogen Energy*, *36*(17), 11116–11127. doi:10.1016/j.ijhydene.2011.05.157

Taheri, H., Ghasemi, H., Rahimi-Kian, A., & Akbari, T. (2011, May). Modeling and optimized scheduling of virtual power plant. In *2011 19th Iranian Conference on Electrical Engineering* (pp. 1-6). IEEE.

Tan, K. M., Ramachandaramurthy, V. K., & Yong, J. Y. (2016). Integration of electric vehicles in smart grid: A review on vehicle to grid technologies and optimization techniques. Renewable and Sustainable Energy Reviews. doi:10.1016/j.rser.2015.09.012

Tang, Y., Lu, J., Wu, B., Zou, S., Ding, W., & Khaligh, A. (2018). An Integrated Dual-Output Isolated Converter for Plug-in Electric Vehicles. *IEEE Transactions on Vehicular Technology*, *67*(2), 966–976. doi:10.1109/TVT.2017.2750076

Tang, Y., Zhang, Q., Li, H., Li, Y., & Liu, B. (2019). Economic Analysis on Repurposed EV batteries in a Distributed PV System under Sharing Business Models. *Energy Procedia*, *158*, 4304–4310. doi:10.1016/j.egypro.2019.01.793

Tawfik, H., Hung, Y., & Mahajan, D. (2007). Metal bipolar plates for PEM fuel cell—A review. *Journal of Power Sources*, *163*(2), 755–767. doi:10.1016/j.jpowsour.2006.09.088

Teixeira, A. C. R., & Sodré, J. R. (2016). Simulation of the impacts on carbon dioxide emissions from replacement of a conventional Brazilian taxi fleet by electric vehicles. *Energy*, *115*, 1617–1622. doi:10.1016/j.energy.2016.07.095

Tesla. (2017). *Powerwall | The Tesla Home Battery*. https://www.tesla.com/en_GB/powerwall

The Cost of Solar Will Drop Another 25% by 2022. (n.d.). https://futurism.com/the-cost-of-solar-will-drop-another-25-by-2022/

Thimmesch, D. (1985). An SCR Inverter with an Integral Battery Charger for Electric Vehicles. *IEEE Transactions on Industry Applications*, *IA-21*(4), 1023–1029. doi:10.1109/TIA.1985.349573

Thingvad, A., Ziras, C., Hu, J., & Marinelli, M. (2017, August). Assessing the energy content of system frequency and electric vehicle charging efficiency for ancillary service provision. In *2017 52nd International Universities Power Engineering Conference (UPEC)* (pp. 1-6). IEEE. 10.1109/UPEC.2017.8231947

Thingvad, A., Calearo, L., Andersen, P. B., Marinelli, M., Neaimeh, M., Suzuki, K., & Murai, K. (2019). *Value of V2G Frequency Regulation in Great Britain Considering Real Driving Data. In 2019 IEEE PES Innovative Smart Grid Technologies Europe*. IEEE.

Thingvad, A., & Marinelli, M. (2018). Influence of v2g frequency services and driving on electric vehicles battery degradation in the nordic countries. *31st International Electric Vehicles Symposium & Exhibition & International Electric Vehicle Technology Conference 2018.*

Tomić, J., & Kempton, W. (2007). Using fleets of electric-drive vehicles for grid support. *Journal of Power Sources*, *168*(2), 459–468. doi:10.1016/j.jpowsour.2007.03.010

Transportation Overview | Center for Climate and Energy Solutions. (n.d.). https://www.c2es.org/energy/use/transportation

Truntič, M., Konjedic, T., Milanovič, M., Šlibar, P., & Rodič, M. (2018). Control of integrated single-phase PFC charger for EVs. *IET Power Electronics*, *11*(11), 1804–1812. doi:10.1049/iet-pel.2017.0663

Tsoleridis, C., Chatzimisios, P., & Fouliras, P. (2016). Vehicle-to-Grid Networks: Issues and Challenges. In Smart grid: networking, data management, and business models (pp. 347-369). CRC Press.

Un, E., & Hava, A. M. (2007). A Near State PWM Method With Reduced Switching Frequency And Reduced Common Mode Voltage For Three-Phase Voltage Source Inverters. *2007 IEEE International Electric Machines & Drives Conference*, 235–240. doi:10.1109/IEMDC.2007.383583

University, B. (2017). *BU-1003: Electric Vehicle (EV) – Battery University*. https://batteryuniversity.com/learn/article/electric_vehicle_ev

Un-Noor, F., Padmanaban, S., & Mihet-Popa, L. (2017). *A Comprehensive Study of Key Electric Vehicle (EV) Components, Technologies, Challenges, Impacts, and Future Direction of Development*. doi:10.20944/preprints201705.0090.v1

Van Schalkwijk, W., & Scrosati, B. (2002). Advances in lithium ion batteries introduction. In *Advances in Lithium-Ion Batteries* (pp. 1–5). Springer. doi:10.1007/0-306-47508-1_1

Vandael, S., Holvoet, T., Deconinck, G., Kamboj, S., & Kempton, W. (2013, October). *A comparison of two GIV mechanisms for providing ancillary services at the University of Delaware. In 2013 IEEE international conference on smart grid communications (SmartGridComm)*. IEEE.

Vasirani, M., Kota, R., Cavalcante, R. L., Ossowski, S., & Jennings, N. R. (2012). Technical Report: Virtual Power Plants of Wind Power Generators and Electric Vehicles. Tech. rep.

Vasirani, M., Kota, R., Cavalcante, R. L., Ossowski, S., & Jennings, N. R. (2013). An agent-based approach to virtual power plants of wind power generators and electric vehicles. *IEEE Transactions on Smart Grid*, *4*(3), 1314–1322. doi:10.1109/TSG.2013.2259270

Viana, C., & Lehn, P. W. (2019). A Drivetrain Integrated DC Fast Charger With Buck and Boost Functionality and Simultaneous Drive/Charge Capability. *IEEE Transactions on Transportation Electrification*, *5*(4), 903–911. doi:10.1109/TTE.2019.2925211

Wang, H. (2003). Stainless steel as bipolar plate material for polymer electrolyte membrane fuel cells. *Journal of Power Sources*, *115*(2), 243–251. doi:10.1016/S0378-7753(03)00023-5

Wang, H., & Turner, J. A. (2007). SnO2:F coated ferritic stainless steels for PEM fuel cell bipolar plates. *Journal of Power Sources*, *170*(2), 387–394. doi:10.1016/j.jpowsour.2007.04.028

Wang, J. (2015). Barriers of scaling-up fuel cells: Cost, durability and reliability. *Energy*, *80*, 509–521. doi:10.1016/j.energy.2014.12.007

Wang, Q., Rongrong, L., & Rui, J. (2016). Decoupling and Decomposition Analysis of Carbon Emissions from Industry: A Case Study from China. *Sustainability*, *8*(10), 1059–1076. doi:10.3390u8101059

Wang, R., Xiao, G., & Wang, P. (2017). Hybrid Centralized-Decentralized (HCD) Charging Control of Electric Vehicles. *IEEE Transactions on Vehicular Technology*, *66*(8), 6728–6741. doi:10.1109/TVT.2017.2668443

Wang, S., Gangammanavar, H., Eksioglu, S. D., & Mason, S. J. (2019). Stochastic Optimization for Energy Management in Power Systems with Multiple Microgrids. *IEEE Transactions on Smart Grid*, *10*(1), 1068–1079. doi:10.1109/TSG.2017.2759159

Wang, Z., Liu, B., Guan, L., Zhang, Y., Cheng, M., Zhang, B., & Xu, L. (2019). A Dual-Channel Magnetically Integrated EV Chargers Based on Double-Stator-Winding Permanent-Magnet Synchronous Machines. *IEEE Transactions on Industry Applications*, *55*(2), 1941–1953. doi:10.1109/TIA.2018.2879869

Wang, Z., Zhang, Y., You, S., Xiao, H., & Cheng, M. (2019). An Integrated Power Conversion System for Electric Traction and V2G Operation in Electric Vehicles With A Small Film Capacitor. *IEEE Transactions on Power Electronics*, 1–1. doi:10.1109/TPEL.2019.2944276

Weakley, S. A. (2012). *Pathways to Commercial Success: Technologies and Products Supported by the Fuel Cell Technologies Program* (PNNL-21804, 1057844; p. PNNL-21804, 1057844). doi:10.2172/1057844

Wilberforce, T., Alaswad, A., Palumbo, A., Dassisti, M., & Olabi, A.-G. (2016). Advances in stationary and portable fuel cell applications. *International Journal of Hydrogen Energy*, *41*(37), 16509–16522. doi:10.1016/j.ijhydene.2016.02.057

Wu, F., & Sioshansi, R. (2019). A stochastic operational model for controlling electric vehicle charging to provide frequency regulation. *Transportation Research Part D, Transport and Environment*, *67*, 475–490. doi:10.1016/j.trd.2018.12.005

Xiang, Y., Hu, S., Liu, Y., Zhang, X., & Liu, J. (2018). Electric vehicles in smart grid: a survey on charging load modelling. IET Smart Grid. doi:10.1049/iet-stg.2018.0053

Xiao, Y., Liu, C., & Yu, F. (2019). An Integrated On-Board EV Charger with Safe Charging Operation for Three-Phase IPM Motor. *IEEE Transactions on Industrial Electronics*, *66*(10), 7551–7560. doi:10.1109/TIE.2018.2880712

Xu, B., Oudalov, A., Ulbig, A., Andersson, G., & Kirschen, D. S. (2016). Modeling of lithium-ion battery degradation for cell life assessment. *IEEE Transactions on Smart Grid*, *9*(2), 1131–1140. doi:10.1109/TSG.2016.2578950

Yang, J. J., Zhao, J. H., Wen, F. S., Xue, Y. S., Li, L., & Lv, H. H. (2014). Development of bidding strategies for virtual power plants considering uncertain outputs from plug-in electric vehicles and wind generators. *Dianli Xitong Zidonghua*, *38*(13), 92–102.

Yang, L., Yu, H., Jiang, L., Zhu, L., Jian, X., & Wang, Z. (2010). Improved anticorrosion properties and electrical conductivity of 316L stainless steel as bipolar plate for proton exchange membrane fuel cell by lower temperature chromizing treatment. *Journal of Power Sources*, *195*(9), 2810–2814. doi:10.1016/j.jpowsour.2009.11.018

Yang, Z., Li, K., & Foley, A. (2015). Computational scheduling methods for integrating plug-in electric vehicles with power systems: A review. *Renewable & Sustainable Energy Reviews*, *51*, 396–416. doi:10.1016/j.rser.2015.06.007

Zecchino, A., Knezović, K., & Marinelli, M. (2017, September). Identification of conflicts between transmission and distribution system operators when acquiring ancillary services from electric vehicles. In 2017 IEEE PES Innovative Smart Grid Technologies Conference Europe (ISGT-Europe) (pp. 1-6). IEEE. doi:10.1109/ISGTEurope.2017.8260127

Zecchino, A., & Marinelli, M. (2018). Analytical assessment of voltage support via reactive power from new electric vehicles supply equipment in radial distribution grids with voltage-dependent loads. *International Journal of Electrical Power & Energy Systems*, *97*, 17–27. doi:10.1016/j.ijepes.2017.10.034

Zecchino, A., Prostejovsky, A. M., Ziras, C., & Marinelli, M. (2019). Large-scale provision of frequency control via V2G: The Bornholm power system case. *Electric Power Systems Research*, *170*, 25–34. doi:10.1016/j.epsr.2018.12.027

Zecchino, A., Thingvad, A., Andersen, P. B., & Marinelli, M. (2019). Test and Modelling of Commercial V2G CHAdeMO Chargers to Assess the Suitability for Grid Services. *World Electric Vehicle Journal*, *10*(2), 21. doi:10.3390/wevj10020021

Zhang, D., Duan, L., Guo, L., Wang, Z., Zhao, J., Tuan, W.-H., & Niihara, K. (2011). TiN-coated titanium as the bipolar plate for PEMFC by multi-arc ion plating. *International Journal of Hydrogen Energy*, *36*(15), 9155–9161. doi:10.1016/j.ijhydene.2011.04.123

Zhang, K., Mao, Y., Leng, S., He, Y., Maharjan, S., Gjessing, S., Zhang, Y., & Tsang, D. H. K. (2018). Optimal Charging Schemes for Electric Vehicles in Smart Grid: A Contract Theoretic Approach. *IEEE Transactions on Intelligent Transportation Systems*, *19*(9), 3046–3058. doi:10.1109/TITS.2018.2841965

Zhang, Y., Fang, J., Gao, F., Song, T., Gao, S., & Rogers, D. J. (2020). Second-Harmonic Ripple Voltage Suppression of Integrated Single-Phase Pulsewidth Modulation Rectifier Charging System for EVs. *IEEE Transactions on Power Electronics*, *35*(4), 3616–3626. doi:10.1109/TPEL.2019.2937745

Zheng, Y., Niu, S., Shang, Y., Shao, Z., & Jian, L. (2019). Integrating plug-in electric vehicles into power grids: A comprehensive review on power interaction mode, scheduling methodology and mathematical foundation. *Renewable & Sustainable Energy Reviews*, *112*, 424–439. doi:10.1016/j.rser.2019.05.059

Zhevago, N. K. (2016). Other methods for the physical storage of hydrogen. In *Compendium of Hydrogen Energy* (pp. 189–218). Elsevier. doi:10.1016/B978-1-78242-362-1.00008-0

Ziras, C., Zecchino, A., & Marinelli, M. (2018, June). Response accuracy and tracking errors with decentralized control of commercial v2g chargers. In 2018 Power Systems Computation Conference (PSCC) (pp. 1-7). IEEE. doi:10.23919/PSCC.2018.8442488

About the Contributors

Mohammad Abdullah is an Associate Professor at the School of Electrical Engineering, Faculty of Engineering, Universiti Teknologi Malaysia (UTM). He is currently the Director at the Centre of Electrical Energy Systems (CEES), Institute of Future Energy (IFE), UTM. He is a Chartered Engineer registered with the Engineering Council, United Kingdom, a Professional Engineer registered with the Board of Engineers Malaysia (BEM) and a Certified Energy Manager registered with the ASEAN Energy Management Scheme (AEMAS). He is also a Senior Member of the Institute of Electrical and Electronics Engineers, USA (SMIEEE) and a Corporate Member of the Institution of Engineers Malaysia (MIEM). He received the bachelors degree (B.Eng) in Electrical & Electronic Engineering from Universiti Tenaga Nasional (Uniten), Malaysia in 2002, masters degree (M.Sc) in Electrical Power Engineering and doctorate degree (Ph.D) from University of Strathclyde, Glasgow, United Kingdom in 2003 and 2008 respectively.

Sadegh Aberoumand does research on energy storage and nanofluids. He is doing PhD at the School of Engineering and Built Environment at Griffith University, Australia.

Zeeshan Ahmad Arfeen is an Assistant Professor at "The Islamia University of Bahawalpur", Pakistan in the Electrical Power Engineering Department. Ph.D. scientist from University Technology Malaysia (UTM) Skudai Johor Bahru, Johor, Malaysia. He is a senior member of IEEE organization and member of IEEE (PES) Power and Energy Society. He is an author of many high indexed Journals and potential reviewer of more than 100 + papers. His research interests include electrical transportations, microgrid, energy management system and sustainable energy.

Furkan Ahmad received the Bachelor of Technology (2012), Master of Technology (2015) and Ph.D. (2019), all in Electrical Engineering, from Aligarh Muslim University (AMU), Aligarh, India. Ahmad was the recipient of the 2016–2017 CSIR Junior Research Fellowship and 2018-2019 Senior Research Fellowship Award. Currently, he working as a post-doctoral fellow at CART, Indian Institute of Technology, Delhi. His research interests include smart microgrids, electric vehicles, energy management, power market operations, and renewable energy integration.

Essam Al Ammar worked as a Power/Software Engineer at Lucent Technologies in Riyadh for two years. In 2003, he received his MS degree from University of Alabama, Tuscaloosa, AL, and Ph.D. degree from Arizona State University in 2007. He is an Associate Professor in Electrical Engineering Department, King Saud University, Riyadh, Saudi Arabia. He was Governor's Advisor in Electricity Co-

Generation Regulatory Authority (ECRA). Also, he was an advisor at Ministry of Water and Electricity (MOWE) and a former Energy Consultant for Riyadh Techno Valley (RTV). He is involved in many local and international committees in different aspects in the area of power, electrical energy and water. He is an opinion columnist in local newspapers. He is a PMP and PBA certified. His current interests include electricity market, high voltage engineering, power system transmission and distribution as well as renewable energy and smart grid. He is a Senior Member of IEEE since 2007 and Saudi Engineering Committee since 1997. He published nearly 120 papers and 20 patents in Energy and Water. Fully interested in the followings: Smart Grid, Renewable Energy, Energy Efficiency, Energy Economics, Electric Vehicles, Energy Regulation, Power System Transients, High Voltage Engineering, Power System Diagnostics, Digital Transformation, Fourth Industrial Revolution (4IR).

Samir Al Sharif is currently working at Taibah University as a faculty member and he is serving as the Director of the Strategic Research Unit and Director of MM Binladin Scientific Chair on O&M. Dr. Shariff has graduated with his BS and MS in Electrical Engineering from Washington University – St. Louis MO in 1994 and 1996 respectively. Dr. Shariff has received his Ph.D. in Electrical Engineering from Wichita State University in 2000 with an honorary award of Phi Kappa Phi (Academic). After that Dr. Shariff has started his practical life by working as a project manager at Saudi Binladin Group (SBG) for the operation and maintenance of the Makkah Harm projects. In 2002 Dr. Shariff has joined Saudi Aramco as a project engineer working with the PMT Department. In 2004 Dr. Shariff moved to the academic life and worked at Taibha University where he served at various academic, research as well as management posts.

Mohamed Alhamad is Manager of Market Operations in GCCIA. He is a member of the GCC Cigre Technical Committee and Chairman of the Electricity Market and Regulatory Committee (SC8). He has authored more than 30 transaction and conference papers. He joined GCC Interconnection Authority in 2012. He worked 3 years in Distribution Planning & Design Section, & 8 years system control engineer at Bahrain Electricity and Water Authority. He worked as an instructor in the University of Bahrain from 1996 to 2001. Al-Hamad received his Ph.D. degree in Electricity Markets in 2016 from the University of Bahrain and received his M.Sc and B.Sc in Electrical Power from the University of Bahrain in 2002 and 1996 respectively. His Ph.D. program covers the Development of the GCC Electrical Power Market and Trading platforms. He is responsible about create Auctioning for the GCCIA Interconnection looking after the Interconnection rights for operations, Energy, and Financial settlements for the scheduled and unscheduled energy.

Ahmed M. Ali is a mechanical engineering student in Faculty of Engineering and Applied Science in Nile University, Egypt. His major interest is in Electric vehicles and their effects on society and environment. Also, he participated in the design and manufacturing of an electric car in one of the national electric rally competitions. His current research interests include electrical vehicles, autonomous cars, and Robotics.

Syed Qaseem Ali got his B. Engg. Degree from NED University of Technology, Karachi, and his MS in Electrical Engineering Power and Control from IllinoisInstitute of Technology, Chicago in 2010. He obtained his Ph.D. in electrical engineering from McGill University in the control and design of integrated battery chargers for electric vehicles in 2018. He joined Quanta Technology as an Advisor in

Technology Engineering. He had worked with OSI International as a Power System consultant in the area of ADMS and DERMS. Earlier he worked with OPAL-RT Technologies as a lead simulation expert for the Distribution and generation team in the Application eXpertise and Electrical Simulation (AXES) division. Before joining Opel-RT he worked as a Graduate Research Assistant at McGill University, Research Engineer at King Saud University, Riyadh. His research interests include microgrid controls, DER integration, design and control of EV chargers and power electronic converters for renewable energy applications.

Hasan K. Alrajhi received a bachelor degree (with honor) from the University of Umm Alqura in Makkah in Saudi Arabia on September 2009, and the M.Sc. and Ph.D. degrees in electrical engineering from the University of Waterloo, Ontario, Canada, in 2014 and 2018, respectively. He is currently an assistant professor with the Electrical Engineering Department, Umm Alqura University, Makkah, SA. His research interests include the areas of smart active distribution systems, control of power electronic, interconnect renewable energies into microgrids, MTDC-VSC, and modelling converters.

Ibrahim Alsaidan received the B.S. degree from Qassim University, Buraydah, Saudi Arabia, in 2008, and the M.S. and Ph.D. degrees from the University of Denver, in 2012 and 2018, respectively. He joined Qassim University, in 2018, where he is currently an Assistant Professor and the Chair of the Electrical Engineering Department. His current research interests include renewable energy and distributed generation, photovoltaic generation forecasting, energy storage planning and operation, and smart grid.

Hossam Hassan Ammar is Instructor of Automatic Control and Mechatronics systems at School of Engineering and Applied Science and Researcher at Smart Engineering Systems Research Center (SESC), Nile University, Egypt. He obtained his B.Sc. 2011 at industrial electronics and control and M.Sc. in Mechatronics and Automatic Control, Egypt in 2016. His main research interest is in the area of Mechatronics and Robotics system modelling, identification, and control. His recent research interest is Renewable Energy solutions and their impacts on the environment.

Mohammad Arqam is finishing his PhD (Mechanical Engineering) in the School of Engineering and Built Environment at Griffith University, Australia. Currently working on a live industry project concerns investigation and development of an electric swashplate compressor for heavy electric vehicle mobile refrigeration and air-conditioning applications. His role in the project is to develop thermodynamic models and the cooling technology required to keep the size of the motor to a minimum. His research interests include computational and experimental heat transfer, computational fluid dynamics, mobile refrigeration and energy systems.

Mehreen Kausar Azam is teaching at the Industrial Management Department at Institute of Business Management (IoBM) at graduate and post graduate level. She is a PhD scholar and currently her major research areas include Resilience, Sustainability and Industry 4.0. She has a number of research papers and a few book chapters published in international journals.

Sanchari Deb is currently associated with Power and Control Laboratory of School of Engineering, University of Warwick as EUTOPIA SIF Postdoctoral Fellow. Earlier, she was associated with Smart e fleet group of VTT Technical Research Centre as ERCIM Fellow. Her research interests broadly covers different aspects of power and energy such as e mobility, charging infrastructure planning, Artificial intelligence applications in power and energy, microgrid planning, distribution network planning, local energy systems with cogeneration, Vehicle Grid integration, optimization, metaheuristics algorithms, tool development for charging station placement, machine learning applications, power system reliability, intelligent transport, autonomous vehicle, quantum computing applications, solar based charging infrastructure for Electric Vehicles (EVs). Sanchari Deb completed her Bachelor of Engineering (BE) in Electrical Engineering from Assam Engineering College (AEC), Guwahati in the year 2014, Master of Engineering from Birla Institute of Technology (BIT) Mesra, Ranchi with specialization as Power Systems in the year 2016, and PhD in Energy from Centre for Energy, Indian Institute of Technology (IIT), Guwahati in the year 2020. Post Phd, she joined Alliance for an Energy Efficient Economy, New Delhi as Research Associate and worked on a project related to Vehicle Grid Integration in Indian aspect. Later, she was associated with Centre for Advanced Research in Electrified Transportation (CARET) of Aligarh Muslim University (AMU) as a postdoctoral researcher for a short duration. In the year 2020, she received European Research Council of Informatics and Mathematics (ERCIM) fellowship and joined VTT Technical Research Centre, Finland. She has published nearly 30 research articles and co-edited a book on Smart Charging. Sanchari Deb serves as the reviewer of several peer reviewed journals of Elsevier, IEEE and Springer.

Thoudam Prabha Devi is heading the research and linkages activities in University of Buraimi and works closely with the management and the ministry of higher education of sultanate of Oman for research and quality assurance. To her profile, there are 5 awarded Ph.Ds., 5 books, 5 international and national Projects, 25 papers in journal of repute including SCOPUS and Web of Sciences and 15 book chapters. She has a special focus on overall planning, implementation and control for the establishment and running of educational setups and universities. She has been sharing her experience by conducting MDPs and FDPs for working executives and Faculty members. She has been awarded with many awards including a Life Time Achievement Award 2019 by International Social Science & Management Welfare Association.

Giulio Ferro received the B.S. degree in industrial engineering, the M.S. degree in power systems engineering and the Ph.D degree in Systems engineering in 2014, 2016, and 2020 respectively, from the University of Genoa (Italy). He was a visiting student at the AAC Lab at MIT, Cambridge, MA. His research work is focused on optimization and management of microgrids, distribution networks and distributed optimization.

Rabia Hassan is currently doing PhD in Engineering Management from University of Engineering and Technology, Taxila. She has earned an MS degree (Engineering Management) from Institute of Business Management. She obtained her B.E Electronics Engineering degree from PAF-Karachi Institute of Economics & Technology. She has been associated with the automotive industry of Pakistan for last five years where her major area was product development.

Geza Joos (F'2004) graduated from McGill University, Montreal, Canada, with an M.Eng. and Ph.D. in Electrical Engineering. He is a Professor in the Department of Electrical and Computer Engineering at McGill University (since 2001). He holds the NSERC/Hydro-Quebec Industrial Research Chair on the Integration of Renewable Energies and Distributed Generation into the Electric Distribution Grid (since

2009) and a Canada Research Chair in Powering Information Technologies (since 2004). His research interests are in distributed energy resources, including renewable energy resources, advanced distribution systems, and microgrids. He was previously with ABB, the Université du Québec, and Concordia University (Montreal, Canada). He has consulting experience in the areas of power electronic converters, with applications to power systems. He is Secretary of CIGRE Study Committee C6 on Distributed Energy Resources and Active Distribution Systems, and active in IEEE Power and Energy Society working groups on power electronic applications to power systems and IEEE Standards Association working groups on distributed energy resources and microgrids. He is a Fellow of CIGRE, the Canadian Academy of Engineering, and the Engineering Institute of Canada.

Wasif Karim is a Design Engineer in Zia Semiconductor, Bangalore. He graduated in Electrical Engineering from Zakir Hussain College of Engineering and Technology, Aligarh Muslim University. He also worked as an Intern in Centre for Advanced Research in Electrified Transportation (CARET), AMU.

Mohd Rizwan Khalid received the B.Tech. and M.Tech. degrees in electrical engineering from Zakir Husain College of Engineering and Technology, Aligarh Muslim University, Aligarh, India, in 2013 and 2016, respectively. He has previously worked with Indian Institute of Technology, Kanpur, India, as project engineer. He is currently working towards the Ph.D. degree in the Department of Electrical Engineering, Aligarh Muslim University, Aligarh, India. His research interests include power electronics converters for renewable energy integration and electric vehicles.

Vinay Kumar is working as a researcher in Queensland Micro-Nanotechnology Centre (QMNC), Griffith University, Australia. His principal research areas are Solar Energy Technology, Thermal Energy Storage, Heat-Transfer, and Micro-fluidics.

Jomana A. Mahmoud is a mechanical engineering student in Faculty of Engineering and Applied Science in Nile University, Egypt. Her passion in life is to spread the knowledge of technology to the kids and try to make them aware with the new discoveries in this field; so, she is currently working in Fantasia for Self-Development and Technology Training. She also interests in the electric vehicles which triggers her to participate in the national electric rally (EVER) competitions in Egypt. According to her concern in electric vehicles, her research was about the microgrids and the effects of electric cars in the future world.

Riccardo Minciardi is Full Professor of Modelling and Simulation at the University of Genoa, and his current research interests include modelling and control of automated manufacturing systems, transport planning in urban systems, risk management, and sustainable planning and optimal control of energy and environmental systems. He has co-authored more than 250 publications in international journals, books, invited chapters, and conference proceedings.

Mamdouh Mohamed Ezzeldin is currently a mechanical engineering student in Faculty of Engineering and Applied Science in Nile University, Egypt, since 2016. Since 2019, he became vice president in a student organization called RPMUS and a supervisor on the racing and technical team patriating in international competitions. He participated in the national electric rally (EVER) competition for designing and manufacturing an electric vehicle. His current research interests include electrical vehicles, autonomous cars, and Robotics.

Luca Parodi is a Ph.D. Student in Systems Engineering at the Department of Informatics, Bioengineering, Robotics and Systems Engineering (DIBRIS), University of Genoa. He received his B.S. degree in Industrial Engineering (2017) and his M.Sc. in Energy Engineering (2019) at the University of Genoa. His research is mainly focused on EV charging and scheduling, and Demand Response models in microgrids and smart buildings.

Isa S. Qamber teaching at University of Bahrain since 1988. Now he is Emeritus Professor University of Bahrain and a member of Bahrain Society of Engineers. He has extensive academic experience in electrical engineering and specialized in Power Systems Reliability and High Voltage. He has supervised masters and PhD-level students. He has served as Dean of Scientific Research, Dean of Applied Studies and Chairman of Electrical and Electronics Engineering Department. Qamber is a member of professional societies such as IEEE (USA), CIGRE (France) and Bahrain Society of Engineers. He is the founder of IEEE Bahrain Section. He has served as a steering committee's member, session chair and a member of technical program committees of many international conferences. He has presented several tutorials and published more than 100 papers in prestigious journals, conference proceedings and book chapter. He is the editor in chief of ALMOHANDIS Magazine and the Honorary Chair of IEEE Bahrain section.

Sohaila Ahmed Rabie is a student at Nile University in the Faculty of Engineering and Applied Sciences, Egypt. She is specifically majoring in Mechanical Engineering, Mechatronics track. Her recent research interests are in the field of renewable energy, energy storage systems, electric vehicles, and battery recycling strategies.

Yasser Rafat is the founding member of CARET.

Michela Robba (M 12') is Associate Professor of Systems Engineering at University of Genoa. The research activity is in the field of optimization and control of smart grids, renewable energy resources, and natural resources management. She is the Italian responsible at EU ESFRI (European Strategy Forum on Research Infrastructures) for the Energy area, and she is a member of the scientific board of the Italian Energy Cluster and Liguria Region Innovation Pole on Energy, Environment and Sustainable Development. Actually, she is lecturer for the courses "Simulation of energy and environmental systems" and "Models and methods for Energy Engineering" at Savona Campus Polytechnic School, University of Genoa. She is author of more than 120 publications in international journals, books and proceedings of international conferences.

Sulabh Sachan received the Ph.D. degree from MNNIT Allahabad, India in 2018 and M.Tech. degree in Power System from the Indian Institute of Technology, Roorkee, India, in 2013. He is an Assistant Professor in the Department of Electrical Engineering, MJP Rohilkhand University Bareilly, India. His research interests include electric vehicle charging discharging and its integration issues in distribution network.

Imran Saleem is Professor and Dean Faculty of Commerce, Aligarh Muslim University India. He is also visiting Professor at IIM Kashipur, and has been Vice Chancellor at Singahania University, Rajasthan India, Head of Management Studies, Jamia Hamdard, New Delhi (India), Head, SIEMAT (World Bank Project), Allahabad (India), Former Consultant, LBSNAA (IAS Academy), Mussoorie (India). His main research interests are entrepreneurial processes, innovation and academic entrepreneurship. He has managed several research and development projects funded by research councils, ministries and government agencies. He publishes regularly in international journals such as International Entrepreneurship and Management Journal, and Journal Global Entrepreneurship Research etc.

Adil Sarwar is working as an Assistant Professor in the Department of Electrical Engineering, AMU, Aligarh. He has teaching and research experience of over 8 years. He is a member of IEEE, USA, and life member of the systems society of India. He has published more than 40 research papers in reputed journals and conferences including IEEE transactions. He has contributed a chapter on multiphase converters in handbook of Power Electronics 4th Ed edited by M. H. Rashid. He is a co-principle investigator in ongoing projects on hybrid converters worth INR 2 Million. His research interest includes Power Converters, Solar PV system, and microcontrollers.

Marzieh Sefid received a bachelor of technology in electronic engineering in 2010 from Islamic Azad University, Dezful, Iran. Master of Technology degree in electrical engineering with specialization in power drive engineering, in 2014, from A.M.U, India. Ph.D. degree in electrical engineering, in 2019, A.M.U, India. His research interests include smart grids, PMU, and renewable energy.

Mahdi Shafaati Shemami received a bachelor of technology in electronic engineering in 2009 from Islamic Azad University, Qazvin, Iran. Master of Technology degree in electrical engineering with specialization in instrumentation and control engineering, in 2013, from A.M.U, India. Ph.D. degree in electrical engineering, A.M.U, India. His research interests include smart grids, electric vehicles, V2X and especially V2H strategies, energy management, and renewable energy.

Gagandeep Sharma has completed his Masters in Science degree at Ontario Tech. University, Oshawa, Canada. His topic of research is "Impact of EV charging on Electrical distribution Grid". He is having 12 years of teaching experience in DAVIET, Jalandhar (India). His field of interest is Power systems, power electronics and renewable energy sources.

Vijay Sood received the Ph.D. degree from the University of Bradford, Bradford, U.K., in 1977., He is currently an Associate Professor at UOIT, Oshawa, Ontario. He has extensive experience in the simulation of HVDC-F ACTS systems and their controllers. He has authored two textbooks on HVDC Transmission. His research focuses on the monitoring, control, and protection of power systems and on the integration of renewable energy systems into the smart grid, Dr. Sood is a Registered Professional Engineer in the province of Ontario, a Life Fellow of the IEEE, Fellow of the Engineering Institute of Canada and Emeritus Fellow of the Canadian Academy of Engineering.

Chu Sun received the B.Sc. and M.Sc. degrees in Electrical Engineering from Beihang University, Beijing, China, in 2012 and 2015 respectively. He obtained his Ph.D. degree in Electrical Engineering from McGill University, Montreal, QC, Canada, in 2021. He is now an assistant professor in the School of Automation Science and Electrical Engineering at Beihang University. His research interests include microgrid and distributed generation, power electronic converter for power system, motor design and drives for electric vehicles.

Sameer Usmani is an environmental researcher and a PhD student at Australian National University, Australia. His research work revolves around renewable energy (biomass, solar), material analysis. He has expertise in simulation software viz. Aspen Plus, UniSim, MATLAB and completed a range of projects for industries. Sameer has several publications in respected journals, proving his competency and expertise with numerical modelling and experimental works.

Mohd Yasir Arafat has completed PhD in Commerce (Entrepreneurship Area) in 2019 and Master of Commerce from Aligarh Muslim University (AMU), Aligarh, India. Currently, he is working as Assistant Professor at Centre for Distance Education, AMU. He has been Ministry of Minority Affairs Scholar, Senior Research Fellow, Maulana Azad National Fellow, Centre for Research and Entrepreneurship Education (CREED) Fellow at Entrepreneurship Development of India (EDII). He is recipient of Best Paper Award and Best Scholar Award. His research interests include electric vehicles entrepreneurship, agricultural entrepreneurship, renewable energy entrepreneurship and entrepreneurship education. He has published papers in learned journals like International Entrepreneurship and Management Journal (Springer), Journal Global Entrepreneurship Research (Springer), International Journal of Intelligent Enterprise (InderSciences), International Journal of Knowledge Management (IGI Global), and Academy of Management Global Proceedings. He is also Lead Editor of a Book "Driving Factors for Venture Creation and Success in Agricultural Entrepreneurship" to be published by IGI Global.

Index

A

Auxiliary Load 95

B

battery degradation 96, 105-106, 112, 205, 256, 264
battery life 8, 11, 48, 96, 98-99, 111-113, 158, 160, 176, 196, 200, 202
Battery Swapping Station 154, 195, 206-207
Blockchain 38-67, 220-225, 229-232

C

capacity fading 99, 101, 104, 111-113
charging level 195-196
charging station 6-10, 20-23, 25, 33-36, 47-48, 50, 52, 58, 60, 63, 102, 114, 116-117, 122, 126, 129, 131-134, 139-140, 144, 151, 153, 195-196, 203, 206-207, 229-230, 261
converters 20, 70-72, 75-76, 79, 88-89, 102, 114, 116-117, 120, 124, 130, 178-179, 185, 189, 191, 193, 198, 200
cryptocurrency 53, 63, 220-221, 232

D

decentralised network 38
decentralization 52
Discrete Event Model 155
Discrete Events 133
Distributed Energy Resources (DER) 38, 43, 48
distributed ledger 40, 61, 63, 220-222

E

electric grid 35, 116, 120, 123-124, 195
Electric Vehicle (EV) 1-9, 11-13, 15, 19-25, 35-40, 42-44, 47, 50, 52, 56-67, 69, 91-97, 102, 105-106, 111-114, 131-132, 153-154, 157-158, 162-164, 166-169, 174-178, 195, 207-211, 214, 220-221, 232, 234, 237, 252, 258, 260-265, 268, 278, 281-282, 291-293
electric vehicle charging 1, 4, 6-8, 20, 22-23, 35-40, 44, 56-59, 61-65, 102, 114, 131-132, 153, 178, 207-208, 214, 220, 232, 258, 261-264, 292
Electric Vehicle Charging Infrastructure 1, 38, 40, 114, 262, 292
Electric Vehicle Station 113
Electrified Transportation 22, 231, 233
emergency backup power 268, 279
emissions 2, 4, 6-7, 19, 39, 112, 134, 136-138, 162-167, 171, 174-175, 179, 195, 197, 233-234, 236-237, 246-247, 256, 259, 263-264, 269, 292
energy sources 4-6, 20, 39, 96-97, 110, 112-113, 117, 134, 161-162, 179, 181, 249, 278-279
energy storage 1-2, 4-5, 7-8, 18, 23, 39, 43, 45, 96-97, 104, 110, 112-113, 123, 137, 153, 162, 174, 177, 195-197, 206-207, 248, 257, 270, 272, 274, 289
energy storage system 2, 7-8, 45, 104, 113, 196-197, 270
entrepreneurship 208-212, 214, 216-219
ESU 1, 289
Ethereum 47, 51, 53, 56-57, 222-225, 231-232
EV Challenges 195

F

Fast Charging Station (FCS) 114, 131
FCEV 1, 12, 234, 239
fuzzy logic inference system 268

G

grid 2-4, 8-9, 17, 19-21, 23-24, 26-27, 29, 34-35, 37-38, 42-46, 48, 52-53, 55-56, 58-60, 62-63, 65-67, 74-80, 82-83, 85-88, 94-97, 105, 109-117, 119-124, 126-127, 130-142, 146-155, 162, 174, 178-179, 190-191, 195-198, 206-207, 253, 256-265, 268,

270-281, 283-285, 288-291

H

harmonics 26, 34, 36, 77, 79, 83, 87, 89, 117-120, 124, 126-128, 130, 178-179, 189, 253
Hydrogen Infrastructure 233, 242

I

impact 3-4, 6, 10, 23, 25, 27-28, 30-37, 58, 60, 64, 80, 88, 90, 96-98, 111, 114-115, 117, 119, 122-124, 128-130, 132, 137-138, 152, 156, 162-163, 166-168, 176, 211-213, 215, 217, 250, 258, 261
Integrated Battery Charger (IBC) 69, 90-91, 93-95
Integrated Chargers 68

L

Li-ion batteries 156, 158, 165
load shedding 262, 268, 270, 272-274, 276, 289-291

M

micro grids 21, 35, 39, 43, 62-63, 96-97, 104-105, 109-111, 113, 133-134, 137, 139, 142, 144, 151, 153-155, 178-179, 206, 260-261, 292
multi-level inverters 178, 191

O

Offboard Chargers 68
on-board chargers 68-69, 114-117
optimization 3-5, 22-23, 46-47, 51, 55, 57, 93, 133-142, 144-145, 149, 151-155, 176, 196, 206, 260-261, 263, 272

P

PEM 233-236, 238-242, 245-250
PHEV 1, 11-12, 20, 44, 114-115, 175, 209, 268-269, 271
photovoltaic arrays (PVA) 268
pilot 39, 50, 252-253, 256-260, 262, 267
plugin electric vehicles 68-69
Plug-in Hybrid Vehicle 208-209
power quality 27, 30, 35-36, 46, 97, 114-115, 117-118, 122, 124, 126, 128, 130-132, 178-179, 181-182, 190-191, 197
Pulse Width Modulation 79, 94, 121, 186

R

recycle 156, 164, 171
regulation 5, 79, 112, 117, 120, 123, 193, 196, 252-253, 256-257, 264
reuse 69, 78, 82-83, 88, 156, 162, 165-167, 173-174, 176

S

smart charging 34, 64-65, 133-134, 195-196, 256, 258-259, 263
smart contract 39, 45, 47-49, 51, 57, 220-221, 223-224, 226-232
Solidity 220-221, 223-224, 227, 229, 232
Special Machines 71, 76, 95
start-up 208-209, 212, 214, 216-218, 234, 241
switched capacitors 178

T

tardiness 133, 140-142, 146-149, 155
three-phase 8-9, 71-79, 82, 84-85, 88-90, 92-95, 116, 124, 182, 185, 194
Torque Cancelation 86, 91, 95
total harmonics distortion 178-179
traction drive 69-72, 74-75, 77, 80-84, 90, 92, 95

U

Unit Template Control (UTC) 114, 121

V

V2H 23, 259, 267-268, 270-273, 275-276, 278-280, 288-291, 293
Vehicle-to-Grid (V2G) 20, 22, 64, 67, 95, 112, 134, 178-179, 195-196, 252, 259, 261-264, 292
Voltage Source Converter (VSC) 89, 116, 120
VPP 34, 252, 258-259

www.ingramcontent.com/pod-product-compliance
Lightning Source LLC
LaVergne TN
LVHW080535300926
849162LV00007B/1614
* 9 7 8 1 7 9 9 8 6 8 5 8 3 *